Handbook of Microbial Artificial Intelligence

This book explores the transformative intersection of artificial intelligence (AI) and microbiology, introducing the concept of microbial artificial intelligence (MAI). Designed to bridge the gap between microbiology and AI, it provides a comprehensive overview of fundamental AI concepts, practical applications, limitations, ethical considerations, and future opportunities in this emerging field.

Featuring contributions from a global team of experts, the book is structured into two distinct sections for a deeper exploration of MAI. The first section introduces the fundamentals of MAI, offering microbiologists a clear understanding of AI principles and their relevance to microbiology. The second section highlights real-world applications of MAI across diverse industries, including healthcare, genomics, metagenomics, drug discovery, antimicrobial resistance, biofuel production, food quality and safety, astrobiology, and forensic science. This section addresses critical issues such as bias in MAI, ethical challenges, and the impact of data collection and interpretation on research outcomes, while proposing strategies to mitigate these challenges.

This essential resource is ideal for students and researchers eager to explore the potential of MAI and understand AI's growing influence on microbiology. It also serves as valuable supplementary reading for AI professionals seeking insights into broader applications of AI across scientific disciplines.

Handbook of Microbial Artificial Intelligence

Applications and Ethical Considerations

Edited by
Sesan Abiodun Aransiola, Kondakindi Venkateswar Reddy, Oluwafemi Adebayo Oyewole, Ebenezer Esenogho, and Naga Raju Maddela

CRC Press
Taylor & Francis Group
Boca Raton London New York

CRC Press is an imprint of the
Taylor & Francis Group, an **informa** business

Designed cover image: Jonathan Pennell

First edition published 2027
by CRC Press
2385 NW Executive Center Drive, Suite 320, Boca Raton FL 33431

and by CRC Press
4 Park Square, Milton Park, Abingdon, Oxon, OX14 4RN

CRC Press is an imprint of Taylor & Francis Group, LLC

ISBN: 978-1-041-21491-5 (hbk)
ISBN: 978-1-041-21493-9 (pbk)
ISBN: 978-1-003-72751-4 (ebk)

DOI: 10.1201/9781003727514

Typeset in Times
by KnowledgeWorks Global Ltd.

Contents

Foreword

The rapid expansion of Artificial Intelligence within the microbial sciences marks a transformative moment in global research and innovation. As microbiologists generate unprecedented volumes of genomic, biochemical, ecological, and clinical data, the imperative for advanced computational systems has never been more critical. Artificial Intelligence now serves as a pivotal enabler—facilitating precision analytics, accelerating discovery, and reshaping the operational capabilities of laboratories, healthcare systems, bioprocess industries, and environmental monitoring platforms worldwide.

The ***Handbook of Microbial Artificial Intelligence: Applications and Ethical Considerations*** provides a timely and comprehensive evaluation of this emerging discipline. The editors bring together a distinguished consortium of contributors whose collective expertise spans microbiology, computational intelligence, environmental science, medical diagnostics, biotechnology, and data ethics. This collective effort yields a technically robust volume, multidisciplinary in scope, and strategically positioned to guide the next generation of innovation in microbial research. What differentiates this handbook is its balanced and integrative architecture. The early chapters offer a foundational grounding in the principles of AI and their intersection with microbiological inquiry, while the subsequent sections examine sector-specific applications that illustrate the expanding utility of Microbial Artificial Intelligence (MAI). These include microbial genomics, metagenomics, antimicrobial resistance prediction, biofuel process optimization, food safety analytics, wastewater monitoring, and advanced forensics. The breadth of coverage reinforces a central insight: MAI is rapidly becoming a core capability in contemporary science, enabling researchers to address longstanding challenges with new levels of accuracy, scalability, and foresight.

Equally consequential is the book's attention to the ethical, methodological, and governance considerations surrounding the use of AI in microbiology. Issues of data quality, algorithmic bias, research integrity, and responsible deployment are treated not as supplementary concerns but as essential components of the discipline. In doing so, the editors underscore the dual imperative of advancing scientific capability while safeguarding societal trust and professional accountability. This volume is poised to become a leading reference for academics, practitioners, policy experts, and emerging scholars seeking a consolidated resource on the theory, practice, and ethics of MAI. Its strategic clarity, scientific depth, and global perspective offer readers a comprehensive roadmap for navigating one of the most consequential technological frontiers of our time. The editors are to be commended for delivering a work of enduring relevance and scholarly value.

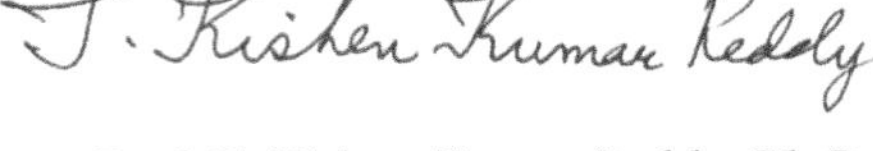

Prof. T. Kishen Kumar Reddy, Ph.D.,
Vice Chancellor, Jawaharlal Nehru Technological University (JNTUH)
Hyderabad, Telangana, 500085, India.
23rd December, 2025.

Preface

The integration of Artificial Intelligence into microbiology represents a fundamental shift in how microbial systems are studied, interpreted, and applied across research and industry. As data volumes expand by next-generation sequencing, digital diagnostics, environmental surveillance, and advanced imaging, traditional analytic frameworks have become increasingly insufficient. The emergence of Microbial Artificial Intelligence (MAI) provides a strategic response to these demands, enabling researchers to interrogate complex datasets with unprecedented speed, depth, and analytical precision.

This handbook originated from the recognition that the field required a unified, multidisciplinary resource that could articulate both the operational potential and the ethical complexities of MAI. Our objective was to develop a reference framework that brings coherence to a rapidly evolving domain, supports informed adoption, and catalyzes responsible innovation. The chapters presented herein reflect the collective insights of international contributors whose expertise spans the microbial sciences, computational modeling, biotechnology, clinical practice, environmental engineering, and digital ethics. The content is organized to guide the reader from conceptual foundations to advanced applications and finally to the ethical and governance issues that must accompany the use of AI in biological research. The first section introduces the intellectual architecture of MAI, tracing its evolution and defining core analytical principles. The second section translates these foundations into practice, demonstrating how AI supports pathogen detection, antimicrobial resistance forecasting, bioenergy modeling, food safety assurance, microbial forensics, and even space-based microbial exploration. The second section addresses the methodological and ethical considerations that underpin responsible adoption, including data integrity, bias mitigation, and the evolving expectations of researchers in an era increasingly influenced by automated tools.

While the scientific opportunities enabled by MAI are extensive, the responsible use of AI remains a central theme throughout this volume. The accelerating integration of computational technologies into microbiology requires a renewed commitment to transparency, data stewardship, and ethical vigilance. We hope that this handbook not only informs the technical advancement of the field but also enhances the quality and reliability of microbial research in an AI-driven era. We are profoundly grateful to the contributors, reviewers, and institutional partners who supported the development of this volume. Their collective expertise and dedication were instrumental in shaping a resource that we believe will be valuable to students, researchers, educators, and professionals seeking to understand and leverage the emerging dynamics of MAI. We trust that readers will find in these pages both the insight and the strategic orientation necessary to navigate the expanding frontier of AI-enabled microbial science.

The chapters were contributed by 60 Academicians/Scientists/Researchers from six different countries (Ecuador, India, Nigeria, USA, Poland, and South Africa) across the globe.

Sesan Abiodun Aransiola, Ph.D.,
University of South Africa
Pretoria, South Africa,
University of Abuja,
Abuja, Nigeria,

Kondakindi Venkateswar Reddy, Ph.D.,
Jawaharlal Nehru Technological University
Hyderabad. Telangana, India

Oluwafemi Adebayo Oyewole, Ph.D.,
Federal University of Technology,
Minna, Nigeria

Prof. Ebenezer Esenogho
University of South Africa
Pretoria, South Africa

Prof. Naga Raju Maddela
Universidad Técnica de Manabí,
Portoviejo, Ecuador

Acknowledgments

We express our profound gratitude to all the contributors of this book, whose expertise, timely input, and unwavering commitment have been instrumental in bringing this book to fruition. The enthusiasm, generosity, and collaborative spirit of each contributor ensured that this work was completed efficiently and to the highest standard. We also extend our sincere appreciation to the anonymous reviewers, whose insightful and constructive feedback significantly enhanced the quality of this book by encouraging the inclusion of additional expert perspectives. Our heartfelt thanks go to the CRC Press team and the production staff, whose guidance, professionalism, and support were critical in ensuring the timely and successful completion of this project. Collaborating with such dedicated professionals has been both an honor and a privilege. Finally, we wish to acknowledge the invaluable support of the management and colleagues at the University of South Africa (South Africa), University of Abuja (Nigeria), Jawaharlal Nehru Technological University (India), Federal University of Technology (Nigeria), and Universidad Técnica de Manabí (Ecuador), whose encouragement and assistance throughout the proposal, preparation, and publication process have been indispensable.

About the Editors

Dr. Sesan Abiodun Aransiola is a lecturer at the Department of Microbiology, University of Abuja, and Philomath University, Abuja, Nigeria. He obtained his B.Tech., M.Tech., and Ph.D. in Environmental Microbiology from the Federal University of Technology, Minna, Nigeria. He has demonstrated his research expertise in the production of vermicasts from vermicomposting of organic wastes to assist plants in the remediation of polluted soil with heavy metals. His expertise spans environmental health, phytoremediation, vermicomposting, marine resources, and bioremediation of pollutants in the soil. Aransiola's research is deeply rooted in innovative solutions for soil and water remediation, with significant contributions to microbial biotechnology and waste management. He is an award-winning researcher of different grants, including TETFUND/DR&D/CE/UNU/ABUJA/IBR/2025/VOL.1. He has over 130 publications, including book chapters, research and review articles of good international repute with a high impact factor. He has co-edited scientific books of global interest, which include *Microbial Biotechnology for Bioenergy, Marine Bioresources: Prospects and Obstacles, Marine Bioresources: Prospects and Obstacles, Marine Microbial Products: Applications and Opportunities. Ecological Interplays in Microbial Enzymology, Prospects for Soil Regeneration and Its Impact on Environmental Protection, Soil Microbiome in Green Technology Sustainability, Marine Bioprospecting for Sustainable Blue-bioeconomy, Vermitechnology: Economic, Environmental and Agricultural Sustainability, White Pollution: Biodiversity and Hazards in Marine Plastisphere, Wetland Ecosystems: Conservation Strategies, Policy Management and Applications. Microbial Biofilms – Applications and Control, Marine Greens: Environmental, Agricultural, Industrial and Biomedical Applications, Environmental Footprint of Bioplastic Additives, Emerging Contaminants in Food and Food Products, Phytoremediation in Food Safety: Risks and Prospects, Biotoxins in Food: Threats and Benefits, Fuel Cells in Environmental Sustainability, Biotechnology of Climate-smart Agriculture in Food Security and Sustainable Farming* with many more in the process. Dr. Aransiola has been a review editor, and as a guest editor, he hosted two special issues with scientific journals including *Research in Microbiology* and *Frontiers on Biofilms – Community, Structure, Applications, and Mitigation, and Effects of Microplastics on Soil Ecosystems.* Previously, before joining academia, he rose to the rank of assistant chief scientific officer at the National Biotechnology Research and Development Agency, Nigeria, where he was involved in the uses of vinasse (a by-product of ethanol) as bio-fertilizer to reclaim polluted soil for agricultural purposes. Dr. Aransiola's academic journey is complemented by robust research, a teaching career (with proper mentoring and supervising both undergraduate and postgraduate students), and active involvement in community services, where he leverages his knowledge to foster sustainability. His contributions continue to impact both academia and the broader environmental sciences community. He is a member of the Nigerian Society for Microbiology, the Association of Environmental Impact Assessment of Nigeria (an Affiliation of the International Association for Impact Assessment), and the American Society for Microbiology.

Kondakindi Venkateswar Reddy received his M.Sc. (Microbiology) (2006–2008) from Sri Krishnadevyaraya University, Anantapur, Andhra Pradesh, M.Tech. (Environmental Management), and Ph.D. (2017) in Environmental Science and Technology from Jawaharlal Nehru Technological University, Hyderabad, Telangana. During his doctoral program in the area of industrial microbiology, he worked on isolation, characterization, and molecular identification of cellulase producing bacterial isolates used in aquaculture wastewater treatment. He has been working as a faculty member in Microbiology and Biotechnology for the past 15 years, teaching undergraduate and postgraduate students. His forte is Industrial microbiology, Biotechnology, and Environmental management. He has successfully completed four AICTE-approved faculty development programs in the areas of Applied Environmental Microbiology, Experimental Biotechnology, Biochemistry, and Cell Biology. He has published 50 book chapters, four books, 50 research papers, and five review articles in reputed journals. He is an active reviewer for multiple scientific publishers. He has actively participated in 25 National/International conferences and presented various research papers at prestigious universities. Currently, he is working as an asst. professor at the Center for Biotechnology, University College of Engineering Science and Technology, Jawaharlal Nehru Technological University, Hyderabad, Telangana, India.

Dr. Oluwafemi Adebayo Oyewole holds a B.Tech. degree in Microbiology (2004) and M.Tech. degree in Microbiology with specialization in Environmental Microbiology (2010) from the Federal University of Technology, Minna, Niger State, Nigeria. He obtained his Ph.D. in Environmental Microbiology (2016) from the University of Portsmouth, UK. He is currently an associate professor in the Department of Microbiology, Federal University of Technology, Minna, Nigeria. His current roles include teaching undergraduate and postgraduate students and research supervision in the areas of Environmental, Productive, and Applied Microbiology. He has been involved in several administrative responsibilities within and outside the university. He has supervised 77 undergraduate, 14 M.Tech., and 11 Ph.D. students. He is also engaging in community development and professional consultancy services outside of the university, and engaged in editorial/academic support services. He has received multiple categories of Awards of Excellence in his field and was listed among the top 500 authors based on scholarly output between 2021 and 2025, according to SCOPUS/SCiVal at Elsevier B.V 2025. He was also named among the top scientists in three fields—microbiology, biochemistry, and agricultural and biological sciences. So far, his publications have produced more than 1982 total citations with an h-index of 20 and i-10 index of 48, according to Google Scholar. He has over 300 research publications in journals, books, conference proceedings, and GenBank submission; won some research grants; has six patents from his research findings; and is a member of professional associations.

Prof. Ebenezer Esenogho (NRF C2 Rated) is a highly accomplished academic and researcher with a distinguished career spanning close to two decades. He holds a Diploma in Computer Engineering (Upper Division, 2002/2003), a Bachelor of Engineering degree in Computer Engineering (Second Class Honours, Upper Division, 2007/2008), and a Master of Engineering in Electronics Engineering (2010/2011) from the University of Benin, Nigeria. Prof. Ebenezer lectured at the University of Benin, where he rose to the rank of Senior Lecturer. From 2013 to 2016, he pursued his Ph.D. in Electronic Engineering at the University of KwaZulu-Natal (UKZN), South Africa, under the Centre of Radio Access and Rural Technology (CRART), a Centre of Excellence. His doctoral studies were funded by leading industry players, including Telkom SA, Eskom, Alcatel, Ericsson, and Huawei. Following his Ph.D., Prof. Ebenezer was awarded the prestigious Global Excellence and Stature (GES) Postdoctoral Research Fellowship at the University of Johannesburg (UJ), where he conducted cutting-edge research in the Fourth Industrial Revolution (4IR) domain at the Institute for Intelligent Systems (IIS) and the Centre for Telecommunications (CfT). After that, he served as a Senior Research Associate at CfT and as Research Coordinator for the Centre for Collaborative Digital Networks. Prior to his deployment on the Diplomatic/Foreign mission to Uganda with the Nigerian Ministry of Foreign Affairs, under the Technical Aid Corps program. Prof. Ebenezer was appointed as an associate professor in the Department of Electrical, Telecommunications, and Computer Engineering at Kampala International University, Uganda, where he contributed to the training of manpower in the East African educational sector. He was later appointed as an expatriate senior lecturer at the University of Botswana, where he supported the research roadmap in Electronics, Telecommunications, Artificial Intelligence, developed new short programs, and contributed to the Botswana Qualifications Authority (BQA) review of undergraduate and postgraduate programs. Currently, Ebenezer Esenogho is appointed a distinguished professor and the head of the Centre for Artificial Intelligence and Multidisciplinary Innovation Studies (CAIMIS) at the University of South Africa (UNISA), Pretoria campus, where he is mentoring over 30 postdoctoral research fellows and hosting over ten research associates. He is also a visiting professor at Mbarara University of Science and Technology (MUST), Uganda, where he supports postgraduate supervision and mentors junior staff members in the field of AI/ML. Prof. Ebenezer is a recipient of numerous grants, scholarships, awards, and fellowships, including the CEPS/Eskom HVDC (2013, 2014), J.W. Nelson Award (2015), GES Postdoctoral Fellowship (2017, 2018, 2019/20), the French/South Africa Institute of Technology (F' SAIT) SARChi Chair (2021), and NRF research grant award (2025). In 2025, he was awarded the prestigious National Research Foundation (NRF) C2 rating for researchers of international repute. His research excellence was recognized with the best oral paper presentation award at the UJ Postdoctoral Fellowship Conference in 2019. A dedicated academic leader, Prof. Ebenezer has served in various leadership roles, which include the first postgraduate representative on the Faculty Management Board of the University of KwaZulu-Natal (2015–2016) and the first postdoctoral research fellow representative in the University of Johannesburg Senate (2018–2020). He was also a UJ/DST/NRF research delegate to the H2020-ESASTAP EU-South Africa STI Cooperation in Vienna, Austria, aimed at strengthening technology, research, and innovation ties between South Africa and the European Union. Prof. Ebenezer has supervised numerous postgraduate students (Postdoc, Ph.D., M.Phil., and Master) and has authored/co-authored several peer-reviewed journal articles and conference papers. He has chaired sessions at international conferences and serves on the editorial board of the KIU Journal of Science, Engineering, and Technology. His expertise as a reviewer has led to his appointment as an external grant proposal assessor for the Australian University Kuwait (Grant No. PN18-15EE-01) and a grant reviewer for the Kuwait Foundation for Advancement Science (KFAS). His research interests include, but are not limited to, Fifth Generation (5G) Wireless Networks, Cognitive Radio Networks, Network Security/Cybersecurity, Smart Grid, IoT/IoE, SDN/SDR, Wireless Sensor Networks, Artificial Intelligence/Machine Learning, Big Data, and Mobile/Cloud Computing. Prof. Ebenezer is a registered Engineer in Nigeria, and a member of the South African Institute of Electrical Engineers (SAIEE), IEEE Region 8, and several other professional bodies. Prof. Ebenezer is deeply committed to service to humanity, having engaged in numerous community development initiatives in South Africa, Uganda, Botswana, and his home country, Nigeria. His contributions to academia, research, and mentorship continue to inspire and shape the future of engineering and technology in Sub-Saharan Africa.

Naga Raju Maddela received his M.Sc. (1996–1998) and Ph.D. (2012) in Microbiology from Sri Krishnadevaraya University, Anantapuramu, India. During his doctoral program in the area of Environmental Microbiology, he investigated the effects of industrial effluents/insecticides on soil microorganisms and their biological activities, and he has been working as a faculty member in Microbiology since 1998, teaching undergraduate and postgraduate students. He worked on Prometeo Investigator (fellowship received from SENESCYT) at Universidad Estatal Amazónica, Ecuador, during 2013–2015, and received a "Postdoctoral Fellowship" (2016–2018) from Sun Yat-sen University, China. He also received external funding from the "China Postdoctoral Science Foundation" in 2017 and internal funding from the "Universidad Técnica de Manabí" in 2020. He participated in national/international conferences and presented research data in China, Cuba, Ecuador, India, and Singapore. Currently, he is working as a full professor at the Facultad de Ciencias de la Salud, Universidad Técnica de Manabí, Portoviejo, Ecuador. He has been actively publishing scientific articles, books (authored and edited), and chapters since 2007. As of now, he has published over 100 articles, 30 books, and 40 book chapters.

Contributors

Olabisi Peter Abioye
Department of Microbiology, School of Life Sciences, Federal University of Technology, Minna, Niger State, Nigeria
Department of Public Health, School of Life Sciences, Federal University of Technology, Minna, Niger State, Nigeria

David Oyeyemi Aboyeji
Department of Microbiology, Federal University of Technology, Minna, Niger State, Nigeria
African Centre of Excellence for Mycotoxins and Food Safety, Federal University of Technology, Minna, Niger State, Nigeria

Samson Olusola Adeboye
Department of Biological Sciences, Covenant University, Ota, Ogun State, Nigeria

Isiaka Adedayo Adelere
Department of Microbiology, Federal University of Technology, Minna, Niger State, Nigeria

Mayowa Agbi
Department of Computer & Information Sciences, Covenant University, Ota, Ogun State, Nigeria

Ganiyu Ayodele Ajibade
Centre for Artificial Intelligence & Multidisciplinary Innovation Studies, Department of Auditing, College of Accounting Sciences, University of South Africa, Pretoria, South Africa
Mathematics Department, Federal University of Technology and Environmental Sciences, Iyin Ekiti, Nigeria

Ebelechukwu Gloria Ajuka
Centre for Artificial Intelligence & Multidisciplinary Innovation Studies, College of Accounting Sciences, University of South Africa, Pretoria, South Africa

Bolaji Oluwatosin Akanbi
Department of Microbiology, Faculty of Science, University of Abuja, Abuja, Nigeria

Yedla Akanksha
Centre for Biotechnology, University College of Engineering Science and Technology, Jawaharlal Nehru Technological University, Hyderabad, Kukatpally, Telangana, India

Shamsudeen Aliyu
Department of Biochemistry and Molecular Biology, Federal University, Birnin Kebbi, Kebbi State, Nigeria

Iisominea Isozo Amos
Department of Computer and Information Sciences (Bioinformatics), College of Science and Technology, Covenant University

Fabricio Andocilla-Oleas
Universidad Israel, Quito, Ecuador

Daniel Abu Anyebe
Department of Biochemistry and Molecular Biology, Federal University, Birnin Kebbi, Kebbi State, Nigeria

Sesan Abiodun Aransiola
Centre for Artificial Intelligence & Multidisciplinary Innovation Studies, College of Accounting Sciences, University of South Africa, Pretoria, South Africa
Department of Microbiology, Faculty of Science, University of Abuja, Abuja, Nigeria

Victoria-Grace Onyekachi Aririguzoh
Department of Biological Sciences, Covenant University, Ota, Ogun State, Nigeria

Ben Asante
Centre for Artificial Intelligence & Multidisciplinary Innovation Studies, College of Accounting Sciences, University of South Africa, Pretoria, South Africa

Helen Shnada Auta
Department of Microbiology, Federal University of Technology, Minna, Niger State, Nigeria
African Centre of Excellence for Mycotoxins and Food Safety, Federal University of Technology, Minna, Niger State, Nigeria

Daniel Oluwatobiloba Balogun
Department of Biological Sciences, Covenant University, Ota, Ogun State, Nigeria

Zion Charles
Federal University of Technology, Minna, Niger State, Nigeria

P. Gnana Deepu
Allele Tech Private Limited, Jeedimetla, Hyderabad, Telangana India

Rodrigo Duarte-Casar
Pontificia Universidad Católica del Ecuador Sede Manabí, Portoviejo, Ecuador

Zawu Yoowa Duyann
Department of Molecular Biology and Bioinformatics, Africa Centre of Excellence for Mycotoxins and Food Safety (ACEMFS), Federal University of Technology, Minna, Niger State, Nigeria

Ajimah Nnabueze Edmund
Centre for Artificial Intelligence & Multidisciplinary Innovation Studies, College of Accounting Sciences, University of South Africa, Pretoria, South Africa

Chiedu Epiphany Edoamodu
Department of Biotechnology, Ivy Tech Community College, Indianapolis, Indiana, USA

Prudence Ejordenore Ehizuenlen
Centre for Artificial Intelligence and Multidisciplinary Studies, College of Accounting Sciences, Department of Auditing, University of South Africa, Pretoria, South Africa

Pius Uagbae Ejodamen
Centre for Artificial Intelligence and Multidisciplinary Studies, College of Accounting Sciences, Department of Auditing, University of South Africa, Pretoria, South Africa

Ebenezer Esenogho
Centre for Artificial Intelligence & Multidisciplinary Innovation Studies, College of Accounting Sciences, University of South Africa, Pretoria, South Africa

Grace Galadima
Federal University of Technology, Minna, Niger State, Nigeria

Bitrus Rose Giwa
Federal University of Technology, Minna, Niger State, Nigeria

Abdurrashid Usman Gulumbe
Department of Pure and Industrial Chemistry, Federal University, Birnin Kebbi, Kebbi State, Nigeria

Taofeek Olajide Hammed
Institute of Microelectronic and Optoelectronic, Warsaw University of Technology, Poland

Jogipeta Harihara
Centre for Biotechnology, University College of Engineering Science and Technology, Jawaharlal Nehru Technological University, Hyderabad, Kukatpally, Telangana, India

Abdulrazaq Izuafa
Department of Biological Sciences, Federal University, Birnin Kebbi, Kebbi State, Nigeria

Joy Oziohu James
Department of Biochemistry, University of Ilorin, Kwara State, Nigeria

B. S. R. S. Sai Kaustubh
Centre for Biotechnology, University College of Engineering Science and Technology, Jawaharlal Nehru Technological University, Hyderabad, Kukatpally, Telangana, India

Oyebode Kazeem
Department of Auditing, School of Accountancy, College of Accounting Sciences, University of South Africa, South Africa

Olamide Lasisi
Department of Microbiology, Faculty of Science, University of Abuja, Abuja, Nigeria

Gafari Lukumon
Centre for Artificial Intelligence & Multidisciplinary Innovation Studies, College of Accounting Sciences, University of South Africa, Pretoria, South Africa

Naga Raju Maddela
Departamento de Ciencias Biológicas, Facultad la Ciencias de la Salud, Universidad Técnica de Manabí, Portoviejo, Manabí, Ecuador

Hussaini Anthony Makun
African Centre of Excellence for Mycotoxins and Food Safety, Federal University of Technology, Minna, Niger State, Nigeria
Department of Biochemistry, Federal University of Technology, Minna, Niger State, Nigeria

Darío Mondavi-Sobby
Pontificia Universidad Católica del Ecuador Sede Manabí, Manta, Ecuador

Hadiza Lami Muhammad
African Centre of Excellence for Mycotoxins and Food Safety, Federal University of Technology, Minna, Niger State, Nigeria
Department of Biochemistry, Federal University of Technology, Minna, Niger State, Nigeria

Josephine Nathaniel
Africa Centre of Excellence for Mycotoxins and Food Safety, Federal University of Technology, Minna, Niger State, Nigeria

Ayomide Iyanuoluwa Olaniyan
Department of Biological Sciences, Covenant University, Ota, Ogun State, Nigeria

Gbotemi Jerry Oni
Department of Molecular Biology and Bioinformatics, Africa Centre of Excellence for Mycotoxins and Food Safety (ACEMFS), Federal University of Technology, Minna, Niger State, Nigeria

Ifeanyi Famous Ossamulu
African Centre of Excellence for Mycotoxins and Food Safety, Federal University of Technology, Minna, Niger State, Nigeria
Department of Biochemistry, Federal University of Technology, Minna, Niger State, Nigeria

Oluwatoyin Comfort Owoyemi
Department of Microbiology, University of Ilorin, Kwara State, Nigeria

Samuel Adeniyi Oyegbade
Department of Biological Sciences, Covenant University, Ota, Ogun State, Nigeria

Oluwafemi Adebayo Oyewole
Department of Microbiology, Federal University of Technology, Minna, Niger State, Nigeria

Juliana Okwena Pondei
Department of Microbiology, Faculty of Science, University of Abuja, Abuja, Nigeria

Kondakindi Venkateswar Reddy
Centre for Biotechnology, University College of Engineering Science and Technology, Jawaharlal Nehru Technological University, Hyderabad, Kukatpally, Telangana, India

Marlene Rojas-Le-Fort
Pontificia Universidad Católica del Ecuador Sede Manabí, Portoviejo, Ecuador

Aishat Olorunkemi Salisu
Department of Microbiology, Faculty of Science, University of Abuja, Abuja, Nigeria

Jamilu Sani
Department of Demography and Social Statistics, Federal University, Birnin Kebbi, Kebbi State, Nigeria

Akinola Solomon
Department of Auditing, School of Accountancy, College of Accounting Sciences, University of South Africa, South Africa

Amos Musa Sumakada
Department of Microbiology, Faculty of Science, University of Abuja, Abuja, Nigeria

Fakuade Olubusayo Victor
Centre for Artificial Intelligence & Multidisciplinary Innovation Studies, College of Accounting Sciences, University of South Africa, Pretoria, South Africa

Tajudeen Olanrewaju Yahaya
Department of Biological Sciences, Federal University, Birnin Kebbi, Kebbi State, Nigeria

Tsado Priscilla Yetu
Federal University of Technology, Minna, Niger State, Nigeria

Nengak Yunana
Federal University of Technology, Minna, Niger State, Nigeria

SECTION ONE

Microbial Artificial Intelligence at a Glance

Microbial Artificial Intelligence (MAI) *Introductory Perspectives* 1

Sesan Abiodun Aransiola, Gafari Lukumon, Ganiyu Ayodele Ajibade, Ebelechukwu Gloria Ajuka, and Ebenezer Esenogho

1.1 INTRODUCTION

Contemporary advancements in artificial intelligence (AI) have paved the way for changes in the field of microbiology by allowing for the quick, sensitive, and comprehensive computation and analysis of biological data. For instance, AI methods like machine learning (ML) and computer vision can be applied in simplifying clinical microbiology diagnoses through analysis of genomic, metagenomic, spectrometric, and imaging data. This improves cost-effectiveness, increases accuracy of diagnoses, and reduces turnaround time (Alsulimani *et al.*, 2024).

In that same manner, AI models capable of analyzing complex sequencing data are being applied to microbiome data analysis, and these models provide more in-depth, non-culture-dependent insights on microbial populations, using AI (Barh, 2020). Microbial artificial intelligence (MAI) comprises the integration of microbial research and AI. MAI aims to replace time-consuming, human-restricted methods of microbial research with scalable AI systems, informed by large datasets, that can adapt to dynamic environments with the use of predictive algorithms (Vamathevan *et al.*, 2019). MAI, therefore, is positioned in between biology and computing, positioned to improve knowledge of ecology, treatment, and diagnosis (Alsulimani *et al.*, 2024).

Simply put, MAI is the application of AI tools to the study of microorganisms. These AI tools are concepts like ML, computer vision, and deep learning. For practical applications, MAI assists researchers in the modeling and understanding of microbes' behavior, interaction, and functioning within different systems (Alsulimani *et al.*, 2024). For instance, these AI-powered methods are capable of sifting through metagenomic data for species identification, prediction of the probability of antibiotic resistance, and could possibly link activities of microorganisms to health or environmental impact (Pillay *et al.*, 2022). AI is currently being applied to different areas of microbiology, like bacteriology, mycology, parasitology, and virology (Tsitou *et al.*, 2024). The application of AI in diagnostics has revolutionized the methods of detection and management of infection (Alsulimani *et al.*, 2024). The

DOI: 10.1201/9781003727514-2

traditional methods of microbe identification are reliant on culture-based methods like microscopy and biochemical assays, but despite their effectiveness, these methods are labor-intensive, time-consuming, and error-prone (Alsulimani *et al.*, 2024).

Technological improvements in areas like digital imaging, mass spectrometry, and next-generation sequencing (NGS) have enabled the generation of complex and extensive microbiological datasets (Gao & Liu, 2024). AI systems are capable of processing these large swaths of data, as they are equipped with massive processing power, and this leads to faster, easier to scale up, and generally more effective diagnostic tools (Alsulimani *et al.*, 2024; Gao & Liu, 2024).

1.2 MACHINE LEARNING IN PATTERN RECOGNITION

ML technology has been usefully applied to pattern recognition within laboratory data. For example, algorithms trained on Matrix-Assisted Laser Desorption Ionization–Time-of-Flight (MALDI-TOF) mass spectrometry profiles can classify a number of microbial species, and this number is increasing (Andreadi *et al.*, 2025). Pillay *et al.* (2022) showed that supervised ML models can increase the accuracy of classification and improve time efficiency when compared to traditional methods.

Convolutional neural networks (CNNs), with the ability to differentiate between bacterial shapes and bacterial structures, assist clinicians in speedy pathogen identification have also made it useful in microscopy imaging (Vunduk *et al.*, 2023). AI has exponentially increased genomic and metagenomic applications in the field of diagnostics. It is now possible to produce terabytes of data, a lot more than can be analyzed manually, by sequencing platforms (Vunduk *et al.*, 2023). This has led to the employment of deep learning models like random forest classifiers and recurrent neural networks (RNNs) by researchers. This has made the identification of genes associated with antibiotic resistance, novel strain identification, and pathogenicity prediction directly from raw sequencing data possible (Vamathevan *et al.*, 2019).

AI-mediated initiatives like these improve diagnostic accuracy, particularly in complex cases of polymicrobial infections where conventional culture techniques may fail (Alsulimani *et al.*, 2024). AI systems can also be applied to clinical decision-making through clinical decision support systems (CDSS). These systems create an intersection between electronic health records and laboratory findings, which allows their algorithms to aid health workers in choosing appropriate antimicrobial therapies for treatment (Lin *et al.*, 2024). In addition to that, these predictive models have the ability to *predict* treatment outcomes that are most likely and identify emerging resistance patterns, thus supporting antimicrobial management efforts (Alsulimani *et al.*, 2024; Lin *et al.*, 2024).

1.3 COMPARISON OF MICROBIAL PRINCIPLES AND AI TECHNIQUES

Microorganisms have developed various interesting ways of survival and interaction with and within their environments over the course of adaptation and evolution. Interestingly, a number of these microorganisms' natural behaviors mimic the modus operandi of AI systems, including their

TABLE 1.1 Parallels between microbial principles and artificial intelligence techniques

MICROBIAL PRINCIPLE	*AI TECHNIQUE*	*REFERENCE*
Adaptation and evolution	Machine learning with adaptive algorithms	Topçuoğlu *et al.* (2020)
Biofilm formation	Neural networks that build complex, interconnected structures	Vunduk *et al.* (2023)
Metabolic pathway networks	Graph-based AI models for mapping and analyzing metabolic relationships	Yuan & Nault (2025)
Rapid diagnostics via pattern recognition	Computer vision and spectral analysis (e.g., MALDI-TOF mass spectrometry)	Andreadi *et al.* (2025)
Pathogen detection by immune response	Pattern recognition and classification methods (e.g., CNNs)	Li & Wu (2021)

N/B: The parallels discussed above do not suggest that microorganisms and AI systems share similar operating mechanisms; rather, it means that AI systems can be engineered to mimic specific microbial strategies.

methods of learning, adapting, and decision-making (Mohseni & Ghorbani, 2024). It becomes possible to paint a picture of the very similar manner in which the fields of AI and microbial principles mirror each other by drawing parallels between them (Table 1.1). This existing similarity also explains why AI has become invaluable to the exploration and understanding of microbiological processes (Barh, 2020).

1.4 AI TECHNIQUES APPLICABLE TO MAI

AI has become popular in nearly all fields of human endeavor. Its application to microbial systems (an emerging area at the intersection of microbiology and AI) is therefore unsurprising (Noecker *et al.*, 2016; Topçuoğlu *et al.*, 2020). Before the advent of AI, which now empowers researchers to perform a powerful computation such as pattern extractions and predictions, research in microbiology was primarily conducted through experimental and laboratory approaches.

Data from microbial research are dynamic because they shift in response to both the environmental factors and the host interactions. They are also high-dimensional; that is, they comprise genomic, proteomic, metabolomic, and ecological variables that are difficult to analyze simultaneously. Additionally, microbial systems are highly interactive, with strong dependencies among microbial species and between microbes and their hosts (Knight *et al.*, 2018). All these challenges require analytical approaches capable of managing large-scale data, identifying hidden relationships, and generating predictive models.

1.4.1 Modeling and Representation of Microbial Data

Consequently, a central task in MAI is the effective modeling of this microbial data, which exists in diverse forms.

Microbial data requires extensive preprocessing and representation. Unfortunately, some raw datasets often contain noise that can only be corrected through the process of normalization. Once this is sorted, feature extraction methods identify the most informative variables while removing redundancy. But because microbial datasets are typically high-dimensional, there is a need to compress them into lower-dimensional spaces without losing essential information using a technique such as the principal component analysis (PCA) (Jolliffe & Cadima, 2016; Tenenbaum *et al.*, 2000).

It is worth mentioning that microbial data can be either structured or unstructured, with each requiring a unique way of processing them. For example, structured data such as abundance tables or metabolite concentrations can be processed directly with conventional ML algorithms. Unstructured data, on the other hand, require a specialized preprocessing pipeline such as sequence encoding or image recognition algorithms because they contain raw DNA sequences, microscopy images, or free-form clinical records.

1.4.2 Graph-Based Modeling of Microbial Systems

Another approach that has proven valuable in microbial research is the use of graph-based models. They come to play since the microbial communities are highly interactive systems, and graph representation can capture the co-occurrence and ecological relationships between species. Graphs uncover emergent properties that are somehow overlooked by traditional statistical methods (Aguiar-Pulido *et al.*, 2016; Faust & Raes, 2012), thereby enabling host–microbiome graph analysis that reveals links between microbial populations and host health.

ML provides a powerful set of tools for analyzing microbial data because they ensure pattern discovery, prediction, and intervention design. But depending on the type of data and objectives of the research, different branches of ML can be applied. For example, supervised learning is commonly used when labeled data are available. Classification models distinguish healthy from diseased microbiomes, while the regression models predict microbial growth or metabolite levels. Unsupervised learning becomes important when labels are limited or absent, as is often the case in microbial studies. Clustering algorithms help uncover natural groupings in microbial communities. This reveals species with shared ecological roles or identifies community shifts under stress. Dimensionality reduction methods such as t-distributed stochastic neighbor embedding (t-SNE) and uniform manifold approximation and projection (UMAP) make it possible to visualize high-dimensional microbial datasets in two or three dimensions.

In microbial contexts, reinforcement learning has been applied to optimize interventions, such as identifying effective drug combinations against pathogens or adjusting conditions for industrial fermentation processes. Deep learning has further advanced MAI through the provision of architectures capable of modeling highly nonlinear and heterogeneous data.

1.4.3 Deep and Hybrid Learning Architectures in MAI

CNNs have been adapted to detect DNA, RNA, and protein motifs, enabling the identification of conserved regions associated with microbial functions or pathogenicity (Min *et al.*, 2017; Zou *et al.*, 2019), which is useful in detecting biomarkers. RNNs, long short-term memory (LSTM) networks, and transformer architectures are suitable for capturing temporal microbial dynamics. These models track microbial dynamics and intervention effects. Graph neural networks (GNNs) extend deep learning to microbial interaction networks.

Autoencoders and generative models, including variational autoencoders (VAEs) and generative adversarial networks (GANs). They are very useful to reduce the dimensionality in multi-omics datasets and generate synthetic microbial profiles, offering strategies to fill gaps where experimental data are incomplete.

Beyond these standard methods, hybrid and advanced approaches expand the potential of MAI. Integrating systems biology models with ML combines mechanistic equations describing microbial processes with data-driven methods, producing more accurate and interpretable models. Multi-omics integration merges genomic, proteomic, metabolomic, and transcriptomic data into unified frameworks, offering a more comprehensive view of microbial systems (Misra *et al.*, 2019; Wang *et al.*, 2009).

Transfer learning is very useful when there are not many datasets with labels. Models trained on one microbial system, for example, the human gut microbiota, can be adapted for others. explainable AI (XAI) methods improve trust because they ensure predictions are interpretable. XAI helps align computational predictions with biological plausibility (Holzinger *et al.*, 2017; Tjoa & Guan, 2020).

Several algorithms are commonly applied in MAI. Support vector machines (SVMs) are effective for classification in small but high-dimensional datasets. Random forests provide robust predictions and help identify important microbial features. Gradient boosting algorithms, such as extreme gradient boosting (XGBoost) and light gradient boosting machine (LightGBM), perform strongly in predicting microbial responses to environmental changes or treatments. Neural network architectures are also widely used. CNNs for sequence-based problems, RNNs and LSTM networks for temporal microbial dynamics, and GNNs for network-based analyses.

1.4.4 Visualization, Interpretability, and Limitation

To turn complicated model outputs into biologically useful information, it is important to observe and interpret them effectively. People often use heatmaps to show how many genes are being used or the presence of microbes, while network visualizations illustrate how microbes communicate and how they also come together at the same time and place. Researchers can also use feature rankings to identify the microbial taxa, genes, or metabolites that have the biggest effect on how well the model works.

Interpretability makes sure that the results of the computer simulations are possible and at the same time useful. Without transparency, predictions are not reliable and so cannot be trusted. So, simplifying the steps for MAI makes the results more reliable and helps experimental research grow.

Despite its promise, MAI faces certain important limitations. First, small datasets increase the risk of overfitting, where models capture noise rather than genuine biological patterns (Hawkins, 2004). Microbial data are also prone to sequencing errors, sampling inconsistencies, and other forms of noise. Data imbalance and sparsity present further challenges, with some species or conditions well represented while others remain under-sampled.

It is also essential to have well-grounded biological interpretations. Although AI models can be very good at making highly accurate prediction, their outputs must connect to known biological mechanisms or inspire testable hypotheses. Without this foundation, predictions may continue to be unclear and not useful to real-world microbiological uses.

1.5 APPLICATIONS OF MICROBIAL ARTIFICIAL INTELLIGENCE IN MEDICINE, ENVIRONMENT, AND BEYOND

The integration of AI with microbial science has uncovered innovative perspectives. Some applications within Microbial-AI interactions consider microorganisms as active participants rather than passive subjects. Microbial systems with added AI or microbial datasets with AI are designed for novel, groundbreaking applications. Each of these areas celebrates innovative applications for microbial-AI

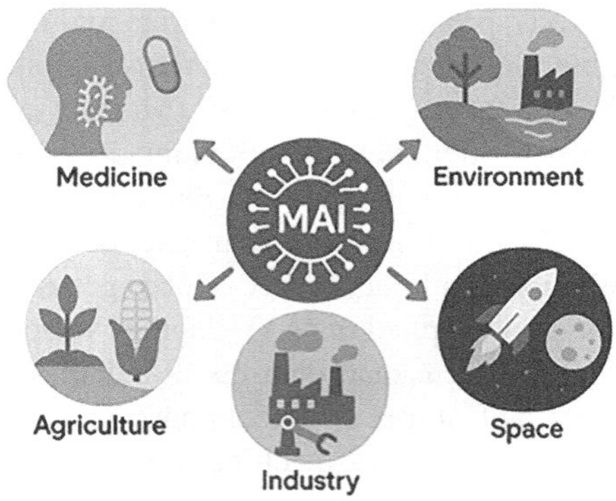

FIGURE 1.1 Applications of microbial artificial intelligence (MAI) across diverse domains, including medicine, environment, agriculture, industry, and space exploration.

interactions. Medicine benefits from fully automated microbial-AI interactions through personalized, sensitive, and specific medicines and radically transformed diagnostics. Microbial-AI interactions provides autonomous, fully automated systems for environmental monitoring and active remediation of pollutants. MAI makes advanced, targeted engineering of microbial consortia for biotechnology and agricultural purposes possible. Thus, this section aims to refine our understanding of the literature survey by focusing on multiple domains. Also, we will discuss various current applications of MAI and its future potential applications.

This section is organized as follows: examination of MAI in the field of medicine and environmental applications, followed by discussions on agriculture and industrial biotechnology, space, and finally addresses emerging future directions. Figure 1.1 displays various applications of MAI across different fields.

1.6 APPLICATIONS OF MAI IN MEDICINE

Using microbial knowledge along with AI-driven models has already started to change the way diagnostics, therapies, and precision health are done in the medical field. These will be discussed in the subsections as follows.

1.6.1 Application of MAI for Drug Discovery

AI is primarily used to help mitigate the global crisis of antimicrobial resistance. AI techniques help speed up the antimicrobial drug discovery process by quickly identifying new candidates and effectively repurposing previously existing drugs (Mulat *et al.*, 2025). These techniques help develop new therapies for resistant pathogens. AI's incredible influence on microbial genomics also helps in quickly annotating pathogens' genomes and accurately predicting antimicrobial resistance genes to help discover new drugs as well as novel diagnostics and therapeutics (Dhaarani & Reddy, 2025). AI is also improving the core biotech tools of foundational synthetic biology by enhancing functional microbial trait annotation and improving protein engineering and novel enzyme design (Huo & Wang, 2024). These improvements contribute to new enhancements in industrial bio-production, AI has advanced microbial bio-production by improving the protein mining, engineering, and control of bioprocesses (Gong *et al.*, 2026).

1.6.2 Diagnostic Biosensors and Biomarker Discovery

Engineered microbial biosensors create a fascinating interface where modified microbes detect disease metabolites, pathogen signals, or inflammation markers and produce measurable signals (e.g., fluorescence, luminescence, and metabolic flux). AI analyzes the signals alongside clinical data and provides a diagnostic result. For example, AI and microbial technology together have been used to identify disease-associated biomarker patterns and to predict microbiome-derived biomarkers for precision health (Tsitou *et al.*, 2024).

On the other hand, metagenomic and multi-omics data from patient microbiomes are processed through ML and deep learning to assess hidden biomarkers. Multiple reviews on deep learning for microbiome analysis describe innovative architectures, such as autoencoders, convolutional, and GNNs, which surpass linear models in multiple classification tasks, for example, disease versus healthy. Such advances allow clinicians to intervene more swiftly after the initial microbial signatures are detected for diseases like inflammatory bowel disease and colorectal cancer. AI improves the results in some of the most data-scarce areas of medical diagnostics. In the field of ophthalmology, a StyleGAN2-ADA model was used to synthetically create high-quality slit-lamp photos from a dataset of only 84 real images, a dataset that was later used to train a deep learning classifier for microbial keratitis. Using self-augmentation for training data increased the classifier's AUROC from 0.76 to 0.93 (Wang *et al.*, 2025). This demonstrates a powerful and viable way to train precise models in situations where there is a deficit of annotated clinical data.

1.6.3 Therapeutics, Microbiome Engineering, and Phage Design

Apart from diagnostics, MAI also enhances precision therapeutics through intervention, modeling, and understanding the dynamics of the host and the microbiome. AI models are used for predicting changes in diet, medication, and drugs that impact the reshaping of microbial networks. With such insight, engineered microbes can respond to the changing microbiome therapeutically and release payloads when they detect certain pathological conditions. An example of this is the AI optimization of strain selection and metabolic and regulatory pathways of microbes to release anti-inflammatory agents in the gut to treat targeted inflammation.

Simultaneously, MAI supports the design of bacteriophage therapy. One of the more persistent challenges in this area is the prediction of phage–host specificity. Deep learning models, particularly CNNs, have shown significant accuracy in predicting phage host ranges using genomic data. Structural embeddings (e.g., PHIStruct) and contrastive learning (CL4PHI) improve predictive accuracy regarding phage–host specificity (Zhang *et al.*, 2023). These advances minimize the necessary empirical trial-and-error that is required to create custom phage cocktails to target resistant pathogens.

1.6.4 Rapid Pathogen Detection and Antimicrobial Discovery

By combining optical imaging, microcolony detection, and AI inference, MAI reduces the time needed for the detection of pathogens in clinical or food matrices. Ma *et al.* (2023) demonstrated the ability to capture and identify bacterial microcolonies in 3 hours using AI-augmented optical imaging. In recent microbiological diagnostics, AI-powered image classification that is based on CNN comprises colony counting, bacterial morphology recognition, and quick antibiotic susceptibility determination (Alsulimani *et al.*, 2024).

With the daily advancements of MAI, there is no doubt that it will transform the discovery landscape within the antibacterial field. Prior studies have highlighted the application of AI in the development of new antibiotics and the utilization of microbiome peptide libraries (D'Urso & Broccolo, 2024). In addition, several researchers have suggested that AI systems are substantial for predictive new chemical synthesis, metabolic pathway optimization, and the design of probiotic interventions. AI systems that utilize large collections of microbial genomes and metabolomes are capable of pinpointing the most valuable candidates for in vitro validation. Therefore, AI systems can streamline the entire discovery process. In the field of medicine, being able to bridge the gap between diagnostics and therapeutics establishes MAI as a unique tool for improving the sensitivity, personalization, and speed of current approaches in a transformative manner.

1.7 APPLICATIONS OF MAI IN ENVIRONMENTAL SYSTEMS

Microbes exist in every major terrestrial and marine ecosystem (Aransiola *et al.*, 2025), and applying AI to harness their abilities yields enormous environmental benefits. MAI's applications in environmental systems include bioremediation and pollutant degradation, ecosystem monitoring and predictive ecology, forensic microbiology, and postmortem examination. We will briefly discuss them below.

1.7.1 Bioremediation and Pollutant Degradation

Bioremediation is one of the most tangible uses of MAI. With the usage of AI models built on enzyme and metabolic databases, one can predict which types of microbes or enzyme variants can break down certain contaminants, such as plastics, hydrocarbons, and heavy metals. For instance, one of the recent review papers describes microbial technologies that are enhanced by AI for targeted diagnostics using microbial biomarkers and therapies that are unique to pollutants In another related research, studies have shown that enzyme functions that are predicted by AI can be used to speed up biocatalyst activities. These biocatalysts can break down pollutants.

In engineered systems, microbes can be used as autonomous cleanup agents. MAI-controlled consortia can be used in identifying pollutant gradients and flexibly adjust metabolic flow to optimize degradation while minimizing ecological disruption. In bioremediation, a dual-phase AI modeling framework was established to initially forecast the succession of microbial community structures, subsequently employing these forecasts, in conjunction with a genetic algorithm, to optimize a bioreactor for the concurrent elimination of multiple pollutants at rates surpassing 90% (Liu *et al.*, 2025). Furthermore, in environmental monitoring, a wireless microbial taste chip solved the critical challenge of sensor selectivity by employing a gated recurrent unit (GRU) deep learning algorithm to classify heavy metal ions with 98.9% accuracy, achieving an ultra-short response time of 48 seconds for real-time water quality monitoring (Wang *et al.*, 2025). These studies demonstrate a progression toward environmental systems that are intelligent and capable of optimizing themselves.

1.7.2 Ecosystem Monitoring and Predictive Ecology

Changes in microbial communities can occur prior to large-scale ecosystem changes, such as a drop in water quality and dysregulation of ecosystems. Integrated microbial biosensors and AI technology allow for real-time monitoring of and surveillance for potential environmentally triggered microbial community changes. These AI patterns and models can identify micro and macro changes within an ecosystem, predict stress response implementations, and predict proximate triggering of an ecosystem dysbiosis. These features of AI in epidemiology and ecological forecasting, as well as pathogen surveillance, are discussed in the editorial focusing on AI in environmental microbiology (Sarrafzadeh *et al.*, 2022).

In predictive activities such as anticipating the onset of algal blooms, the spread of pollutants, and other transitions within the ecological systems, AI uses time-series data of microbes in predictive ecology. In achieving this, MAI works as a sentinel network for the health of the planet.

1.7.3 Forensic Microbiology and Postmortem Analysis

MAI also extends into forensic microbiology. ML is being used to study the microbial succession on decomposing bodies and the surrounding soil. For instance, AI-powered microbiome analysis was used to estimate the postmortem interval and infer the location of the crime scene (Przymus, 2025). This combines metagenomic sequencing with deep learning to estimate time and space variables (Przymus, 2025).

Moreover, the method works well with different substrates, whether skin, organs, or soil, and can use metabolomic and proteomic data to improve estimates. This shows that MAI can penetrate new areas of study to anticipate and estimate ventures.

1.8 APPLICATIONS IN AGRICULTURE AND INDUSTRIAL BIOTECHNOLOGY

In the intersection of ecology and human systems, there are many applications of MAI in agriculture and industrial biotechnology.

1.8.1 Positive Impact of Microbiome Engineering on Crop Productivity

The condition of crops and their output is significantly influenced by the soil and the root microbiome. The MAI allows for effective integrations of microbial consortia that help without compromise on any specific focus, such as nutrient acquisition, drought-resilience, and pathogen tolerance. AI strategies can also be used as predictive models on root, soil, and plant microbiome interactions in relation to certain crop phenotypes and yield outcomes for specific crop-genotype and environment combinations.

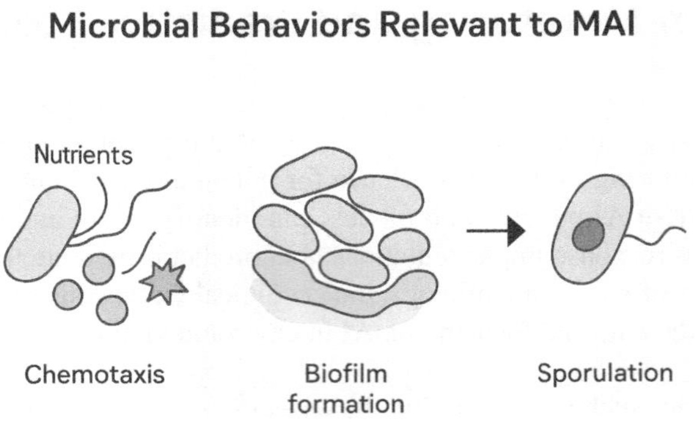

FIGURE 1.2 Microbial behaviors relevant to microbial artificial intelligence (MAI).

Engineered microbes can also be deployed with variable controls and with certain responsiveness, where they release growth-promoting microbes when plants exhibit certain stress signals. This encourages productive plant growth and allows for the reduction of progressive chemical control and pesticide overuse in the field. This promotes sustainable agricultural practice (Figure 1.2).

1.8.2 Biomanufacturing and Synthetic Biology

The self-optimizing bioprocesses, where AI links and integrates with certain synthetic and bioprocesses to guide the production of microbes for industrial applications, widen the functional scope of microbial bioprocesses. Applied AI has been considered in relation to microbiomes and in the larger scope of microbiology (Mohseni *et al.*, 2024; Wang *et al.*, 2024). This includes the application of AI in predictive metabolic regulation, genome prediction, and path optimization systems (Table 1.2).

In synthetic biology, AI aids in designing robust control metabolic pathways in organisms to allow them to control certain metabolic processes as they shift between different industrially relevant metabolic pathways. That is, turning on/off pathways based on crowding, resource availability, or feedback control. Following the integration of techniques such as ML and protein language models, some methods like traditional trial-and-error are being transformed into a predictive and automated design-build-test-learn (DBTL) framework (Gong *et al.*, 2026). This is a component of a bigger transition that is taking place. Therefore, MAI is responsible for transforming microbial biotechnology from static, manually regulated systems into bioprocesses that are adaptable and learning.

TABLE 1.2 Summary of applications of microbial artificial intelligence (MAI) across domains

DOMAIN	*REPRESENTATIVE USE CASES*
Medicine	Diagnostic biosensors; microbiome-based therapies; phage therapy; antimicrobial discovery
Environment	Bioremediation; pollutant degradation; ecosystem monitoring; forensic microbiology
Agriculture	Soil microbiome engineering; crop yield optimization; biocontrol against pathogens
Industry	Synthetic biology; biomanufacturing; biofuel production; enzyme optimization

1.9 BIOLOGICAL SYSTEMS AND PROCESSES UNDERLYING MICROBIAL ARTIFICIAL INTELLIGENCE (MAI)

MAI is an evolving concept that draws insight from microbial systems to develop computational models, algorithms, and adaptive problem-solving strategies. Microbes, being among the oldest and most diverse forms of life on Earth, have evolved a wide range of processes for survival in dynamic environments. These processes include highly adaptive behaviors, intricate communication networks, and collective decision-making processes. The study of such biological systems provides a foundation for decoding microbial intelligence into artificial systems (Lanka *et al.*, 2025). This section, therefore, explores the biological principles that underlie MAI with emphasis on microbial behavior, communication, and adaptation.

Microbes are single-celled organisms that exhibit prominent behavioral complexity. They exhibit carefully regulated processes that maximize survival and reproduction, as seen in chemotaxis, biofilm formation, and sporulation (Figure 1.2). During chemotaxis, the ability of microbes to move toward nutrients and away from toxins, the *Escherichia coli* detects chemical gradients using transmembrane receptors that trigger flagellar movement. This sensing and navigation mechanism reflects how ML algorithms use input data to make predictions and decisions (Zajdel *et al.*, 2018). Microbes continuously adjust their movement in search of optimal conditions just as algorithms adapt their pathways based on reward signals. Also in biofilm formation, microbes adhere to surfaces and secrete extracellular polymeric substances that create a protective matrix for the microbes (Penesyan *et al.*, 2021). They tend to organize into subgroups in the form of division of labor with specialized roles such as nutrient acquisition, structural maintenance, and defense against antibiotics within the biofilm (West & Cooper, 2016). This resembles a distributed computing, where multiple processors handle different tasks while contributing to the stability of the system. According to Freire *et al.* (2024), some microbes, when faced with starvation or stress, especially the *Bacillus subtilis*, undergo sporulation, producing highly resistant spores, which are not a random response but a calculated survival decision based on multiple environmental signals. Sporulation highlights a microbial form of conditional reasoning: "if resources are depleted, then initiate dormancy." In computational terms, it mirrors conditional branching in algorithms, ensuring resilience and long-term survival.

Another interesting means by which microbes act as intelligent systems is communication. Communication, a language of life, lies at the heart of microbial intelligence. Microbes exchange information to coordinate their behaviors, adapt collectively, and interact with hosts across different kingdoms of life. One of the most studied microbial communication systems is quorum sensing (QS). The QS is a population density-dependent element of bacterial composition (Hense & Schuster, 2015). The microbes measure their population density through the discharge of signaling molecules known as Autoinducers (Hense & Schuster, 2015). They display collective actions such as virulence, luminescence, or biofilm formation within a certain limit (Juszczuk-Kubiak, 2024). This act is similar to the decision-making threshold in AI systems, where a model implements a response after gathering sufficient evidence. Microbes also participate in horizontal gene transfer (HGT) through conjugation, transformation, and transduction processes, which permit them to successfully update their "knowledge base" while exchanging their genetic information, such as antibiotic resistance genes (Sakagianni *et al.*, 2024). The HGT resembles the distributed learning in AI, where information is shared between nodes for enhanced system performance. In addition, microbial communication often extends to plants, animals, and humans, as observed during the interaction between gut microbiota and the human immune system, which influences neurological pathways through chemical indicators (Fadhil *et al.*, 2025).

These cross-kingdom interactions between microbes and other living organisms (plants, animals, and humans) reflect the universal nature of communication capable of incorporating diverse and heterogeneous data sources.

Microbes gain insight from the environment as they adapt to rapid change, which is hurled by a high amount of mutation and short doubling time, enabling them to withstand selective pressures such as antibiotics or temperature fluctuations (Guo *et al.*, 2025). This evolutionary process resembles the genetic algorithms in AI, where results evolve over multiple recurrences toward optimal outcomes. Microbes have the capacity to change their form and function without modifying their genetic code. This is known as phenotypic plasticity. A typical example of such is *Candida albicans*, which changes between yeast and filamentous growth forms subject to its environment (Chen *et al.*, 2020). This flexibility is similar to the reconfiguration process based on incoming data in dynamic neural networks. The most perplexing aspects of microbial adaptation are the development of resistance mechanisms. For instance, bacteria yield enzymes that counteract antibiotics, thrust toxins out of their cells, or modify molecular targets while averting drug binding. These strategies represent forms of biological "learning" that improve the survival of these microbes. The MAI can model resilience by developing adaptive algorithms capable of fighting against emerging challenges, such as cybersecurity threats.

The behaviors, communication methods, and adaptive strategies of microbes bring about rich descriptions and outlines for AI development. In MAI, these biological concepts are abstracted into computational outlines, enabling algorithm designs that demonstrate adaptability, scalability, and robustness. For example, the idea of swarm intelligence algorithms, such as ant colony optimization and particle swarm optimization, was derived from these biological concepts (Selvi & Umarani, 2010). In the dimension of microbial models, it gives an opportunity to design algorithms that incorporate quorum-sensing-inspired communication or HGT-inspired information sharing (Beltrán-Velasco & Clemente-Suárez, 2025; Islam *et al.*, 2025). In environment monitoring, MAI could be used to develop flexible monitoring systems that respond vigorously to pollution changes, mimicking microbial flexibility. In addition, distributed sensor networks could integrate quorum-sensing principles to prompt collective responses when pollution limits are surpassed.

The study of microbial systems as foundations for MAI highlights the importance of interdisciplinary collaboration between microbiology, computer science, and systems engineering. Future research should explore how microbial intelligence can be formalized into computational models and tested in real-world applications, from environmental monitoring to health informatics. In conclusion, the emergence of MAI has caused a paradigm transformation in AI development. Rather than creating systems using only top-down logic, researchers can mimic the bottom-up, emergent intelligence of microbial collectives, which provides scalable, resilient, and sustainable solutions that are in line with the complexity of natural ecosystems.

1.10 CHALLENGES AND FUTURE GUIDELINES IN MICROBIAL ARTIFICIAL INTELLIGENCE (MAI)

The rapid incorporation of AI into the field of microbiology has resulted in the development of some fascinating new discoveries. Despite the fact that this integration is encountering certain difficulties due to the fact that MAI is currently constrained by some obstacles. A few examples of these issues are the problems that are associated with data, the computational opacity that is associated with biosafety concerns, and global disparities. Consequently, in order to guarantee an equal and appropriate deployment, every effort that leads to the future must not only emphasize the technical requirements but also specify the governance and ethics frameworks that will be necessary.

1.10.1 Data Scarcity, Bias, and Quality Concerns

In practice, datasets that are extensive, varied, and thoroughly annotated are generally required by AI-driven microbiological models. The world of microbes, on the other hand, is characterized by an extremely high variety, the majority of which are still subject to inadequate sampling. Microbiome datasets have been proven to be biased toward the human gut and a small number of model species, according to a number of studies. On the other hand, there is not enough research that has been conducted on the soil, marine, and harsh environments (Gilbert *et al.*, 2018; Pasolli *et al.*, 2019). Because of this imbalance, systematic biases are created. This makes it more difficult to generalize MAI models when they are applied to a variety of different scenarios. Other challenges that are associated with this include variances in sequencing platforms, annotations of metadata, and preparations for bioinformatics investigations. As a result of this, reproducibility becomes more difficult (Knight *et al.*, 2018). In addition, the lack of data and problems with its quality are significant obstacles to the implementation of MAI.

1.10.2 Complexity of Microbial Interactions

Microbial systems are characterized by their complexity as a result of the interaction of several components acting on a variety of scales. These components include a wide variety of genes, pathways, and communities. The AI systems of today are quite good at recognizing patterns. Yet, they frequently fail to recognize emergent characteristics and causal linkages. The usage of only deep learning modeling of microbiome–host interactions is extremely difficult due to the presence of feedback mechanisms that span the immunological, metabolic, and ecological systems (Noe & Clementi, 2017; Quince *et al.*, 2017). As a consequence of this, the combination of mechanistic models and ML has been proposed as the solution to the problem at hand. This is because it has the ability to generate predictions that are not only interpretable but also generalizable (Bikel *et al.*, 2015; Hasin *et al.*, 2017).

1.10.3 Algorithmic Opacity and Interpretability

Deep learning has been instrumental to recent advances in microbiome analyses and drug discovery, yet its "black box" quality remains a major limitation in high-stakes fields such as medicine. Decision makers across clinics, regulations, and environmental policymaking expect interpretability in addition to accuracy. AI-generated insights into microbiology lack explainability and therefore trust (Samek *et al.*, 2017; Topol, 2019). Hence, designing XAI systems focused on microbial systems to demonstrate the reasoning used in the predictions made and the microbial features or pathways involved remains a key objective.

1.10.4 Ethical, Biosafety, and Privacy Risks

Microbial engineering, which is led by AI, opens up new avenues of possibility. However, it also simultaneously elevates the probability of adverse outcomes. An example of this would be the possibility that designed microbial consortia could generate unintended HGT during environmental restoration. This could potentially destabilize ecosystems. Additionally, medical procedures involving the microbiome have the potential to generate immunological consequences that are not intended (Trump, 2017). Also, personal microbiome data may leak some important personal information, such as health issues, lifestyle choices, and geographic origin (Franzosa *et al.*, 2015). Considering the facts above, MAI requires ethical biosecurity frameworks. These may include, but are not limited to, the following: biosafety precautions, data privacy, and informed consent.

1.10.5 Global Inequities in Access and Capacity

Since MAI requires a huge amount of computational and experimental resources, there is a disparity between laboratories that are well-resourced in countries with high incomes and those that are in countries with high biodiversity but a lack of infrastructure. Apart from the scientific imbalances that may grow, these under-represented habitats have microbial diversity, which is required for the development of new therapeutics and the resilience of the environment (Curd, 2016). This is a particularly troubling fact. For this reason, future frameworks need to have an emphasis on fair access, international cooperation, and the construction of infrastructure in developing countries in order to make MAI research more accessible.

1.11 FUTURE GUIDELINES AND DIRECTIONS

When considering the future, some notable principles can be termed essential. Firstly, the interoperability of microbial datasets calls for the adoption of findable, accessible, interoperable, reusable (FAIR) data practices (Wilkinson *et al.*, 2016). Secondly, the creation of digital twins of microbial ecosystems can safely spawn in silico ecosystems for intervention testing prior to real-world implementation (Björnsson *et al.*, 2020). Moreover, the addition of more data layers in the multi-omic spectrum (genomics, transcriptomics, and metabolomics) will enhance predictive capabilities (Hasin *et al.*, 2017). Throughout the process of system development, the proactive incorporation of ethics-by-design will ensure that emerging systems are biosafety-compliant and that data privacy is properly protected. Lastly, the extension of MAI to space and extreme environment demonstrates the vastness of its potential, involving microbial life-support systems and bioproduction in outer space habitats (Menezes *et al.*, 2015).

1.12 CONCLUSION

The integration of AI into microbial research marks a transformative step in the study of complex biological systems. Traditional microbiological methods, while invaluable, are often limited in handling the scale, diversity, and dynamism of microbial data. AI techniques overcome these barriers through advanced tools for classification, prediction, visualization, and interpretation, thereby opening new pathways for discovery and application. From supervised and unsupervised learning approaches to deep learning, graph-based methods, and hybrid techniques, AI provides flexible frameworks capable of addressing the multifaceted nature of microbial systems. These approaches enable not only the extraction of meaningful insights from high-dimensional data but also the modeling of microbial interactions, the integration of multi-omics layers, and the design of adaptive interventions. Equally important are visualization and interpretability tools, which ensure that AI-generated knowledge remains transparent and biologically grounded. While challenges such as data sparsity, overfitting, and the need for robust biological validation remain, the future of MAI is promising. As computational and experimental methods continue to advance in parallel, MAI is poised to play a central role in medicine, environmental management, biotechnology, and beyond, ultimately bridging data-driven insights with real-world microbial applications.

REFERENCES

Aguiar-Pulido, V., Huang, W., Suarez-Ulloa, V., Cickovski, T., Mathee, K., & Narasimhan, G. (2016). Metagenomics, metatranscriptomics, and metabolomics approaches for microbiome analysis: Supplementary issue: Bioinformatics methods and applications for big metagenomics data. *Evolutionary Bioinformatics, 12*, EBO-S36436. https://doi.org/10.4137/EBO.S36436

Alsulimani, A., Akhter, N., Jameela, F., Ashgar, R. I., Jawed, A., Hassani, M. A., & Sajad, A. D. (2024). The impact of artificial intelligence on microbial diagnosis. *Microorganisms, 12*(6), 1051–1051. https://doi.org/10.3390/microorganisms12061051

Andreadi, A., Tsivelekidou, E., Dermitzakis, I., Theotokis, P., Gargani, S., Meditskou, S., & Manthou, M. E. (2025). Innovations in MALDI-TOF mass spectrometry: Bridging modern diagnostics and historical insights. *PubMed, 20*(1), 20251136–20251136. https://doi.org/10.1515/biol-2025-1136

Aransiola, S. A., Bhatnagar, S., & Maddela, N. R. (Eds.). (2025). *Marine Bioresources: Prospects and Obstacles*. Elsevier. https://www.sciencedirect.com/book/9780443363207/marine-bioresources

Barh, D. (2020). *Artificial Intelligence in Precision Health: From Concept to Applications*. Academic Press.

Beltrán-Velasco, A. I., & Clemente-Suárez, V. J. (2025). Harnessing gut microbiota for biomimetic innovations in health and biotechnology. *Biomimetics, 10*(2), 73.

Bikel, S., Valdez-Lara, A., Cornejo-Granados, F., Rico, K., Canizales-Quinteros, S., Soberón, X., Del Pozo-Yauner, L., & Ochoa-Leyva, A. (2015). Combining metagenomics, metatranscriptomics, and viromics to explore novel microbial interactions: Towards a systems-level understanding of the human microbiome. *Computational and Structural Biotechnology Journal, 13*, 390–401.

Björnsson, B., Borrebaeck, C., Elander, N., Gasslander, T., Gawel, D. R., & Gustafsson, M. et al (2020). Digital twins to personalize medicine. *Genome Medicine, 12*(1), 4.

Chen, H., Zhou, X., Ren, B., & Cheng, L. (2020). The regulation of hyphae growth in *Candida albicans*. *Virulence, 11*(1), 337–348.

Curd, E. E. (2016). Microbial community diversity, function, and succession in California's Mediterranean habitats. (Doctoral dissertation, University of California, Los Angeles). eScholarship.

D'Urso, F., & Broccolo, F. (2024). Applications of artificial intelligence in microbiome analysis and probiotic interventions—An overview and perspective based on the current state of the art. *Applied Sciences, 14*(19), 8627.

Dhaarani, R., & Reddy, M. K. (2025). Progressing microbial genomics: Artificial intelligence and deep learning driven advances in genome analysis and therapeutics. *Intelligence-Based Medicine, 11*, 100251.

Fadhil, S. A., Radif, M., & Alrammahi, A. H. (2025). Modeling neuroelectrical-microbiome crosstalk: AI-driven insights into gut-brain bioelectrical signaling. *Journal of Applied Data Sciences, 6*(3), 1953–1967.

Faust, K., & Raes, J. (2012). Microbial interactions: From networks to models. *Nature Reviews Microbiology, 10*(8), 538–550. https://doi.org/10.1038/nrmicro2832

Franzosa, E. A., Huang, K., Meadow, J. F., Gevers, D., Lemon, K. P., Bohannan, B. J. M., & Huttenhower, C. (2015). Identifying personal microbiomes using metagenomic codes. *Proceedings of the National Academy of Sciences, 112*(22), E2930–E2938.

Freire, V., Condón, S., & Gayán, E. (2024). Impact of sporulation temperature on germination of bacillus subtilis spores under optimal and adverse environmental conditions. *Food Research International, 182*, 114064.

Gao, Y., & Liu, M. (2024). Application of machine learning based genome sequence analysis in pathogen identification. *Frontiers in Microbiology, 15*, 1474078. https://doi.org/10.3389/fmicb.2024.1474078

Gilbert, J. A., Jansson, J. K., & Knight, R. (2018). The earth microbiome project: Successes and aspirations. *BMC Biology, 12*, 69.

Gong, X., Zhang, J., Gan, Q., Ruan, W., Liu, T., & Yan, Y. (2026). Leveraging artificial intelligence for efficient microbial production. *Bioresource Technology*, *439*, 133286.

Guo, Y. W., Liu, Y., Huang, P. C., Rong, M., Wei, W., Xu, Y. H., & Wei, J. H. (2025). Adaptive changes and genetic mechanisms in organisms under controlled conditions: A review. *International Journal of Molecular Sciences*, *26*(5), 2130.

Hasin, Y., Seldin, M., & Lusis, A. (2017). Multi-omics approaches to disease. *Genome Biology*, *18*(1), 83.

Hawkins, D. M. (2004). The problem of overfitting. *Journal of Chemical Information and Computer Sciences*, *44*(1), 1–12. https://doi.org/10.1021/ci0342472

Hense, B. A., & Schuster, M. (2015). Core principles of bacterial autoinducer systems. *Microbiology and Molecular Biology Reviews*, *79*(1), 153–169.

Holzinger, A., Biemann, C., Pattichis, C. S., & Kell, D. B. (2017). What do we need to build explainable AI systems for the medical domain? arXiv preprint arXiv:1712.09923.

Huo, D., & Wang, X. (2024). A new era in healthcare: The integration of artificial intelligence and microbial. *Medicine in Novel Technology and Devices*, *23*, 100319.

Islam, M. R., Summers, A. O., & Arpinar, I. B. (2025). *Knowledge Graph-Based Framework for Detecting Horizontal Gene Transfer Events Driving Antimicrobial Resistance. bioRxiv*, 2025-06. https://doi.org/10.1101/2025.06.09.658534

Jolliffe, I. T., & Cadima, J. (2016). Principal component analysis: A review and recent developments. *Philosophical Transactions of the Royal Society A: Mathematical, Physical and Engineering Sciences*, *374*(2065), 20150202. https://doi.org/10.1098/rsta.2015.0202

Juszczuk-Kubiak, E. (2024). Molecular aspects of the functioning of pathogenic bacteria biofilm based on quorum sensing (QS) signal-response system and innovative non-antibiotic strategies for their elimination. *International Journal of Molecular Sciences*, *25*(5), 2655.

Knight, R., Vrbanac, A., Taylor, B. C., Aksenov, A., Callewaert, C., Debelius, J., & Dorrestein, P. C. (2018). Best practices for analysing microbiomes. *Nature Reviews Microbiology*, *16*(7), 410–422. https://doi.org/10.1038/s41579-018-0029-9

Lanka, S., Mounika, K., Anitha, K., & Pandrangi, S. L. (2025). Computational and Systems Biology Approaches in Biofilm Research. In *Omics Approaches in Biofilm Research: Perspectives and Applications* (pp. 391–405). Cham: Springer Nature Switzerland.

Li, D., & Wu, M. (2021). Pattern recognition receptors in health and diseases. *Signal Transduction and Targeted Therapy*, *6*(1), 1–24. https://doi.org/10.1038/s41392-021-00687-0

Lin, T. H., Chung, H. Y., Jian, M. J., Chang, C. K., Lin, H. H., Yu, C. M., Perng, C. L., Chang, F. Y., Chen, C. W., Chiu, C. H., & Shang, H. S. (2024). Artificial intelligence-clinical decision support system for enhanced infectious disease management: Accelerating ceftazidime-avibactam resistance detection in klebsiella pneumoniae. *Journal of Infection and Public Health*, *17*(10), 102541. https://doi.org/10.1016/j.jiph.2024.102541

Liu, Z., Lei, J., Yang, R., Cheng, L., Du, Y., Zhang, Y., Wang, J., & Liu, Y. (2025). An artificial intelligence modeling framework based on microbial community structure prediction enhances the pollutant removal efficiency of the algae-bacteria granular sludge system. *Journal of Environmental Management*, *392*, 126648.

Ma, L., Yi, J., Wisuthiphaet, N., Earles, M., & Nitin, N. (2023). Accelerating the detection of bacteria in food using artificial intelligence and optical imaging. *Applied and Environmental Microbiology*, *89*(4), e01828-22.

Menezes, A. A., Montague, M. G., Cumbers, J., Hogan, J. A., & Arkin, A. P. (2015). Grand challenges in space synthetic biology. *Journal of the Royal Society Interface*, *12*(113), 20150803.

Min, S., Lee, B., & Yoon, S. (2017). Deep learning in bioinformatics. *Briefings in Bioinformatics*, *18*(5), 851–869. https://doi.org/10.1093/bib/bbw068

Misra, B. B., Langefeld, C., Olivier, M., & Cox, L. A. (2019). Integrated omics: Tools, advances and future approaches. *Journal of Molecular Endocrinology*, *62*(1), R21–R45. https://doi.org/10.1530/JME-18-0055

Mohseni, P., & Ghorbani, A. (2024). Exploring the synergy of artificial intelligence in microbiology: Advancements, challenges, and future prospects. *Computational and Structural Biotechnology Reports*, *1*, 100005–100005. https://doi.org/10.1016/j.csbr.2024.100005

Mulat, M., Banicod, R. J. S., Tabassum, N., Javaid, A., Kim, T.-H., Kim, Y.-M., & Khan, F. (2025). Application of artificial intelligence in microbial drug discovery: Unlocking new frontiers in biotechnology. *Journal of Microbiological Methods*, *237*, 107232.

Noe, F., & Clementi, C. (2017). Collective variables for the study of long-time kinetics from molecular trajectories: Theory and methods. *Current Opinion in Structural Biology*, *43*, 141–147.

Noecker, C., Eng, A., Srinivasan, S., Theriot, C. M., Young, V. B., Jansson, J. K., & Borenstein, E. (2016). Metabolic model-based integration of microbiome taxonomic and metabolomic profiles elucidates mechanistic links between ecological and metabolic variation. *MSystems*, *1*(1), 10–1128. https://doi.org/10.1128/msystems.00013-15

Pasolli, E., Asnicar, F., Manara, S., Zolfo, M., Karcher, N., & Armanini, F. et al (2019). Extensive unexplored human microbiome diversity revealed by over 150,000 genomes from metagenomes. *Cell*, *176*(3), 649–662.e20.

Penesyan, A., Paulsen, I. T., Kjelleberg, S., & Gillings, M. R. (2021). Three faces of biofilms: A microbial lifestyle, A nascent multicellular organism, and an incubator for diversity. *NPJ Biofilms and Microbiomes*, *7*(1), 80.

Pillay, S., Calderón-Franco, D., Urhan, A., & Abeel, T. (2022). Metagenomic-based surveillance systems for antibiotic resistance in non-clinical settings. *Frontiers in Microbiology*, *13*, 1066995. https://doi.org/10.3389/fmicb.2022.1066995

Przymus, P., Rykaczewski, K., Martín-Segura, A., Truu, J., Carrillo De Santa Pau, E., Kolev, M., Naskinova, I., Gruca, A., Sampri, A., Frohme, M., & Nechyporenko, A. (2025). Deep learning in microbiome analysis: A comprehensive review of neural network models. *Frontiers in Microbiology*, *15*, 1516667.

Quince, C., Walker, A. W., Simpson, J. T., Loman, N. J., & Segata, N. (2017). Shotgun metagenomics, from sampling to analysis. *Nature Biotechnology*, *35*(9), 833–844.

Sakagianni, A., Koufopoulou, C., Koufopoulos, P., Kalantzi, S., Theodorakis, N., Nikolaou, M., & Feretzakis, G. (2024). Data-driven approaches in antimicrobial resistance: Machine learning solutions. *Antibiotics*, *13*(11), 1052.

Samek, W., Wiegand, T., & Müller, K.-R. (2017). Explainable artificial intelligence: Understanding, visualizing and interpreting deep learning models. arXiv preprint arXiv:1708.08296.

Sarrafzadeh, M.-H., Mansouri, S. S., Zahiri, J., Mussatto, S. I., & Asgharnejad, H. (2022). Editorial: Artificial intelligence in environmental microbiology. *Frontiers in Microbiology*, *13*, 944242.

Selvi, V., & Umarani, R. (2010). Comparative analysis of ant colony and particle swarm optimization techniques. *International Journal of Computer Applications*, *5*(4), 1–6.

Tenenbaum, J. B., Silva, V. D., & Langford, J. C. (2000). A global geometric framework for nonlinear dimensionality reduction. *Science*, *290*(5500), 2319–2323. https://doi.org/10.1126/science.290.5500.2319

Tjoa, E., & Guan, C. (2020). A survey on explainable artificial intelligence (XAI): Toward medical XAI. *IEEE Transactions on Neural Networks and Learning Systems*, *32*(11), 4793–4813.

Topçuoğlu, B. D., Lesniak, N. A., Ruffin, M. T. IV, Wiens, J., & Schloss, P. D. (2020). A framework for effective application of machine learning to microbiome-based classification problems. *MBio*, *11*(3), 10–1128. https://doi.org/10.1128/mbio.00434-20

Topol, E. J. (2019). High-performance medicine: The convergence of human and artificial intelligence. *Nature Medicine*, *25*(1), 44–56.

Trump, B. D. (2017). Synthetic biology regulation and governance: Lessons from TAPIC for the United States, European Union, and Singapore. *Health Policy*, *121*(11), 1139–1146.

Tsitou, V. M., Rallis, D., Tsekova, M., & Yanev, N. (2024). Microbiology in the era of artificial intelligence: Transforming medical and pharmaceutical microbiology. *Biotechnology & Biotechnological Equipment*, *38*(1). https://doi.org/10.1080/13102818.2024.2349587

Vamathevan, J., Clark, D., Czodrowski, P., Dunham, I., Ferran, E., Lee, G., Li, B., Madabhushi, A., Shah, P., Spitzer, M., & Zhao, S. (2019). Applications of machine learning in drug discovery and development. *Nature Reviews Drug Discovery*, *18*(6), 463–477. https://doi.org/10.1038/s41573-019-0024-5

Vunduk, J., Klaus, A., Lazić, V., Kozarski, M., Radić, D., Šovljanski, O., & Pezo, L. (2023). Artificial neural network prediction of antiadhesion and antibiofilm-forming effects of antimicrobial active mushroom extracts on food-borne pathogens. *Antibiotics*, *12*(3), 627–627. https://doi.org/10.3390/antibiotics12030627

Wang, D., Sklar, B., Tian, J., Gabriel, R., Engelhard, M., McNabb, R. P., & Kuo, A. N. (2025). Improving artificial intelligence-based microbial keratitis screening tools constrained by limited data using synthetic generation of slit-lamp photos. *Ophthalmology Science*, *5*(3), 100676.

Wang, X., Wang, T., & Liu, Y. (2024). Artificial intelligence for microbiology and microbiome research. arXiv preprint arXiv:2411.01098.

Wang, Y., Tang, F., Liu, B., Wu, Y., Zhang, R., & Ren, H. (2025). A bioinspired microbial taste chip with artificial intelligence-enabled high selectivity and ultra-short response time. *Biosensors and Bioelectronics*, *277*, 117264.

Wang, Z., Gerstein, M., & Snyder, M. (2009). RNA-Seq: A revolutionary tool for transcriptomics. *Nature Reviews Genetics*, *10*(1), 57–63. https://doi.org/10.1038/nrg2484

West, S. A., & Cooper, G. A. (2016). Division of labour in microorganisms: An evolutionary perspective. *Nature Reviews Microbiology*, *14*(11), 716–723.

Wilkinson, M. D., Dumontier, M., Aalbersberg, I. J., Appleton, G., Axton, M., & Baak, A. et al (2016). The FAIR guiding principles for scientific data management and stewardship. *Scientific Data*, *3*, 160018.

Yuan, K., & Nault, R. (2025). Application of a metabolic network-based graph neural network for the identification of toxicant-induced perturbations. *Toxicological Sciences*, *206*(1), 19–29. https://doi.org/10.1093/toxsci/kfaf065

Zajdel, T. J., Nam, A., Yuan, J., Shirsat, V. R., Rad, B., & Maharbiz, M. M. (2018, July). Applying machine learning to the flagellar motor for biosensing. In *2018 40th Annual International Conference of the IEEE Engineering in Medicine and Biology Society (EMBC)* (pp. 1–4). IEEE.

Zhang, Y. Z., Liu, Y., Bai, Z., Fujimoto, K., Uematsu, S., & Imoto, S. (2023). Zero-shot-capable identification of phage–host relationships with whole-genome sequence representation by contrastive learning. *Briefings in Bioinformatics*, *24*(5), bbad239.

Zou, J., Huss, M., Abid, A., Mohammadi, P., Torkamani, A., & Telenti, A. (2019). A primer on deep learning in genomics. *Nature Genetics*, *51*(1), 12–18. https://doi.org/10.1038/s41588-018-0295-5

Fundamentals and Principles of Artificial Intelligence in Microbiology

2

Samson Olusola Adeboye, Samuel Adeniyi Oyegbade, Victoria-Grace Onyekachi Aririguzoh, Joy Oziohu James, Ayomide Iyanuoluwa Olaniyan, Oluwatoyin Comfort Owoyemi, and Oluwafemi Adebayo Oyewole

2.1 INTRODUCTION

The field of microbiology has advanced from merely identifying pathogenic microbes to a deeper system-level comprehension of highly complicated microbial communities and host–environment interactions. Microbiota and microbiome concepts are core to the advancement in this domain and offer a framework for comprehending how biological functions are shaped by microbes in humans, animals, and ecosystems. Microbiota encompasses assembling all microbes dwelling in a definite environment within a defined time period. On the other hand, the microbiome entails the collection of gene products and genomes of these microbes, looking into the dynamics surrounding the interactions and functional potential (Sorbara and Pamer, 2022). The microbiome supports maturation of the immune system, pathogen protection, and physiological homeostatic maintenance. These microbial communities are active inhabitants taking core roles in development, host physiology, disease susceptibility, and immunity in animals and humans. The host–microbiota relationship is formed via metabolic incorporation, long-term co-evolution, co-regulation, and co-development, which influence disease resilience and health consequences. Dysbiosis is imbalance in interactions connected to a vast array of conditions, such as inflammatory conditions, infectious diseases, and metabolic syndromes. Microbiomes' importance goes beyond human health. Several microbes are spread between the environment, wildlife, animals, and humans, establishing biological links between ecological health, veterinary, and clinical human medicine.

DOI: 10.1201/9781003727514-3

Diverse commensals and bacterial pathogens can colonize diverse host species, indicating the interlinked nature of health maintenance and disease transmission across boundaries of species (Gilbert et al., 2018). Consequently, changes in human-linked microbiomes can have cascading influences on animals, especially in environments where there is close coexistence, as observed in urban ecosystems and agricultural systems. Environmental health is largely affected by microbial communities. Atmospheric, water, and soil microbiomes support the stability of the ecosystem and regulate climate processes. Climate change, pollution, antimicrobial contamination, and land-use changes can substantially reshape microbial community structures and functions, with far-reaching consequences for both host health and ecosystems (Zhu et al., 2023).

These observations enhance the comprehension that host health, microbial health, and environmental health are jointly connected. This interrelatedness is common.

In the twentieth century, Karl Friedrich Meyer expanded on the "One Health" concept, pointing to the dependence of the health of the ecosystem, animals, and humans and paving the way for networking across veterinary, medical, and environmental sciences (Pospischil, 2015). The One Health concept gained substantial relevance as environmental degradation poses global concerns. Computational biology's and AI's sporadic growth in recent years is transforming microbiological research and providing potential tools for analysis of enormous microbiome datasets, modeling interactions between microbe-host-environment interactions and prediction of disease outcomes (Dakal et al., 2025). Incorporating AI-propelled strategies into microbiology demonstrates potential for advancement of surveillance, interventions, and diagnostics within the One Health Framework (Sahoo et al., 2025)

This chapter aims to investigate the core principles of microbiome science and microbiota within the purview of modern microbiology while pointing out how AI can be applied to better incorporate and manage the complicated interactions that exist between the environment, host, and microbes under the One Health Framework.

2.2 RISING ROLE OF AI IN LIFE SCIENCES

Scientific developments that artificial intelligence (AI) applications have recently become an essential focus of modern innovation facilitate innovation throughout the biohierarchies, including molecular and ecological (Xu et al., 2021). It is critical to analyze sequencing data in large volumes generated (Jin et al., 2024). Through working with extremely large sets of biological data, techniques such as deep learning (DL), machine learning (ML), and data mining can help scientists learn more about the complexity of life and its interaction with its environment. This technological revolution is particularly evident in the biomedical industry. AI makes great contributions to disease prevention, individual therapy, and new medication development, besides being more accurate and effective in the diagnosis of diseases (Huang et al., 2024). The new outlook and approach that AI brings to treatment plans will enable future healthcare to develop approaches that are more personalized and precise. For example, AI can analyze a large amount of genetic information in genomics to find mutations that cause disease and potential targets of treatment (Cheng et al., 2023). AI has the potential to accelerate drug discovery by predicting the potential efficacy and side effects of potential drugs. Moreover, AI will be able to enhance our understanding of complex biological processes (Keyl et al., 2023).

2.3 BASICS OF ARTIFICIAL INTELLIGENCE

The first field to introduce AI was called by John McCarthy during the Dartmouth Conference in the summer of 1956. According to McCarthy (1990), AI was concerned with theoretical methods and techniques, applied systems, and simulations of human intelligence or its extensions. The advent of AI has also served to propel the microbiology field and introduce a new line of thinking throughout the field (Barredo Arrieta et al., 2019). Microbiology has expanded to encompass vast scales and dimensions and is currently applied to such fields as systems biomedicine and systems ecology. The latter is made possible with the help of big data, automation, modeling, and AI. ML is a special field of AI that was introduced by a man named Arthur Samuel, who worked at Bell Labs, IBM, and Stanford in 1959. ML seeks to identify trends in big and diverse data. The easiest method of doing so is to process the data using algorithms to identify trends and use the trends to predict or make decisions regarding actual events (Jordan and Mitchell, 2015). ML is taught a lot of data to train an algorithm to perform some tasks, unlike conventional software, which is programmed to perform certain tasks (Domingos, 2012). ML can reveal the behavior patterns of microorganisms about one another or their hosts by integrating various data types and exploring complex microorganism groups across various scales. It is also used in realizing the roles of the microorganisms and increasing the utilization of microbial technology. It is achieved through preprocessing of large data (to cut down the data) and finding meaningful information in big data. DL is a recent form of ML, which operates on a deep network comprising multiple processing units. These units are optimally trained on the basis of models acquired during the gradient descent, which are useful in interpreting patterns of data on a high level (LeCun et al., 2015). ML may be applied in AI, and DL facilitates this. A field that involves the integration of computer science and computational linguistics is natural language processing (NLP). It aims at converting speech and written communication into data that are analyzable. It is possible to interpret the meaning of a document using NLP or even make a response that can sound like a human being is speaking. It is accomplished with the integration of the approaches of linguistics, statistics, and AI.

2.4 OVERVIEW OF ALGORITHMS RELEVANT TO BIOLOGICAL DATA

Random Forest is an ensemble learning method that creates many decision trees during training and generates the class mode (classification) or mean prediction (regression) of each tree independently. It integrates multiple decision trees to improve classification accuracy and durability, particularly for high-dimensional data. Support Vector Machine (SVM) is a supervised ML method that may be used for both classification and regression tasks. The mechanism of operation involves finding the hyperplane that performs the best in dividing a dataset into classes. SVM is effective in high-dimensional spaces and beneficial when working with small to medium-sized, yet complicated datasets. The artificial neural network (ANN) model was modeled after the structure and function of the human brain; in a similar way that the human brain has a large network of interconnected neurons, ANNs demonstrate clusters of interconnected nodes. ANNs are capable of learning and mimicking complex relationships between inputs and outputs, even if the real relationship is nonlinear (Figure 2.1).

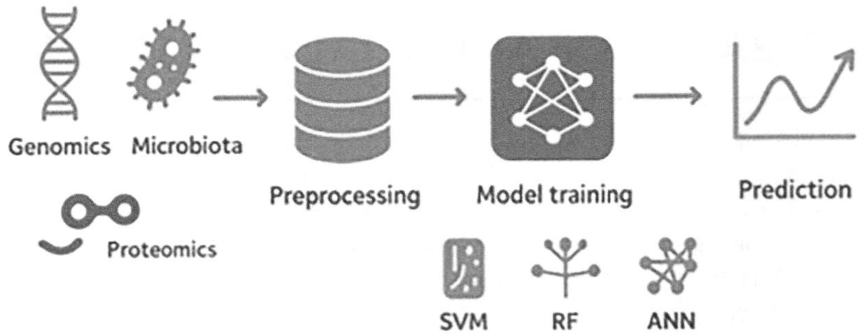

FIGURE 2.1 General workflow of a machine learning pipeline in microbiological research, showing stages from multi-omics data acquisition and preprocessing to model training and predictive analysis.

2.5 DISTINCTION BETWEEN SUPERVISED, UNSUPERVISED, AND REINFORCEMENT LEARNING

Supervised learning involves the use of labeled datasets to develop algorithms. In microbiology, this method is useful in a goal such as predicting microorganisms' behavior or disease progression. Adaptive boosting classifiers are made to identify antibiotic resistance among various bacterial species (Davis et al., 2016). Unsupervised learning is the ability to find latent patterns in complex microbiological data with the use of unlabeled data analysis. The interaction-based learning type is the essential element of reinforcement learning (RL), a special type of ML algorithm (Kaelbling et al., 1996; Keerthi and Ravindran, 1994). To reap the expected rewards, we constantly perform the interaction-based activity in an attempt to understand whether it is linked to the goal and produce rewards or punishment. This is referred to as automatic scoring and escalation.

2.6 DIGITAL DATA IN MICROBIOLOGY

Some of the omics approaches commonly used in multi-omics research include the following technologies. The field of omics, known as genomics, examines all of the genes (the genome) that make up an individual's DNA, along with their variations and how those genes interact with one another in a given environment (Kesheri et al., 2016). Proteomic analysis is the study of all the proteins that a cell or tissue produces at a specific moment in time. This analysis can help design specific medications or vaccines to prevent cervical cancer by revealing information about posttranslational modifications, protein expression levels, and protein-protein interactions (Srivastava and Singh, 2013). The methodical study of all chemical reactions involving metabolites is known as metabolomics. The study of all the little molecules (metabolites) that a cell or tissue produces at a specific moment can reveal information about therapeutic targets, disease biomarkers, and metabolic pathways (Galande et al., 2014).

Phenotypes are observable characteristics of cells. The term "phenotype" has become more and more popular over the last few years and now refers to any cell feature, such as the quantity of mRNA from a certain gene that is present. In clinical microbiology, the primary results are usually images. The images are then analyzed and interpreted. Such images include those of slides, which are observed through a microscope, the appearance of microbes on an agar plate rich in nutrients, and their reactions in susceptibility tests with various antibiotics. Although these images are primarily examined by human observers and subjected to manual analysis, there is a progressive inclination to incorporate digital tools in laboratories. The said transition facilitates the application of computer software for the analysis and interpretation of these images.

Data-driven approaches are now integral to the realm of biomedical research. The way in which researchers manage and disseminate data has a direct impact on the quality of their research and the pace at which new concepts can be formed (Leonelli, 2012). Since biomedical data is usually very intricate and delicate, it requires professional management with extreme care. Unfortunately, the lack of proper training among researchers is a barrier to their full utilization of the data as well as to AI's assistance. Thus, there is an urgent need to develop a comprehensive framework that imparts data skills in the full research cycle. This framework should help prepare data for use with AI and ML. AI-ready data means the data is ready to be used in a way that is both effective and ethical, covering things like training models, classifying data, and making predictions that are clear and responsible.

2.7 AI IN MICROBIAL DIAGNOSTICS

Older ways of detecting microorganisms and their characteristics are commonly associated with great challenges, such as slow and laborious culture procedures, challenges in handling samples, and inefficiently measuring antimicrobial susceptibility (Ahmad et al., 2024). AI is a pioneering technology that is changing the process of identifying and characterizing pathogens (Chaturvedi et al., 2024). AI is providing quicker, more precise, and more accessible diagnostic solutions through automation and enhancement of the traditional laboratory processes, and it is proving to save time needed to make a diagnosis (Ashraf et al., 2025; Naugler and Church, 2019).

The main basis of the picture-based diagnostics in modern times is computer vision, a sub-branch of AI that allows the machine to make sense of the image (Akhtar, 2025; Hammad and Ahmad, 2025; Kalita et al., 2024). Using deep-learning networks, especially convolutional neural networks (CNNs), AI models can analyze visual data of different types, such as stained slides, culture plates, and microscope images, to automate tasks that had previously depended on the manual skills of a microbiologist (Krichen, 2023). This automation is much faster and ensures objectivity.

The counting of colonies, which is a classical method of measuring the growth of microorganisms, is a complicated and tedious procedure, particularly in cases of overlapping colonies or volumes of plates (Hossainpour and Mahmoudi, 2024). AI-based systems employ computer vision and instance segmentation to efficiently outline and count individual colonies even in dense plates. This causes quicker and more predictable outcomes among batches (Ahmad et al., 2023). The interpretation of the microscopy slides is another area of AI revolutionization (Bhattiprolu, 2025). The traditional analysis needs high-skilled professionals and is highly inter-operator variable. AI offers a standardized real-time Gram-stained sample analysis, thus making the analysis less subjective and more accurate in its diagnoses (Ali et al., 2025). A tuberculosis smear microscopy system that is a combination of T-SPOT and computed tomography (CT) analysis and powered by a DL-based system (Ying et al., 2022). The system takes several minutes per slide to seconds by using a deep neural network classifier to analyze slides and identify the slide type, which is much faster and more accurate in diagnosis than traditional techniques (Desruisseaux et al., 2024). The use of AI in image processing can be measured by the speed improvement that it offers as quantifiable and by the data reliability that is offered by the AI as qualitative.

New AI-enabled biosensors that can be used to conduct the diagnostics of resources with limited capabilities have also been developed (Kovur et al., 2025). AI has been a paradigm shift to point-of-care

(PoC) diagnostics (Daramola et al., 2021; Hagos et al., 2025; Olatunji et al., 2024). The intelligent brain is the AI that transforms information on the PoC sensors into sophisticated, actionable diagnostic information. The two in combination facilitate a fresh category of diagnostic knowledge that is mobile, cheap, and scalable (Mardani and Nasiri, 2025). Light AI is a firm that is transforming smartphones into portable diagnostic labs by implementing AI algorithms to analyze pictures and videos (Sampson, 2025). The imaging devices (SIDs) using smartphones are broadly applicable and have a variety of biomedical uses (Banik et al., 2021). This new solution is able to address the drawbacks of bare-bones PoC devices, allowing testing to be conducted using a ubiquitous consumer product, fulfilling the requirements of quick and convenient testing (Wang et al., 2023).

2.8 AI IN INDUSTRIAL AND ENVIRONMENTAL MICROBIOLOGY

Valuable products are produced by the use of living organisms, which is known as bioprocessing, and it is an activity that is influenced by many factors such as cell lines, media compositions, and process conditions (Gopalakrishnan et al., 2024). Conventionally, this has been optimized by trial and error, which is time-consuming and is an inefficient method. This is getting enhanced by AI and ML, which analyze datasets of bioprocesses to identify patterns and foresee the most advantageous conditions of operation in silico (Richter et al., 2025).

The RL algorithms are being actively applied to tuning the parameters of the bioreactor dynamically during fermentation in real-time (Rajasekhar et al., 2024). It will be able to optimize variables such as temperature, pH, and agitation by monitoring the process and learning new knowledge to improve metabolite yield and process failure (Augustin et al., 2024). The implementation of real-time adjusting RL algorithms has contributed to the reduction of fermentation cycles. This has led to bioreactor failure by 60 percent and 300 percent yields increase in certain applications, which have achieved huge efficiency and productivity improvements (Priyadharshini et al., 2025). There are continuous learning smart fermentation platforms, which enhance food ingredient production with minimal human intervention.

Food safety apps can use AI in the detection of pathogens (Onyeaka et al., 2024). The same AI-based vision systems have been used in industrial quality control, where they check surfaces of foods or the indicators of spoilage on conveyor belts in real-time and raise alarms about contamination or defects without having to be inspected by hand (Chhetri, 2024). The transformative effect of AI is also applied to the food supply chain, making food safety farm to fork. The AI-based algorithm models, along with machine vision and sensor networks, can effectively detect spoilage of foods, pesticides, and microbial pollution (Li et al., 2022; Yin et al., 2025). Moreover, it can be an end-to-end traceability system, which is a secure and transparent means to track food quality and safety throughout the entire supply chain because AI has been integrated (Raki et al., 2024).

2.9 AI IN CLINICAL AND PUBLIC HEALTH MICROBIOLOGY

Typically, disease monitoring will be a very reactive and frequently imperfect process that will always be behind when it comes to the fast-spreading modern pathogens (Gardy and Loman, 2018). AI has been predicted to be a game-changing tool in public health monitoring because it processes

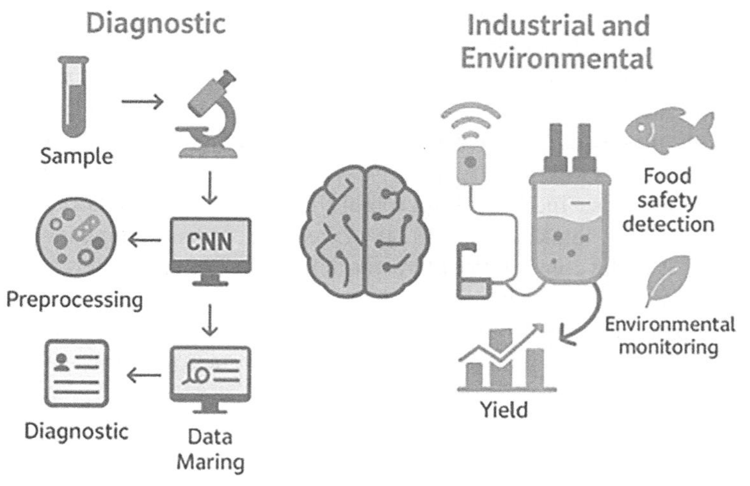

FIGURE 2.2 Applications of artificial intelligence in diagnostic, industrial, and environmental microbiology. AI-driven tools support automated microbial detection, image-based diagnostics, process optimization in bioreactors, food safety monitoring, and environmental quality assessment.

different kinds of both structured and unstructured data simultaneously. The government of the United States has been an early adopter in this regard, using AI to analyze and process vast amounts of non-traditional data, for example, aerial photographs, social networks, and electronic health records, to increase the speed and accuracy of its surveillance models (Figure 2.2) (CDC, 2025). The implementation of AI-based warning systems, such as BlueDot and EPIWATCH, has proven that they can spot possible epidemics sooner than the official declaration. These platforms were important in giving early warning signs of further analysis by scanning their reports of the population, news reports, and social media. Hybrid models, which are faster and scalable than AI but rely on conventional epidemiological frameworks, are the most effective systems that can produce a robust and adaptable system (Gawande et al., 2025). An example of this kind of approach is the Pandemic LLM framework, in which disease prediction is reformulated as a text-based reasoning problem (Coulson, 2025; Ruga et al., 2024). This enables the model to conglomerate non-numerical, real-time, and complex information (textual health policies as well as genomic surveillance information) and predict more powerfully compared to traditional models. This change in the data that is regarded as pertinent to surveillance allows transitioning toward a reactive response of the populace to proactive preparedness. AI is also used to monitor the disease and predict an outbreak by evaluating the spatiotemporal data and clinical trends to identify the way the infectious disease will go (Salam, 2024; Zeng et al., 2021). Such ability is essential to the officials in charge of the health sector, as it allows them to predict future trends and use resources more efficiently in case of an emerging health crisis (Erwin and Brownson, 2017; Morr and Ali-Hassan, 2019). This changes the mental health of the populace health and clinical practice of posing the question of "What is happening now?" to "What is the next thing that is likely to happen?", completely transforming the fight against the health crises of the new century (Coffman and Sandhu, 2018; Siddiqui et al., 2025). The concept of transitioning from reactivity to proactiveness and preventive actions is one of the major goals of AI use in clinical microbiology (Babanejaddehaki et al., 2025). One of the main areas that predictive analytics are making a significant impact in is the growing worldwide menace of antimicrobial resistance (AMR). It is possible to train ML models based on historical trends in laboratory data and patient clinical data to predict the emergence and diffusion of AMR trends (Ali et al., 2023; Ji et al., 2021). This prediction ability enables the clinicians to make proactive therapy in the prescription of the most effective antibiotics.

AI has also made a substantial impact in the personalized medicine sector. The human microbiome is a very intricate system that plays a major role in the overall health of the person, and it affects different body processes like digestion, immunity, metabolism, and even brain functioning (Gomaa, 2020; Ogunrinola et al., 2020). This is where the technology has the potential to make custom diets and therapies per individual microbial fingerprint, thus the collaboration of the three experts (Saxena et al., 2024; Shukla et al., 2024). AI technology is taking us to the time of personalized medicine, and it is the computational power behind it. The AI models have even shown that they are capable not only to analyze but also to discover hidden patterns in the large biological data, for instance, the sequencing of microbiomes, genetic tests, metabolomics, and real-time data from wearable devices, and eventually predict the exact person's reaction to the specific type of food or therapy (Dakal et al., 2025).

2.10 ETHICAL AND REGULATORY CHALLENGES OF AI IN MICROBIOLOGY

In addition to the diagnostics, AI-related tools have become useful in synthetic biology and gene editing. One of them is that AI can suggest new gene circuits or metabolic pathways that can facilitate the engineering of microorganisms. Even though all these can be used in therapies and industrial microbiology, they also have a grave ethical issue. The potential dangers are unintended environmental effects, bioterrorism abuse, and unintended discharges of harmful engineered organisms. Such opportunities require serious control and ethical deliberation of AI usage in microbial sciences (Mohseni and Ghorbani, 2023). In addition to the dangers of the physical abuse of engineered organisms, the ethical consideration also relates to the nature of the management and protection of digital biological-related information. Microbiological data applied to AI frequently include sensitive patient data, including genomic sequences, histories of diagnosis, and antibiotic resistance. The mismanagement of such data poses a serious risk, such as cyberattacks, data leakages, and infringement of patient privacy. The ethical issue of safe data handling and secrecy regulations is, therefore, a pressing ethical issue in AI-operated microbiology (Mohseni and Ghorbani, 2023). Yet patient privacy is not the sole issue in this question, the second issue is just as significant as the first one, the reasonability of AI models themselves (Figure 2.3).

AI models have a critical ethical concern in microbiology, and that is bias. Haruna et al. (2025) note that biased data may be induced by skewed datasets, like an excess of pathogens or a population of a specific geographic area. Although these models might work under the conditions that they are trained on, there is a chance of inaccurate results in underrepresented environments and the resulting occurrence of diagnostic errors and unreliable predictions of AMR. This brings disparities in the provision of healthcare, especially in low-resource areas that have access to limited microbiological datasets. In addition to the technical limitations, such biases have even more significant ethical and societal consequences since they might cause global inequalities in the handling of infectious diseases and putting doubts on the trust in AI-based diagnostic technologies.

On equity and AI-enabled microbiology tools, Mohseni and Ghorbani (2023) also included that the majority of advances in AI models and computational microbiology models are based in high-income countries, where these countries possess better infrastructure, money, and technical skills. In comparison, areas with lower and middle income tend to have fewer computational resources, strong datasets, and trained professionals to develop or deploy AI successfully. It creates a digital divide

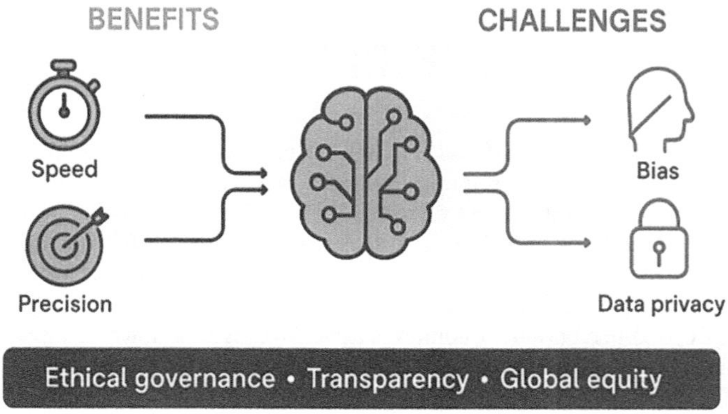

FIGURE 2.3 Ethical and regulatory framework of AI in microbiology. The diagram highlights key benefits (speed, precision, and data-driven insight) and challenges (bias, data privacy, and equity), emphasizing the importance of ethical governance, transparency, and global inclusivity.

within the field of microbiology, as resource-intensive labs and healthcare services can utilize AI to find pathogens faster, to treat them more personally, and to monitor AMR, while poorly resourced regions may be left behind. The disparity not only limits global benefits of AI but can also widen existing health inequities since regions with the greatest infectious disease burden may be the least able to access AI-powered tools. They further indicated the potential cost of prohibitive hardware, software, and licensing needed to have such advanced AI systems, which could be too expensive for small labs. They emphasized the significance of open-source AI platforms, collaborative global data-sharing models, capacity building, including training initiatives, infrastructure assistance, and policy-level interventions, so that AI tools can be made more affordable and accessible in low-income regions to narrow inequalities.

Informed consent is also one of the foundations of ethical medical practice, and the increase in AI in microbiology opens new challenges. Traditionally, consent is described as efforts to explain diagnostic procedures, risks, and anticipated outcomes in patient-understandable terms. With AI-assisted diagnostics, though, patients might not know when their samples are being analyzed by ML or when clinical decisions are being made with the help of ML. Ethically, patients should be informed that AI is part of the process, what role it plays such as supporting pathogen identification or resistance prediction, and what limitations exist, such as potential errors, algorithmic bias, or data privacy risks. It is not feasible to explain the DL models to patients in their entirety, but it is important to be transparent by providing meaningful information such as the fact that the system learns from past data and may be less reliable for underrepresented populations (Zhang and Jiang, 2024).

2.11 CONCLUSION

AI is changing microbiology for the better with data-driven discoveries and the ability to forecast the behavior of microbial systems. Applying AI to microbiology research is a game-changer in terms of precision, effectiveness, and creativity. In order to bring this potential into reality, there should be excellent data that will support the ethical and interdisciplinary issues of providing science and public health with meaningful, trustworthy, interpretable, and safe results.

REFERENCES

Ahmad, F., Khan, M. U. G., Tahir, A., & Masud, F. (2023). Deep ensemble approach for pathogen classification in large-scale images using patch-based training and hyper-parameter optimization. *BMC Bioinformatics*, *24*(1), 273. https://doi.org/10.1186/s12859-023-05398-7

Ahmad, S., Lohiya, S., Taksande, A., Meshram, R. J., Varma, A., Vagha, K., Ahmad, S. Jr, & Varma, A. Sr (2024). A comprehensive review of innovative paradigms in microbial detection and antimicrobial resistance: Beyond traditional cultural methods. *Cureus*, *16*(6), e61476.

Akhtar, Z. B. (2025). Beyond perception: A comprehensive investigation into the advancements, challenges & ethical dimensions of AI and computer vision. *Real-World AI Systems*, *1*(1), 1–27. https://doi.org/10.64797/rwas.v1i1.9577

Ali, H. (2023). Quantum computing and AI in healthcare: Accelerating complex biological simulations, genomic data processing, and drug discovery innovations. *World Journal of Advanced Research and Reviews*, *20*(02), 1466–1484. https://doi.org/10.30574/wjarr.2023.20.2.2325

Ali, M., Benfante, V., Basirinia, G., Alongi, P., Sperandeo, A., Quattrocchi, A., Giannone, A. G., Cabibi, D., Yezzi, A., Di Raimondo, D., Tuttolomondo, A., & Comelli, A. (2025). Applications of artificial intelligence, deep learning, and machine learning to support the analysis of microscopic images of cells and tissues. *Journal of Imaging*, *11*(2), 59. https://doi.org/10.3390/jimaging11020059

Ali, T., Ahmed, S., & Aslam, M. (2023). Artificial intelligence for antimicrobial resistance prediction: Challenges and opportunities towards practical implementation. *Antibiotics*, *12*(3), 523. https://doi.org/10.3390/antibiotics12030523

Ashraf, A. A., Rai, S., Alva, S., Alva, P. D., & Naresh, S. (2025). Revolutionizing clinical laboratories: The impact of artificial intelligence in diagnostics and patient care. *Diagnostic Microbiology and Infectious Disease*, *111*(4), 116728. https://doi.org/10.1016/j.diagmicrobio.2025.116728

Augustin, M. A., Hartley, C. J., Maloney, G., & Tyndall, S. (2024). Innovation in precision fermentation for food ingredients. *Critical Reviews in Food Science and Nutrition*, *64*(18), 6218–6238. https://doi.org/10.1080/10408398.2023.2166014

Babanejaddehaki, G., An, A., & Papagelis, M.. (2025). Disease outbreak detection and forecasting: A review of methods and data sources. *ACM Transactions on Computing for Healthcare*. https://doi.org/10.1145/3708549

Banik, S., Melanthota, S. K., Arbaaz, Vaz, J. M., Kadambalithaya, V. M., Hussain, I., Dutta, S., & Mazumder, N. (2021). Recent trends in smartphone-based detection for biomedical applications: A review. *Analytical and Bioanalytical Chemistry*, *413*(9), 2389–2406. https://doi.org/10.1007/s00216-021-03184-z

Barredo Arrieta, A., Tabik, S., García López, S., Molina Cabrera, D., Herrera Triguero, F., & Díaz Rodríguez, N. A. (2019). Explainable Artificial Intelligence (XAI): Concepts, taxonomies, opportunities and challenges toward responsible AI.

Bhattiprolu, S. (2025). AI-driven microscopy: From classical analysis to deep learning applications. *Methods in Microscopy*, *2*(2), 153–163. https://doi.org/10.1515/mim-2024-0033

CDC. (2025, September 18). *CDC's Vision for Using Artificial Intelligence in Public Health*. Data Modernization. https://www.cdc.gov/data-modernization/php/ai/cdcs-vision-for-use-of-artificial-intelligence-in-public-health.html

Chaturvedi, N., Yadav, M. K., & Sharma, M. (2024). Chapter eleven—Applications of artificial intelligence and machine learning in microbial diagnostics and identification. In A. Srivastava & V. Mishra (Eds.), *Methods in Microbiology* (Vol. *55*, pp. 213–230). Academic Press. https://doi.org/10.1016/bs.mim.2024.05.013

Cheng, J., Novati, G., Pan, J., Bycroft, C., Žemgulytė, A., Applebaum, T., & Avsec, Ž (2023). Accurate proteome-wide missense variant effect prediction with AlphaMissense. *Science*, *381*(6664), eadg7492.

Chhetri, K. B. (2024). Applications of artificial intelligence and machine learning in food quality control and safety assessment. *Food Engineering Reviews*, *16*(1), 1–21. https://doi.org/10.1007/s12393-023-09363-1

Coffman, J., & Sandhu, A. (2018). Using state health improvement planning to guide the way: The future of identifying national trends in public health priorities and emerging public health issues. *Journal of Public Health Management and Practice*, *24*(5), 495. https://doi.org/10.1097/PHH.0000000000000845

Coulson, A. (2025, June 18). AI tool for forecasting infectious disease risk. *BioTechniques*. https://www.biotechniques.com/computational-biology/ai-tool-for-forecasting-infectious-disease-risk/

Dakal, T. C., Xu, C., & Kumar, A. (2025). Advanced computational tools, artificial intelligence and machine-learning approaches in gut microbiota and biomarker identification. *Frontiers in Medical Technology, 6*. https://doi.org/10.3389/fmedt.2024.1434799

Daramola, O., Nyasulu, P., Mashamba-Thompson, T., Moser, T., Broomhead, S., Hamid, A., Naidoo, J., Whati, L., Kotze, M. J., Stroetmann, K., & Osamor, V. C. (2021). Towards AI-enabled multimodal diagnostics and management of COVID-19 and comorbidities in resource-limited settings. *Informatics*, *8*(4), 63. https://doi.org/10.3390/informatics8040063

Davis, J. J., Boisvert, S., Brettin, T., Kenyon, R. W., Mao, C., Olson, R., & Stevens, R. (2016). Antimicrobial resistance prediction in PATRIC and RAST. *Scientific Reports*, *6*(1), 27930.

Desruisseaux, C., Broderick, C., Lavergne, V., Sy, K., Garcia, D.-J., Barot, G., Locher, K., Porter, C., Caza, M., & Charles, M. K. (2024). Retrospective validation of MetaSystems' deep-learning-based digital microscopy platform with assistance compared to manual fluorescence microscopy for detection of mycobacteria. *Journal of Clinical Microbiology*. https://doi.org/10.1128/jcm.01069-23

Domingos, P. (2012). A few useful things to know about machine learning. *Communications of the ACM*, *55*(10), 78–87.

Erwin, P. C., & Brownson, R. C. (2017). Macro trends and the future of public health practice. *Annual Review of Public Health*, *38*, 393–412. https://doi.org/10.1146/annurev-publhealth-031816-044224

Galande, S. H., Kanchan, S., & Kesheri, M. (2014). *Drug Discovery and Design: A Bioinformatics Approach*. LAP Lambert Academy Publishing.

Gardy, J. L., & Loman, N. J. (2018). Towards a genomics-informed, real-time, global pathogen surveillance system. *Nature Reviews Genetics*, *19*(1), 9–20. https://doi.org/10.1038/nrg.2017.88

Gawande, M. S., Zade, N., Kumar, P., Gundewar, S., Weerarathna, I. N., & Verma, P. (2025). The role of artificial intelligence in pandemic responses: From epidemiological modeling to vaccine development. *Molecular Biomedicine*, *6*(1), 1. https://doi.org/10.1186/s43556-024-00238-3

Gilbert, J. A., Blaser, M. J., Caporaso, J. G., Jansson, J. K., Lynch, S. V., & Knight, R. (2018). Current understanding of the human microbiome. *Nature Medicine*, *24*(4), 392–400.

Gomaa, E. Z. (2020). Human gut microbiota/microbiome in health and diseases: A review. *Antonie Van Leeuwenhoek*, *113*(12), 2019–2040. https://doi.org/10.1007/s10482-020-01474-7

Gopalakrishnan, A. V., Suresh, H., Mariappan, G., Kanagaraja, A., & Raman, P. (2024). Industrial microbial bioprocess development: A comprehensive overview. In P. Verma (Ed.), *Industrial Microbiology and Biotechnology: A New Horizon of the Microbial World* (pp. 89–154). Springer Nature. https://doi.org/10.1007/978-981-97-6270-5_4

Hagos, D. H., Aryal, S. K., Ymele-Leki, P., & Burge, L. L. (2025). AI-driven multimodal colorimetric analytics for biomedical and behavioral health diagnostics. *Computational and Structural Biotechnology Journal*, *27*, 2219–2232. https://doi.org/10.1016/j.csbj.2025.05.015

Hammad, M., & Ahmad, S. (2025). Machine learning for image processing in healthcare. In *Techniques and Applications of Biomedical AI* (pp. 131–182). IGI Global Scientific Publishing. https://doi.org/10.4018/979-8-3373-0548-6.ch005

Haruna, A. E., Alhaji, N. B., Adama, J. Y., Monday, O. M., Muhammed, H. L., & Makun, H. A. (2025). Challenges and applications of artificial intelligence in tackling infectious diseases and combating antimicrobial resistance. *BioRes Scientia*, *7*(3), 1–14. https://doi.org/10.59657/2837-4681.brs.25.137

Hossainpour, H., & Mahmoudi, H. (2024). Revolutionizing microbial infection diagnosis: The role of artificial intelligence. *Iranian Journal of Medical Microbiology*, *18*(2), 66–79. https://doi.org/10.30699/ijmm.18.2.66

Huang, J., Zhong, X. F., & Gao, Y. Z. (2024). New antibiotic against multi-drug resistant bacteria. *The Innovation Life*, *2*(1), 100057.

Huang, T., Xu, H., Wang, H., Huang, H., Xu, Y., Li, B., & Qiao, J. (2023). Artificial intelligence for medicine: Progress, challenges, and perspectives. *The Innovation Medicine*, *1*(2).

Ji, L., Senyi, D., & Le, Z. (2021). A review of artificial intelligence applications for antimicrobial resistance. *Biosafety and Health*. https://mednexus.org/doi/10.1016/j.bsheal.2020.08.003

Jin, W. Y., Peng, J., Dai, J., Tang, R., Guo, J. X., Zhao, H., & Gao, Y. Z. (2024). Bacterial load in meconium. *Imeta*, *3*(1), e173.

Jordan, M. I., & Mitchell, T. M. (2015). Machine learning: Trends, perspectives, and prospects. *Science*, *349*(6245), 255–260.

Kaelbling, L. P., Littman, M. L., & Moore, A. W. (1996). Reinforcement learning: A survey. *J. Artif. Intell. Res*, *4*, 237–285. 10.1613/jair.301

Kalita, A. J., Boruah, A., Das, T., Mazumder, N., Jaiswal, S. K., Zhuo, G.-Y., Gogoi, A., Kakoty, N. M., & Kao, F.-J. (2024). Artificial intelligence in diagnostic medical image processing for advanced healthcare applications. In A. Gogoi & N. Mazumder (Eds.), *Biomedical Imaging: Advances in Artificial Intelligence and Machine Learning* (pp. 1–61). Springer Nature. https://doi.org/10.1007/978-981-97-5345-1_1

Keerthi, S. S., & Ravindran, B. (1994). A tutorial survey of reinforcement learning. *Sadhana Acad. Proc. Eng. Sci*, *19*, 851–889.

Kesheri, M., Sinha, R. P., & Kanchan, S. (2016). Advances in soft computing approaches for gene prediction: A bioinformatics approach. In *Medical Imaging in Clinical Applications: Algorithmic and Computer-Based Approaches* (pp. 383–405). Cham: Springer International Publishing.

Keyl, P., Bischoff, P., Dernbach, G., Bockmayr, M., Fritz, R., Horst, D., & Klauschen, F. (2023). Single-cell gene regulatory network prediction by explainable AI. *Nucleic Acids Research*, *51*(4), e20–e20.

Kovur, P., Kovur, K. M., Rayat, D. Y., & Wishart, D. S. (2025). PoC sensor systems and artificial intelligence—Where we are now and where we are going? *Biosensors*, *15*(9), 589. https://doi.org/10.3390/bios15090589

Krichen, M. (2023). Convolutional neural networks: A survey. *Computers*, *12*(8), 151. https://doi.org/10.3390/computers12080151

LeCun, Y., Bengio, Y., & Hinton, G. (2015). Deep learning. *Nature*, *521*(7553), 436–444.

Leonelli, S. (2012). Introduction: Making sense of data-driven research in the biological and biomedical sciences.

Li, Y., Man, S., Ye, S., Liu, G., & Ma, L. (2022). CRISPR-cas-based detection for food safety problems: Current status, challenges, and opportunities. *Comprehensive Reviews in Food Science and Food Safety*, *21*(4), 3770–3798. https://doi.org/10.1111/1541-4337.13000

Liu, Z., DeSantis, T. Z., Andersen, G. L., & Knight, R. (2008). Accurate taxonomy assignments from 16S rRNA sequences produced by highly parallel pyrosequencers. *Nucleic Acids Research*, *36*(18), e120–e120.

Ma, L. C., Zhao, H. Q., Wu, L. B., Cheng, Z. L., & Liu, C. (2023). Impact of the microbiome on human, animal, and environmental health from a One Health perspective. *Science in One Health*, *2*, 100037.

McCarthy J. (2004). What is artificial intelligence. http://www-formal.stanford.edu/jmc/whatisai.html

Mardani, M., & Nasiri, M. J. (2025). *The Role of Artificial Intelligence in the Diagnosis of Infectious Diseases: A Transformative Shift in Global Health* (Editorial 20; Issue 2, p. e161813). Archives of Clinical Infectious Diseases. https://brieflands.com/articles/archcid-161310#abstract

Mohseni, P., & Ghorbani, A. (2023). Exploring the synergy of artificial intelligence in microbiology: Advancements, challenges, and future prospects. *Computers in Biology and Medicine*, *165*, 107380. https://doi.org/10.1016/j.compbiomed.2023.107380

Morr, C. E., & Ali-Hassan, H. (2019). *Analytics in Healthcare: A Practical Introduction*. Springer.

Naugler, C., & Church, D. L. (2019). Automation and artificial intelligence in the clinical laboratory. *Critical Reviews in Clinical Laboratory Sciences*. https://www.tandfonline.com/doi/abs/10.1080/10408363.2018.1561640

Ogunrinola, G. A., Oyewale, J. O., Oshamika, O. O., & Olasehinde, G. I. (2020). The human microbiome and its impacts on health. *International Journal of Microbiology*, *2020*(1). https://doi.org/10.1155/2020/8045646

Olatunji, A. O., Olaboye, J. A., Maha, C. C., Kolawole, T. O., & Abdul, S. (2024). Revolutionizing infectious disease management in low-resource settings: The impact of rapid diagnostic technologies and portable devices. *International Journal of Applied Research in Social Sciences*, *6*(7), 1417–1432.

Onyeaka, H., Akinsemolu, A., Miri, T., Nnaji, N. D., Emeka, C., Tamasiga, P., Pang, G., & Al-sharify, Z. (2024). Advancing food security: The role of machine learning in pathogen detection. *Applied Food Research*, *4*(2), 100532. https://doi.org/10.1016/j.afres.2024.100532

Pospischil, A. (2015). Human and animal health on three continents—a biography of the early life of Karl Friedrich Meyer (1884–1974). *Pathogens and Disease*, *73*(6), ftv039.

Priyadharshini, D., Muthuvel, I., Saraswathy, S., Kavitha, P. S., & Jegadeeswari, V. (2025). Precision to plate: AI-driven innovations in fermentation and hyper-personalized diets. *Frontiers in Nutrition*, *12*. https://doi.org/10.3389/fnut.2025.1659511

Rajasekhar, N., Radhakrishnan, T. K., & Mohamed, S. N. (2024). Reinforcement learning based temperature control of a fermentation bioreactor for ethanol production. *Biotechnology and Bioengineering*, *121*(10), 3114–3127. https://doi.org/10.1002/bit.28784

Raki, H., Aalaila, Y., Taktour, A., & Peluffo-Ordóñez, D. H. (2024). Combining AI tools with non-destructive technologies for crop-based food safety: A comprehensive review. *Foods*, *13*(1), 11. https://doi.org/10.3390/foods13010011

Richter, J., Wang, Q., Lange, F., Thiel, P., Yilmaz, N., Solle, D., Zhuang, X., & Beutel, S. (2025). Machine learning-powered optimization of a CHO cell cultivation process. *Biotechnology and Bioengineering*, *122*(5), 1153. https://doi.org/10.1002/bit.28943

Ruga, T., Vocaturo, E., & Zumpano, E. (2024). On the Role of LLM to Forecast the Next Pandemic. *2024 IEEE International Conference on Bioinformatics and Biomedicine (BIBM)*, 6567–6573. https://doi.org/10.1109/BIBM62325.2024.10822835

Sahoo, A., Singh, J., Alam, K., Alruwaili, N. K., Aodah, A., Almalki, W. H., Almujri, S. S., Alrobaian, M., Barkat, M. A., Singh, T., & Lal, J. A. (2025). Medical artificial intelligence: Opportunities and challenges in infectious disease management. *Current Medicinal Chemistry*, *33*, 1616–1634.

Salam, U. A. (2024). Real-time disease surveillance: AI in tracking and controlling outbreaks. *Emerging Trends in Medicine*, *1*(1), 9–13.

Sampson, M. (2025, August 5). Light AI turning smartphones into portable diagnostic labs for infectious disease detection. *BioTuesdays*. https://biotuesdays.com/2025/08/05/light-ai-turning-smartphones-into-portable-diagnostic-labs-for-infectious-disease-detection/

Saxena, R., Sharma, V., Saxena, A. R., & Patel, A. (2024). Harnessing AI and gut microbiome research for precision health. *Journal of Artificial Intelligence General Science (JAIGS) ISSN:3006-4023*, *3*(1), 74–88. https://doi.org/10.60087/jaigs.v3i1.68

Shukla, V., Singh, S., Verma, S., Verma, S., Rizvi, A. A., & Abbas, M. (2024). Targeting the microbiome to improve human health with the approach of personalized medicine: Latest aspects and current updates. *Clinical Nutrition ESPEN*, *63*, 813–820. https://doi.org/10.1016/j.clnesp.2024.08.005

Siddiqui, S., Khan, A. A., Khan Khattak, M. A., & Sosan, R. (2025). Public health surveillance and resource optimization. In S. Siddiqui, A. A. Khan, M. A. Khan Khattak & R. Sosan (Eds.), *Connected Health Insights for Sustainable Development: Integrating IoT, AI, and Data-Driven Solutions* (pp. 121–140). Springer Nature Switzerland. https://doi.org/10.1007/978-3-031-81433-4_7

Sorbara, M. T., & Pamer, E. G. (2022). Microbiome-based therapeutics. *Nature Reviews Microbiology*, *20*(6), 365–380.

Srivastava, U., & Singh, G. (2013). Comparative homology modelling for HPV type 16 E 7 proteins by using MODELLER and its validations with SAVS and ProSA web server. *J Comput Intell Bioinform*, *6*(1), 27.

Wang, B., Li, Y., Zhou, M., Han, Y., Zhang, M., Gao, Z., Liu, Z., Chen, P., Du, W., Zhang, X., Feng, X., & Liu, B.-F. (2023). Smartphone-based platforms implementing microfluidic detection with image-based artificial intelligence. *Nature Communications*, *14*(1), 1341. https://doi.org/10.1038/s41467-023-36017-x

Xu, Y., Liu, X., Cao, X., Huang, C., Liu, E., Qian, S., & Zhang, J. (2021). Artificial intelligence: A powerful paradigm for scientific research. *The Innovation*, *2*(4), 100179.

Yin, B., Tan, G., Muhammad, R., Liu, J., & Bi, J. (2025). AI-powered innovations in food safety from farm to fork. *Foods*, *14*(11), 1973. https://doi.org/10.3390/foods14111973

Ying, C., Li, X., Lv, S., Du, P., Chen, Y., Fu, H., Du, W., Xu, K., Zhang, Y., & Wu, W. (2022). T-SPOT with CT image analysis based on deep learning for early differential diagnosis of nontuberculous mycobacteria pulmonary disease and pulmonary tuberculosis. *International Journal of Infectious Diseases*, *125*, 42–50. https://doi.org/10.1016/j.ijid.2022.09.031

Zeng, D., Cao, Z., & Neill, D. B. (2021). Chapter 22—Artificial intelligence-enabled public health surveillance—From local detection to global epidemic monitoring and control. In L. Xing, M. L. Giger & J. K. Min (Eds.), *Artificial Intelligence in Medicine* (pp. 437–453). Academic Press. https://doi.org/10.1016/B978-0-12-821259-2.00022-3

Zhang, X., & Jiang, R., 2024. Ethical and legal challenges of medical AI on informed consent: China as an example. *Medicine, Health Care and Philosophy*. https://pubmed.ncbi.nlm.nih.gov/38240080/

Zhu, Y. G., Zhu, D., Rillig, M. C., Yang, Y., Chu, H., Chen, Q. L., & Gillings, M. (2023). Ecosystem microbiome science. *Mlife*, *2*(1), 2–10.

3 Artificial Intelligence and Microbial Science Integration
Technological Frameworks

Ben Asante, Gafari Lukumon, Fakuade Olubusayo Victor, and Ebenezer Esenogho

3.1 INTRODUCTION

The field of microbiological science is now more influenced by data than ever before. Advances in the next generation of sequencing, high-resolution imaging, and multiple-omics profiling now allow researchers to observe the composition of microbial communities at scales that were previously unattainable, from the gene content and expression patterns to the metabolite flow and host–microbe interactions (Sherwani et al., 2025). This increase in progress has also led to a practical limit. Many microbial datasets are large, complex, and context-sensitive. They differ at the sampling sites, the sequencing platform, the laboratory protocols, and the quality of the metadata, which makes it difficult to cross-study and make biological inferences using conventional approaches alone (Rozera et al., 2025; Zhou, 2025).

AI provides a powerful response to the question because it can learn patterns from complex data, recognize nonlinear relations, and operate across different data types. In microbiome and genetic research on microorganisms, machine learning (ML) and deep learning (DL) are increasingly employed to predict the relatedness of microorganisms to disease, to model the dynamics of the ecosystem, and to support the generation of hypotheses in a faster manner than manual analysis can match (Fonseca et al., 2025).

Additionally, natural language processing (NLP) is becoming paramount to the extraction of structured knowledge from the increasing literature; this will facilitate the automated mapping of entities and relationships associated with the microbial function, pathways, and clinical outcomes (Hong et al., 2025). These benefits, however, are not derived solely from algorithms. They rely on specific technological digital data management, preprocessing, computational flexibility, data reproducibility, and workflow reproducibility for microbial data to be used and trusted (Nguyen et al., 2025; Sherwani

DOI: 10.1201/9781003727514-4

et al., 2025). The need for strong and well-designed frameworks is especially clear in research on antimicrobial resistance (AMR). Antibiotic resistance is also a systems problem that includes changes to the genetics of pathogens, clinical decision-making, surveillance, and regulation. There is a growing body of work showing that predicting antibiotic resistance and artificial intelligence (AI) models predictive of antibiotic resistance have improved clinical decision-making at the level of individual patient treatment.

However, these works also indicate the most common problems, which include poor data, a narrow scope of testing, and models that lack interpretability. All of these factors make the technology less transferable for use in clinical practice (Kherabi, 2024; Li et al., 2024). The findings within the scope of DL concerning antibiotic resistance remain true to the ideas that performance improves with the integration of models and the appropriate design and consistency of features of the datasets, complemented with validation that considers the actual clinical or public health settings (López-Cortés et al., 2024). Thus, it is necessary to include the different technological frameworks in the section on AI and microbial science to instill the ability to ethically extract, analyze, and utilize data. This pertains to the capture and integration of primary data and metadata, multi-omics harmonization, workflow automation, and the tiered data redistribution.

Moreover, it comprises frameworks that enhance transparency, reproducibility, and ethical governance (Aswini et al., 2025; Nguyen et al., 2025). Defining these processes explains why this chapter is devoted to the examination of the technological frameworks that allow the use of AI in microbial science. The focus is on how data are collected and processed, how different omics datasets are combined, how scalable computing systems are used, and how these approaches are applied to the detection and treatment of AMR.

3.2 MACHINE LEARNING, DEEP LEARNING, AND NEURAL NETWORK

The potential for AI in microbial science is optimized when two components are combined: powerful AI technologies, namely, ML, DL, NLP, computer vision (CV), and predictive analysis (PA), as well as strong technological frameworks (data pipelines, preprocessing, omics integration, cloud-based infrastructure, and automation of workflows). Now, let us review the technologies involved with AI.

ML refers to the pattern recognition and prediction made possible within data by a non-formal and rule-based approach. ML is carried over to the microbial sciences, where supervised learning (pre-labeled training data) permits the execution of functions such as predicting antibiotic resistance or recognizing dysbiosis. The unsupervised learning reveals clusters of microbial communities associated with properties, whether metabolic composition or taxonomic type. Representation learning and deep neural networks add to this complexity hierarchy, automatically extracting features from raw data (sequence or expression), reducing the need for manual feature engineering (Vidanagamachchi & Waidyarathna, 2024).

Recent emerging studies display high-fidelity prediction of future events and offer interpretable results. LP-Micro delivers an interpretable method for outcome prediction in diseases: it flags the taxa associated with incident disease from both simulated and real-world datasets and improves prediction accuracies over simpler models (Dai et al., 2024). MIOSTONE, a model that applies a taxonomy-encoding neural network similar to the hierarchy structure in microbial species, accurately predicts traits of disease while at the same time offers an interpretation regarding the attribution of features across species, genus, family, and so forth (Jiang et al., 2025).

A standard pipeline in microbiome research typically starts with data collection (usually longitudinal, multi-omics). Then moves to preprocessing (most often quality control, read trimming, normalization, and handling missing data). And further to feature engineering (k-mer frequencies, relative abundances, and taxonomic profiles). Dimensionality reduction (PCA, autoencoders, and taxonomy-based aggregation) is mainly used to remove noise. The models can be random forest or support vector machine (SVM) or deep neural network, or some hybrid model that is taxonomy aware, as in the case of MIOSTONE. The evaluation involves completely independent test sets that are still in the possession of the testator. The metrics are accuracy and F1-scores and AUROC and AUPRC. Shapley-based or feature-association methods are used to determine which taxa/features contribute to predictions (Asher & Bashan, 2024; Dai et al., 2024; Jiang et al., 2025).

Despite these achievements, major challenges remain. When datasets contain many more features than samples (the "small-n, large-p" problem), overfitting is a risk. DL models often behave like black boxes unless interpretability methods are explicitly integrated. Generalization across cohorts with different species assemblages or sequencing technologies can fail, as shown in MIOSTONE's benchmarking (Jiang et al., 2025) and LP-Micro's simulations. Also, ensuring reproducibility demands not only rigorous pipeline documentation but also standardized preprocessing, shared datasets, and transparency about data source, normalization, and transformation choices (Asher & Bashan, 2024; Teixeira et al., 2024).

3.3 NATURAL LANGUAGE PROCESSING FOR LITERATURE MINING

Usually referred to as NLP, it is a subset of AI that involves teaching computers to understand, interpret, and produce language that humans can understand. Its utility in microbial science has been growing. The number of scientific publications, reports from clinics, and records within databases is continuously growing and has, therefore, made it untenable to only manually read and review. NLP tools can mitigate this issue by converting unstructured text into analyzable datasets. Bridging of relevant, yet previously uncorrelated concepts, knowledge consolidation, and improved experimental design are a few of the multiple ways NLP tools assist in addressing the problem (Henao et al., 2024; Wang, Wang, & Liu, 2024; Wang, Ly et al., 2024). In microbial science, the applications and the concentration of studies involving NLP continue to increase. One of the most common is named entity recognition (NER), which textually extracts and identifies units like genes, microbial species, diseases, and biochemical pathways. Then the units are related to each other through relation extraction to demonstrate their interconnectedness. NLP's capability significantly goes beyond literature review and includes genome annotation, facilitating large-scale meta-analyses and study findings from various reports. This helps in improving reproducibility and the removal of duplicated and, therefore, wasted efforts (Pope et al., 2025; Sharma & Sengar, 2025).

An NLP workflow begins with document retrieval through APIs like PubMed or PubMed Central. Then, to prepare and clean the text, it goes through certain processes like tokenization, lemmatization, and removal of stop words. To improve the precision of entity recognition, ontologies from the domain of interest, like gene ontologies or microbial taxonomy, are used. Models such as BioBERT and SciBERT or other custom language models on microbial data are used to perform adaptation for tasks such as entity recognition, text classification, and summarization. The outcomes are presented in knowledge graphs, which help the users to discover and exploit the relationships enabling the formation of new ideas and research questions in a systematic manner (Henao et al., 2024; Pope et al., 2025; Wang et al., 2024).

Conversely, NLP is fraught with problems. Biological terminology is inherently complex, species are frequently renamed or reclassified, which complicates automated extraction. Large, annotated corpora specific to microbial science are still scarce, especially with relation extraction or factuality annotation (Hong et al., 2025). Training and inference for transformer models also demand substantial computational resources, which may not be available in under-resourced laboratories (Sharma & Sengar, 2025). Concerns about interpretability and transparency are also significant for researchers who need clear evidence of how extracted associations are derived and in what context. Addressing these issues will require publicly accessible annotated datasets, community-maintained ontologies and taxonomies, open-source tools, and integrating NLP outputs directly into downstream pipelines so that extracted textual entities/relations feed into ML or predictive models.

3.4 PREDICTIVE ANALYTICS FOR MICROBIAL BEHAVIOR AND ECOSYSTEM MODELING

Predictive analytics leverages past and current data to forecast future behavior in microbial systems, an essential approach for understanding complex and dynamic systems influenced by evolutionary history, host-dependence, and interspecies competition. Traditional statistical models often fail to capture nonlinear dynamics. Predictive analytics, powered by methods such as kNN, deep neural networks, and transformer architectures, offers greater precision (Asher & Bashan, 2024; Wang et al., 2024). This capability is vital for addressing global challenges like AMR, climate-driven changes in microbial communities, and optimizing industrial processes.

Uses of predictive analytics in microbial science are diverse. In healthcare, these models are increasingly used to predict the onset or spread of AMR using genetic markers, patient treatment histories, and epidemiological data (Vidanagamachchi & Waidyarathna, 2024). In environmental microbiology, frameworks forecast how microbial communities respond to climate change, pollution, or nutrient shifts, supporting ecosystem restoration and conservation (Turjeman et al., 2025). Industrial biotechnology benefits where predictions guide fermentation or bioprocess conditions to maximize yield and stability while reducing waste (Bejerano-Blázquez & Familiar-Cabero, 2025). Also, host–microbiome investigations apply predictive analytics to estimate effects of diet, medication, or host genotypes on microbiome composition and function (Wang, Wang, & Liu, 2024).

Recent studies illustrate what is possible with rigorous pipelines. Model-free prediction of microbiome compositions introduces a k-nearest-neighbors regression algorithm that uses only species presence/absence to predict steady-state abundances (Asher & Bashan, 2024). It outperforms null models and several neural network models in both simulated ecological dynamics and real metagenomic datasets when test samples share similar species assemblages.

A typical predictive analytics pipeline begins with gathering complex, multi-omics, longitudinal datasets enriched with environmental or host metadata. Preprocessing steps include standardization, batch correction, and feature engineering to harmonize inputs. Dimensionality reduction (e.g., autoencoders and taxonomy-aware embedding layers) helps reduce noise and computational burden. Predictive modeling uses a mix of methods: classical ML (random forests and gradient boosting), simpler model-free methods such as the Asher & Bashan kNN approach; more complex architectures such as taxonomy-adaptive neural networks (e.g., MIOSTONE); and recurrent neural networks, or transformer-based models. Evaluation includes held-out test data, external cohort validation, and metrics such as AUROC, AUPRC, F1-score, and calibration/uncertainty estimates (Jiang et al., 2025; Lu et al., 2025).

The obstacles confronting these advances include data heterogeneity, limited longitudinal or high-resolution time series datasets, and environmental variability, which diminish model generalizability (Rozera et al., 2025). DL and complex architectures often suffer from lack of interpretability (i.e., "black box" models) unless specifically designed to yield feature attributions, as MIOSTONE does. Ethical concerns around data ownership, patient privacy, and misuse also complicate deployment, especially in sensitive clinical contexts (Vidanagamachchi & Waidyarathna, 2024).

To optimize predictive analytics' promise, the field must invest in pipelines that include uncertainty quantification, accessible annotated datasets, cross-cohort external validations, taxonomy-aware model architectures, and explainable AI methods. With these in place, predictive analytics can become a foundational element of scientifically based and ethical microbiology, enhancing the speed, accuracy, and relevance of discoveries.

3.5 TECHNOLOGICAL FRAMEWORK FOR INTEGRATION

While AI technologies such as ML, NLP, CV, and predictive analytics have begun transforming microbial science, their full potential remains locked unless embedded within rigorous technological frameworks. The integration of these tools into microbial research requires carefully designed pipelines that manage data acquisition, processing, storage, and analysis so as to ensure scalability, reproducibility, and equitable access. Microbial data are inherently large, multimodal, and often fragmented spanning genomics, transcriptomics, proteomics, metabolomics, and environmental measurements, and stored in disparate formats with limited metadata. Without systematic frameworks, data silos and inconsistent preprocessing impede model performance and interpretability (Mujahid et al., 2025; Rozera et al., 2025).

Emerging infrastructures such as findable, accessible, interoperable, reusable (FAIR) data management practices, relational metadata systems, cloud/distributed architectures, and standardized workflow automation are helping to address these issues. Adherence to principles like FAIR is increasingly mandated by funders and institutions and shown to improve data reuse, transparency, and computational readiness (Mujahid et al., 2025).

3.6 DATA ACQUISITION AND PREPROCESSING

Effective data acquisition and preprocessing form the bedrock of AI-enabled microbial science. Without carefully designed pipelines, issues such as inconsistent metadata, poor data quality, and fragmented storage systems can undermine downstream modeling, regardless of how sophisticated the AI techniques are.

A recent study by Mujahid et al. (2025) offers practical guidance: they describe challenges such as the prevalence of data "silos" in nonstandard formats, missing metadata, and inadequate relational databases, which make even basic cross-study comparison difficult. Their framework outlines how to structure metadata at every experimental step (sample source, library prep, sequencing platform, etc.), use relational databases, and enforce FAIR principles across multimodal datasets.

A canonical pipeline for data acquisition and preprocessing in microbial AI should include:

1. **Standardized experimental design and metadata capture**, employing relational metadata systems and documenting all relevant variables (sample source, time, environment, and processing steps) as per Mujahid et al. (2025).
2. **Quality control and filtering** of raw data (e.g., removing reads with low quality, removing contaminant reads, and eliminating low-prevalence features or pathways).
3. **Normalization and transformation**, including adjusting for sequencing depth, batch effects, and variability across modalities.
4. **Feature extraction and dimensionality reduction**, for example, generating relative abundances, k-mers, pathway counts, applying methods like PCA or autoencoders to reduce noise.
5. **Validation cohorts and external testing**, to ensure generalizability and reproducibility (as seen in Serrano-Gómez et al., 2025).

Other challenges include ensuring enough sample diversity across cohorts to avoid overfitting; agreeing on threshold values (e.g., prevalence and abundance) for inclusion/exclusion; dealing with missing or inconsistent metadata; and having the infrastructure (cloud storage and versioned relational databases) and staff training necessary for standardization.

3.7 INTEGRATION OF OMICS DATA WITH AI PLATFORMS

Integrating omics datasets, genomics, transcriptomics, proteomics, and metabolomics poses both a critical challenge and an opportunity in microbial science. Each layer offers unique insight: genomics shows potential, transcriptomics reveals expression, proteomics highlights functional proteins, and metabolomics reflects biochemical activity. When analyzed separately, these omics provide useful but partial views. When combined and processed through AI, they enable systems-level understanding, biomarker discovery, and more accurate predictive modeling (Serrano-Gómez et al., 2025; Shi et al., 2021).

Recent successes show what can be achieved. In the area of multi-omics integration, Shi et al. (2021) defined both functional and health signatures in the intestine through a combination of metagenomics, metabolomics, and transcriptomics analysis. Their results indicated that pathways associated with disease changes were altered by technical variability within each discipline. A study by Serrano-Gómez et al. (2025) documented about 20 species and specific metabolic pathway alterations in Crohn's Disease in a combined omics model, which achieved an external validation AUC of approximately 0.94 using combined omics data. Another research by Wang et al. (2025) used integrated ML on multiomics data from colorectal cancer patients to generate prognostic subtypes (MCMLS) with very high potential in survival prediction and response to therapy, harmonization, feature extraction, and informed model selection, delivering both accuracy and biology. This is possible if emphasized; however, therein lies proof of concept that such results can also be achieved in the pursuit of pure accuracy.

A common route many omics researchers take include:

- Data harmonization: standardizing formats, normalizing across platforms (e.g., correcting for batch effects), aligning feature definitions (gene IDs and metabolites).

- Feature extraction/Embedding: transforming raw counts or signals into embeddings or summary features (pathway scores and module abundances).
- Fusion strategies: early fusion (concatenating features) versus late fusion (separate models combined), or module/deep-learning approaches that preserve layer-specific structure (e.g., graph neural networks).
- Model building and validation, including cross-cohort or external validation and performance metrics (AUC, precision, and recall); interpretability methods (SHAP and feature attribution) are increasingly used to assess which omics layers drive predictions.

Nonetheless, key challenges include high dimensionality versus limited sample sizes, missing data, heterogeneous modalities, batch effects, and inconsistent metadata (Lu et al., 2025; Nam et al., 2024; Ouhmouk et al., 2025). Furthermore, interpretability is often sacrificed: many multi-omics models yield predictions but give little insight into mechanistic pathways. To overcome these issues, strategies such as knowledge-guided integration, modular architectures, open ontologies, and explainable AI methods are employed.

3.8 CLOUD COMPUTING AND DISTRIBUTED ARCHITECTURES FOR LARGE-SCALE MICROBIAL DATASETS

The rising scale of microbial science from multi-omics datasets to high-resolution imaging makes local computation increasingly inadequate. Computational and storage demands for datasets in the terabyte to petabyte range force researchers to seek scalable architectures. Cloud computing and distributed systems offer a path forward, enabling resource elasticity, reproducible workflows, and collaboration across institutions.

A compelling example is HiOmics, a cloud-based platform with nearly 300 plugins built to analyze and visualize large-scale omics data (Lu et al., 2025). It employs Docker containers for reproducibility, leverages WDL + Cromwell for portable workflow descriptions, uses object storage with public cloud providers, and provides tools such as Data Check to verify and convert data formats. These features allow researchers to run complex omics pipelines without needing extensive infrastructure locally.

Another major system is IBM's functional genomics platform (formerly OMXWare) (Seabolt et al., 2022). It contains over 200,000 curated bacterial genomes and annotations for tens of millions of genes, proteins, and domains (over 239 million domains), linked via relational database architecture and accessible via APIs/SDKs. This platform enables querying across huge biological entity spaces and supports rapid cross-genome, functional queries (e.g., genotype-phenotype associations) that would be impractically slow on local hardware.

Despite the promise of cloud and distributed architectures, significant obstacles still stand in the way of their broad adoption across the microbial sciences. For many institutions, especially those in low- and middle-income regions, the cost of cloud compute, and long-term data storage remains prohibitive, often exceeding budgets available for research infrastructure (Banimfreg, 2023). Even when cloud services are accessible, transferring multi-terabyte datasets to or from distant data centers can introduce latency and network charges that hamper project timelines and inflate expenses.

Security and privacy concerns are highly relevant to human-associated microbiological data. Necessary measures needed in such a situation include compliance with regulations, access control, and enforcement of encryption and auditing on the data. These measures could be implemented locally but are difficult to implement in a global setting.

Scalable cloud infrastructures with reproducible containerized workflows, shared resources, and adherence to the rules of metadocumentation and reporting microbial science can begin to overcome these hurdles. This integration will enable a digital delivery of high-performance computing, break down computational barriers, and speed the AI-driven discovery of new substances in a transparent, fair, and uncompromising way on data.

3.9 WORKFLOW AUTOMATION AND AI-ASSISTED LABORATORY SYSTEMS

As the number of microbial research projects grows in both scope and complexity, the manual labor involved in sample handling, culture monitoring, and data acquisition is becoming problematic, causing delay, variation, and error. To address this issue, the application of automated workflows combined with AI-based systems is essential to ensure consistency, reproducibility, and speed. Platforms composed of robots plus laboratory information management systems (LIMS) having AI-based decision support are capable not only to increase experimental throughput but also to wastage efforts (Mencacci et al., 2023)

In reality, almost all practical installations of an AI-driven automation system comprise a robotic technology-based automation system, such as liquid handling robots, specimen sorters, plate readers, integrated with an AI feedback loop. One of the systems that Copan's WASPLab fully automates is all classes: incubation, imaging, and AI analysis to standardize pre-analytic sample preparation and reduce turnaround time.

Overcoming the high cost of robotic hardware and difficulties of integrating it into instruments and the lack of ethical accountability concerns necessitates open standards for the robots' APIs, accessible modular platforms for resource-limited laboratories, and AI systems that are transparent and traceable in their decision-making.

By integrating automated robotic systems with AI, standardized documentation, and larger computational and data frameworks (e.g., the reputation of such next-generation laboratories would presumably be steadier and swifter in responding to intramolecular development for addressing global challenges).

3.10 TECHNOLOGICAL FRAMEWORKS: AI AND MICROBIAL SCIENCE INTEGRATION

Microbes are to life on Earth what the central nervous system is to the body. This is so because they are part of the most fundamental components. This analogy highlights the foundational role microbes play in ecological and biological systems worldwide, just as the central nervous system coordinates the body's essential functions (Bertrand et al., 2015). Microbes are present in nearly every environment. They are found even in deep-sea vents and the human gut and play major roles such as nutrient cycling and disease causation (Nwogwu et al., 2015). Because of this rare diversity and their importance, microbial science has played a major role and intersects with other areas such as biology, medicine, agriculture, and environmental studies, to mention but a few (Frey-Klett et al., 2011). In essence, understanding microbes goes far beyond cataloging species. Additionally, it involves understanding their interactions within larger systems.

Over the past decades, as technology has evolved, research in life sciences has been transformed rapidly because it is not static. Moving forward, sequencing, imaging, and highly powered experiments have generated vast datasets, which in turn make it possible to study microbial communities in more detail (Crowther et al., 2024; Orozco, 2024; Zhou, 2025).

Like other fields, AI (e.g., ML and DL) has now been widely applied, and this has been seen as a solution to these challenges (Roy et al., 2024). In microbial science, it is very useful in many areas. For example, it can be used to develop early statistical models, including but not limited to DL frameworks. It can also handle large datasets and identify patterns (Suster et al., 2024).

3.11 FOUNDATIONS OF MICROBIAL SCIENCE RELEVANT TO AI INTEGRATION

Microbes and microorganisms in general hold the greatest reservoir of biodiversity on our planet. For example, bacteria alone make up several million species present in every environment, many of which are still unexplored and undescribed. No doubt, their genetic diversity surpasses that of plants and animals combined. Most of the time, viruses, which are often overlooked, are the most abundant biological entities on Earth, adding another layer of ecological complexity. At the same time, fungi, although less numerous, also play their roles in decomposition and symbiosis (Wang et al., 2015).

This genetic diversity extends to structure and function, not just species. Microbes can be found as single cells and also exist in colonies. They can recycle essential elements such as carbon, nitrogen, and sulfur. They also produce vitamins and enzymes needed for both human and animal health, as well as degrade pollutants that accumulate in ecosystems. Due to this, microbial communities, that is, the microbiome, are known to be responsible for physiological imbalance, influence digestion, mood, and cognition. Even beyond humans, they are useful in agriculture and industry. They are used to drive fermentation, fix nitrogen, and support biotechnological processes (Jiang et al., 2019).

Based on this immense functional diversity, integrating AI into microbial research is one of the reasons that makes it compelling. But to deeply understand how microbes interact, adapt, and evolve is not in any way a joke. It requires rigorous analysis of complex systems where countless variables, irrespective of their number, change concurrently. For example, from a single human gut study, one can generate many useful datasets, such as genomic data, metabolite profiles, dietary logs, and health parameters, all together. Before AI frameworks that can model nonlinear relationships, detect and check interactions that might go unnoticed, traditional analysis suffered from handling such multidimensional data (Rozera et al., 2025)

3.12 DATA TYPES IN MICROBIAL RESEARCH

Modern microbial research leverages the combination of biology and data science. This field draws on multiple data types, which form a block of microbial life, each having its own analytical challenges. Genomic data is the mother of modern microbiology. Different layers, such as whole-genome sequencing, metagenomics, and metatranscriptomics, help scientists identify species related to microbial community composition and explore functional genes across and within environments. These genome datasets are numerous and rigorous. For example, a single sequencing run can provide several million short reads that may require assembly, annotation, and comparison.

Although progress has been made in algorithmic procedures, since the data is huge, the computational aspect is also rigorous. Therefore, for proper assembly, prediction of gene function, and even inference of ecological roles from incomplete or uncleaned data, AI models are powerful and useful (Roy et al., 2026).

Phenotypic data bring the genetic picture to life. They are useful in capturing observable traits, which include growth rates, motility, morphology, resistance patterns, and host interactions. Because of high-content imaging and automated techniques, phenotypic information has grown exponentially, and interpreting it poses challenges. Thanks to DL approaches, especially those drawn from CV. Currently, it is possible to detect subtle phenotypic variations that were very difficult to observe with the human eye (Mutka et al., 2016; Pratapa et al., 2021).

In addition, we have metabolic and biochemical data. To reveal active pathways and ease metabolic exchanges between species, metabolomics, and proteomics can trace the flow of molecules through microbial systems. But these datasets also suffer from several shortcomings. They are dense, high-dimensional, and very sensitive to environmental conditions. Clustering and network inference, which are AI-based techniques, can reveal co-regulated pathways and disclose hidden metabolic interactions (Badia-i-Mompel et al., 2023). But beyond these types, environmental and clinical metadata provide important contexts that shape microbial composition and behavior, such as temperature, pH, nutrient availability, host health, and treatment conditions. Unfortunately, these data are often incomplete, unmatched, and inconsistent. This weakness can be improved with ML models designed for heterogeneous data because they can reveal complex interactions through the combination of structured and unstructured information (Wang et al., 2024). In essence, all these different datasets are important in microbial life, but they can only be made better with AI computational frameworks, without which the integration of genomics, phenomics, metabolomics, and environmental data, which are fundamental to microbial research, will be inaccessible.

3.13 CHALLENGES IN MICROBIAL DATA ANALYSIS WITHOUT AI TOOLS

Before the advent of AI and before it became prominent, microbial data analysis relied mostly on statistical and rule-based approaches. These methods, such as regression models, clustering, and principal component analysis (PCA), were effective and powerful for smaller datasets but suffered serious limitations with the scale and diversity of modern microbial data (Zhou, 2025):

1. **Data volume:** Scientists now generate data that go far beyond the capacity of traditional computation frameworks (Crowther et al., 2024), which might take time and resources and require good storage, speed, and adaptive algorithms to process.
2. **Complexity and heterogeneity:** Unlike traditional models, which often assume linearity and independence among several variables of consideration (Suster et al., 2024), microbial datasets combine different data types. This can cause important interactions to be obscured and oversimplified, posing significant challenges.

3. **Hidden patterns and nonlinear dynamics:** As mentioned, microbial data are nonlinear. A small change in genetics or environment can alter the composition and function of the community. Regression models often miss such effects and feedback loops, but AI models such as neural networks (Zhou, 2025) have solved this problem, making it more flexible in spotting nonlinearities.
4. **High dimensionality:** There is high dimensionality and interpretability. Standard dimensionality reduction approaches like PCA may become obsolete because they obscure some meaningful biological variations (Zhou, 2025). However, AI-based methods such as manifold learning and graph representation make the data richer and more structurally preserved.
5. **Integration across modalities:** One might want to combine genomic, phenotypic, metabolic, and environmental data into a single framework. But conventional pipelines provide fragmented insights (Crowther et al., 2024) because they treat these datasets separately. AI approaches such as multimodal neural networks can integrate such data, enabling discovery across scales (Suster et al., 2024).

3.14 APPLICATION OF AI IN ANTIMICROBIAL RESISTANCE DETECTION AND TREATMENT

Antibiotics have been a great source of cure for bacterial infections since their discovery. But, unfortunately, sometimes some users of this medicine abuse it (Ali et al., 2023). This behavior mostly enables them to grow resistant to the medicine. Globally, AMR has become a daunting problem that needs a powerful solution to tackle (Ali et al., 2023; Pise & Raskar, 2025). In dealing with this problem, scientists conducted extensive research and came up with a combating tool. The effective method recommended by these scientists is the application of AI (Chang & Chen, 2022; Chen et al., 2022; Didelot & Volz, 2018; Didelot & Pouwels, 2019; Moran et al., 2020). AI has the ability to process data and information and transform them into the knowledge that facilitates data analysis, moving far beyond the capabilities of the minds of human beings. AI analyzes data to reveal the patterns and mechanisms relating to how resistance to antibiotics is developed by pathogens. In addition, the AI uses the gene contents and genomic resistance composition to predict pathogens' resistance to antibiotics.

3.15 METHODS FOR DIAGNOSES

The available methods for diagnosing antibiotics are twofold. They are using whole-genome sequencing for susceptibility testing and antibiotic susceptibility testing. The whole-genome sequencing method can be used to diagnose antibiotic resistance levels with significant accuracy. Furthermore, the method can assist to diagnose antibiotic resistance level in the user constantly. The antibiotic susceptibility testing is an ineffective method, but it is used to quantify AMR level. The shortcoming of this method is its inability to offer the explanation of the mechanism of AMR (Gordon-Rodriguez et al., 2022; Raban et al., 2022). Now, in what ways does AI assist in diagnosing and remediating antibiotic-resistant patients?

The AI models developed predict bacterial susceptibility candidates by searching through existing enormous chemical libraries. For example, the DL models analyze images of bacterial cells or genomic sequences to rapidly detect resistance markers. AI assists to diagnose antibiotic-resistant patients faster and more accurately by processing genome sequences to identify genetic changes that occur in the patient. In some cases, patients are misdiagnosed. But with the assistance of AI, misdiagnosis is reduced to the barest minimum. According to Li et al. (2024), AI technology is used to predict the resistance of pathogens to different antibiotics based on gene content and genomic composition. The selection of targeted DL models in this application is important. Specifically, the recurrent neural network (RNN) and the convolutional neural networks (CNNs) are useful for extracting resistance features and the processing of genetic sequence data. AI algorithms reduce false negatives and false positives when testing patients supposedly resistant to antibiotics. In this sense, patients can be treated with great accuracy.

AMR is identified using the ML and DL algorithms.

3.16 SUPPORT VECTOR MACHINE AND RANDOM NETWORK

Besides the DL and the ML, there are other tools that are used by AI to solve the AMR problems. The tools are SVM and random network (NR). The objective of the SVM is to maximize the distance between the hyperplane and the closest data. It splits data points into different classes by the application of the optimal hyperplane. The optimal hyperplane may be a line in 2D, a plane in 3D, or a multidimensional boundary. Furthermore, the SVM is able to separate nonlinear data. It uses a kernel trick to change the data into a high dimension for separation.

Again, AMR is diagnosed employing two techniques in clinical microbiology. The first one is the classical culture-based antimicrobial susceptibility testing (AST). The second one is whole-genome sequencing for antimicrobial susceptibility testing (WGS-AST). According to Sakagianni et al. (2023), although the former approach is simpler and easier to use, it typically requires a day or more to produce the results, which significantly lengthens the empirical antibiotic regimen and raises the possibility of treatment failure due to ineffective therapy or the threat of antibiotic resistance caused by broad-spectrum antibiotics. AI is employed to assess massive chemical compounds to identify new antibiotics. AI is used to analyze the data belonging to a particular patient. Gathering accurate information about the patient can assist to treat the patient in a fast and accurate manner. The treatment of antimicrobial-resistant patient can be optimized by analyzing the particular patient data and how the one exhibits the pattern of resistance. Furthermore, application of ML and DL can assist in the development of new drugs to combat antimicrobial drug resistance.

Table 3.1 shows the various ML algorithms and their applications in drug screening and delivery.

3.17 CONCLUSION

The chapter described the types of AI that are available to microbial scientists, including ML, NLP, CV, and predictions. The tools presented in the chapter contribute to the expansion of the field of microbial research beyond the descriptive level to predictive, integrative, and practical science. Furthermore, the chapter showed the mechanisms of addressing AMR by humans. It showed that AMR is claiming innocent lives globally. And in trying to solve this problem, medical scientists apply AI. The chapter presented that AI has proven to be an effective tool to treat this problem (Cánovas-Segura et al., 2016).

TABLE 3.1 Machine learning algorithms and their applications in drug screening and discovery

ALGORITHM	*TYPE*	*CORE STRATEGY/PRINCIPLE*	*APPLICATION IN DRUG SCREENING*
SVM (Support vector machine)	Supervised learning	Margine maximization through convex quadratic programming optimization	Predicting active compounds and molecular properties
KNN (K-nearest neighbors)	Supervised learning	Classification based on the most similar K neighbors in feature space	Handling data scarcity and classification in early stages of drug screening
RF (Random forest)	Ensembled learning	Majority voting from multiple decision trees	Predicting drug activity based on genomic characteristics and chemical properties
AE (Autoencoder)	Unsupervised learning	Reproducing input at output nodes through hidden layer transformation	Predicting drug-target interactions, model initialization, and feature dimensionality reduction for drug similarity evaluation
CNN (Convolutional neural network)	Deep learning	Layered architecture including convolutional, pooling, and fully connected layers	Predicting toxicity of compounds and analyzing experimental drug test image data

In summary, AI has the ability to diagnose and provide an effective and fastest way to treat a patient who is plagued by an AMR problem.

It is worth noting that standardized data framework adoption is recommended for microbial integrations. To begin with, FAIR data principles are followed by microbial scientists to develop community standards regarding metadata in order to improve the quality and consistency of data across studies and platforms. And due to the planetary nature of microbial systems, it is recommended that international collaborations intended for sharing datasets, processes, and methods that ensure equally informed access to scientifically based microbial technology for all participants should be put in place. In conclusion, it is recommended that AI be extended to solve other medicine resistance problems that are found difficult to deal with.

REFERENCES

Ali, T., Ahmed, S., & Aslam, M., (2023). Artificial Intelligence for Antimicrobial Resistance Prediction: Challenges and Opportunities towards Practical Implementation. pp. 523. https://doi.org/10.3390/antibiotics 12030523 (accessed: 21/09/2025).

Asher, E. E., & Bashan, A. (2024). Model-free prediction of microbiome compositions. Microbiome, 12(1), 17.

Aswini, R., Saranya, B., Gayathri, K., & Karthikeyan, E. (2025). Revolutionizing infectious disease surveillance: Multi-omics technologies and AI-driven integration. Decoding Infection and Transmission, 3, 100061.

Badia-i-Mompel, P., Wessels, L., Müller-Dott, S., Trimbour, R., Ramirez Flores, R. O., Argelaguet, R., & Saez-Rodriguez, J. (2023). Gene regulatory network inference in the era of single-cell multi-omics. Nature Reviews Genetics, 24(11), 739–754.

Banimfreg, B. H. (2023). A comprehensive review and conceptual framework for cloud computing adoption in bioinformatics. Healthcare Analytics, 3, 100190.

Bejerano-Blázquez, & Familiar-Cabero, M. (2025). On the application of artificial intelligence and cloud-native computing to clinical research information systems: A systematic literature review. Information, 16(8), 684.

Bertrand, A., Kostine, M., Barnetche, T., et al. (2015). Immune-related adverse events associated with anti-CTLA-4 antibodies: Systematic review and meta-analysis. BMC Medicine, 13, 211. https://doi.org/10.1186/s12916-015-0455-8

Bertrand, E. M., Allen, A. E., Dupont, C. L., Norden-Krichmar, T. M., Bai, J., Valas, R. E., ... & Armbrust, E. V. (2015). Influence of cobalamin scarcity on diatom molecular physiology and identification of a cobalamin acquisition protein. Proceedings of the National Academy of Sciences, 112(26), 8451–8456. https://doi.org/10.1073/pnas.1505569112

Cánovas-Segura, B., Campos, M., Morales, A., Juarez, J. M., & Palacios, F. (2016). Development of a clinical decision support system for antibiotic management in a hospital environment. Programming Artificial Intelligence, 5, 181–197.

Chang, A., & Chen, J. H., (2022), BSAC vanguard series: Artificial intelligence and antibiotic stewardship. Journal of Antimicrobial Chemotherapy, 77, 1216–1217. doi: 10.1093/jac/dkac096 (accessed: 10/09/2025).

Chen, Y., Chen, X., Liang, Z., Fan, S., Gao, X., Jia, H., Li, B., Shi, L., Zhai, A., & Wu, C. (2022). Epidemiology and prediction of multidrug resistant bacteria based on hospital level. Journal of Global Antimicrobial Resistance, 29, 155–162, doi: 10.1016/j.jgar.2022.03.003 (accessed: 26/09/2025).

Crowther, T. W., Rappuoli, R., Corinaldesi, C., Danovaro, R., Donohue, T. J., Huisman, J., & van Galen, L. G. (2024). Scientists' call to action: Microbes, planetary health, and the sustainable development goals. Cell, 187(19), 5195–5216.

Dai, L., Sheng, B., Chen, T., Wu, Q., Liu, R., Cai, C., & Jia, W. (2024). A deep learning system for predicting time to progression of diabetic retinopathy. Nature Medicine, 30(2), 584–594.

Didelot, X., & Volz, E. M. (2018). Modeling the growth and decline of pathogen effective population size provides insight into epidemic dynamics and drivers of antimicrobial resistance. Systematic Biology, 67(4), 719–728. https://doi.org/10.1093/sysbio/syy007

Didelot, X., & Pouwels, K. B. (2019). Machine-learning-assisted selection of antibiotic prescription. Nature Medicine, 25(7), 1033–1034. https://doi.org/10.1038/s41591-019-0517-0

Fonseca, D. C., da Rocha Fernandes, G., & Waitzberg, D. L. (2025). Artificial intelligence and human microbiome: A brief narrative review. Clinical Nutrition Open Science, 59, 134–142.

Frey-Klett, P., Burlinson, P., Deveau, A., Barret, M., Tarkka, M., & Sarniguet, A. (2011). Bacterial-fungal interactions: Hyphens between agricultural, clinical, environmental, and food microbiologists. Microbiology and Molecular Biology Reviews, 75(4), 583–609.

Gordon-Rodriguez, E., Quinn, T. P., & Cunningham, J. P. (2022). Data augmentation for compositional data: Advancing predictive models of the microbiome. arXiv. https://doi.org/10.48550/arXiv.2205.09906

Henao, J., Giraldo-Betancur, A., Poblano-Salas, C. A., Espinosa-Arbelaez, D. G., Corona-Castuera, J., Forero-Sossa, P. A., & Diaz-Rebollar, R. (2024). On the role of substrate in hydroxyapatite coating formation by cold spray. Coatings, 14(10), 1302.

Hong, G., Hindle, V., Veasley, N. M., Holscher, H. D., & Kilicoglu, H. (2025). DiMB-RE: Mining the scientific literature for diet-microbiome associations. Journal of the American Medical Informatics Association, 32(6), 998–1006.

Jiang, Y., Aton, M., Zhu, Q., et al. (2025). Modeling microbiome-trait associations with taxonomy-adaptive neural networks. Microbiome, 13, 87. https://doi.org/10.1186/s40168-025-02080-3

Jiang, D., Armour, C. R., Hu, C., Mei, M., Tian, C., Sharpton, T. J., & Jiang, Y. (2019). Microbiome multi-omics network analysis: statistical considerations, limitations, and opportunities. Frontiers in Genetics, 10, 995.

Kherabi, Y. (2024). Machine learning to predict antimicrobial resistance: Future applications in clinical practice. Infectious Diseases Now, 54(3), 104864.

Li, Y., Cui, X., Yang, X., Guangqia Liu, G., & Zhang, J. (2024). Artificial intelligence in predicting pathogenic microorganisms' antimicrobial resistance: Challenges, progress, and prospects. Frontiers in Cellular and Infection Microbiology. 14:1482186. doi: 10.3389/fcimb.2024.1482186 (accessed: 20/09/2025).

López-Cortés, X. A., Manríquez-Troncoso, J. M., Hernández-García, R., & Peralta, D. (2024). MSDeepAMR: Antimicrobial resistance prediction based on deep neural networks and transfer learning. Frontiers in Microbiology, 15, 1361795.

Lu, J., Zhang, W., He, Y., Jiang, M., Liu, Z., Zhang, J., Zheng, L., Zhou, B., Luo, J., He, C., Shan, Y., Zhang, R., Fan, K., Fang, B., & Wan, C. (2025). Multi-omics decodes host-specific and environmental microbiome interactions in sepsis. Frontiers in Microbiology, 16, 1618177. https://doi.org/10.3389/fmicb.2025.1618177

Mencacci, A., De Socio, G. V., Pirelli, E., Bondi, P., & Cenci, E. (2023). Laboratory automation, informatics, and artificial intelligence: Current and future perspectives in clinical microbiology. Frontiers in Cellular and Infection Microbiology, 13, 1188684. https://doi.org/10.3389/fcimb.2023.1188684

Moran, E., Robinson, E., Green, C., Keeling, M., & Collyer, B. (2020). Towards personalized guidelines: Using machine-learning algorithms to guide antimicrobial selection. Journal of Antimicrobial Chemotherapy, 75, 2677–2680.

Mujahid, M., Wang, C., Farooq, M. U., Feng, L., Qiu, Y., Li, J., & Shao, J. (2025). Synergistic optimization of membrane distillation-reverse electrodialysis for sustainable desalination and salinity gradient power generation. Separation and Purification Technology, 360, 130953.

Mutka, A. M., Fentress, S. J., Sher, J. W., Berry, J. C., Pretz, C., Nusinow, D. A., & Bart, R. (2016). Quantitative, image-based phenotyping methods provide insight into spatial and temporal dimensions of plant disease. Plant Physiology, 172(2), 650–660.

Nam, S. W., Nguyen, T. H., Tran, D. T., Dinh, V. A., Ta, T. T. N., Dong, C. L., & Lee, J. H. (2024). Molecularly engineered potential of d-orbital modulated iron-bridged delaminated MBene for rechargeable Zn–air batteries. Energy & Environmental Science, 17(18), 6559–6570.

Nguyen, C., Nguyen, T., Trivitt, G., Capaldo, B., Yan, C., Chen, Q., & Meerzaman, D. (2025). Modular and cloud-based bioinformatics pipelines for high-confidence biomarker detection in cancer immunotherapy clinical trials. PLoS One, 20(8), e0330827.

Nwogu, T. P., Azubuike, C. C., & Ogugbue, C. J. (2015). Enhanced bioremediation of soil artificially contaminated with petroleum hydrocarbons after amendment with *Capra aegagrus hircus* (goat) manure. Biotechnology Research International, 2015, 1–9.

Orozco, H. (2024). The hidden role of microorganisms in ecosystem health. Ukrainian Journal of Ecology, 14(6), 31–33.

Ouhmouk, M., Nhilil, M., Ait Benhassou, H., & Abik, M. (2025). AI-Powered Insights into Genomic Variants. In International Conference on Advanced Intelligent Systems for Sustainable Development (AI2SD 2024): AI2SD Global Submit Symposium Serie on Health, Energy, Environment and Agriculture, Volume 1 (Vol. 1402, p. 67). Springer Nature.

Pise, G. S., & Raskar, I. (2025). Decoding AMR AI's Role in Understanding Pathogen Evolution. IGI Global Scientific Publishing, doi: 10.4018/979-8-3693-7550-1.ch003 (accessed: 18/09/20250).

Pope, Q., Varma, R., Tataru, C., David, M. M., & Fern, X. (2025). Learning a deep language model for microbiomes: The power of large-scale unlabeled microbiome data. PLoS Computational Biology, 21(5), e1011353.

Pratapa, A., Doron, M., & Caicedo, J. C. (2021). Image-based cell phenotyping with deep learning. Current Opinion in Chemical Biology, 65, 9–17.

Raban, M. Z., Gates, P. J., Gamboa, S., Gonzalez, G., & Westbrook, J. I. (2023). Effectiveness of non-interruptive nudge interventions in electronic health records to improve the delivery of care in hospitals: A systematic review. Journal of the American Medical Informatics Association, 30(7), 1313–1322.

Roy, C., Whetsell, M., & Frederickson, K. (2009). The Roy adaptation model and research. Nursing Science Quarterly, 22(3), 209–211.

Roy, D., Banerjee, S., Ansari, A., Goswami, S., Raval, N. P., Adhikary, K., & Ghosh, T. S. (2026). Artificial intelligence (AI) in microbiome-directed biotherapeutics development. Progress in Molecular Biology and Translational Science, 220, 87–102.

Roy, G., Prifti, E., Belda, E., & Zucker, J. D. (2024). Deep learning methods in metagenomics: A review. Microbial Genomics, 10(4), 001231.

Rozera, T., Pasolli, E., Segata, N., & Ianiro, G. (2025). Machine learning and artificial intelligence in the multi-omics approach to gut microbiota. Gastroenterology, 169(3), 487–501.

Sakagianni, A., Koufopoulou, C., Feretzakis, G., Kalles, D., Verykios, V. S., Myrianthefs, P., & Fildisis, G. (2023). Using machine learning to predict antimicrobial resistance—A literature review. Antibiotics, 12, 452. https://doi.org/10.3390/antibiotics12030452 (accessed: 22/9/2025).

Seabolt, E. E., Roellig, D. M., & Konstantinidis, K. T. (2022). Functional genomics platform: A cloud-based platform for studying microbial life at scale. IEEE/ACM Transactions on Computational Biology and Bioinformatics, 19(2), 940–952.

Serrano-Gómez, G., Yañez, F., Soler, Z., Pons-Tarin, M., Mayorga, L., Herrera-deGuise, C., & Manichanh, C. (2025). Microbiome multi-omics analysis reveals novel biomarkers and mechanisms linked with CD etiopathology. Biomarker Research, 13(1), 85.

Sharma, P., & Sengar, A. (2025). Trends and insights in renewable energy research: A comprehensive bibliometric analysis (2000–2023). International Journal of Energy Sector Management, 19(4), 751–771.

Sherwani, M. K., Ruuskanen, M. O., Feldner-Busztin, D., Nisantzis Firbas, P., Boza, G., Móréh, Á., & Lahti, L. (2025). Multi-omics time-series analysis in microbiome research: A systematic review. Briefings in Bioinformatics, 26(5), bbaf502.

Shi, X., Zheng, Y., Lei, Y., Xue, W., Yan, G., Liu, X., & Wang, J. (2021). Air quality benefits of achieving carbon neutrality in China. Science of the Total Environment, 795, 148784.

Suster, C. J., Pham, D., Kok, J., & Sintchenko, V. (2024). Emerging applications of artificial intelligence in pathogen genomics. Frontiers in Bacteriology, 3, 1326958.

Teixeira, B., Afonso, C., Severo, M., & Oliveira, A. (2024). Are the EAT-lancet dietary recommendations associated with future cardiometabolic health?-Insights from the generation XXI cohort from childhood into early adolescence. The American Journal of Clinical Nutrition, 120(6), 1344–1353.

Turjeman, S., Rozera, T., Elinav, E., Ianiro, G., & Koren, O. (2025). From big data and experimental models to clinical trials: Iterative strategies in microbiome research. Cell, 188(5), 1178–1197.

Vidanagamachchi, S. M., & Waidyarathna, K. M. G. T. R. (2024). Opportunities, challenges and future perspectives of using bioinformatics and artificial intelligence techniques on tropical disease identification using omics data. Frontiers in Digital Health, 6, 1471200.

Wang, X.-W., Wang, T., & Liu, Y.-Y. (2024). Artificial intelligence for microbiology and microbiome research. arXiv. https://doi.org/10.48550/arXiv.2411.01098

Wang, W., Lv, Q., Yu, W., Hong, W., Qi, J., Wang, Y., & Tang, J. (2024). Cogvlm: Visual expert for pretrained language models. Advances in Neural Information Processing Systems, 37, 121475–121499.

Wang, W. L., Xu, S. Y., Ren, Z. G., Tao, L., Jiang, J. W., & Zheng, S. S. (2015). Application of metagenomics in the human gut microbiome. World Journal of Gastroenterology: WJG, 21(3), 803.

Wang, X. W., Wang, T., & Liu, Y. Y. (2024). Artificial Intelligence for Microbiology and Microbiome Research. arXiv preprint arXiv:2411.01098.

Zhou, X. (2025). Enhancing pathogen identification through AI-assisted metagenomic sequencing. Frontiers in Microbiology, 16, 1634194.

Intersection of Artificial Intelligence and Microbiology

4

Oyebode Kazeem and Akinola Solomon

4.1 INTRODUCTION

Advances in artificial intelligence (AI) have considerably broadened the scope of microbiology, revolutionizing techniques for microbe classification, genomics, and drug discovery. Traditionally, the study of microbes was based on hands-on methods such as deciphering genomic sequences, which were time-consuming and susceptible to human error. With the development of high-performance computers, cloud technologies, and large-scale databases, AI can now easily automate several tasks, offering unprecedented speed and accuracy in data interpretation (Gershell & Atkins, 2003; Lee & Lee, 2000). Furthermore, AI algorithms can now recognize disease-related patterns in genetic sequences and suggest relevant therapeutic actions.

AI algorithms are becoming increasingly important in finding disease-related patterns in genomic sequences, predicting antimicrobial resistance (AMR), and developing personalized treatment plans. For instance, machine learning models have been helping to detect pathogens and predict mutation trends by analyzing intricate genomic, proteomic, or metabolomic datasets, while deep learning techniques enhance image-based diagnostics by automatically detecting the presence of pathogens from microscopy images or medical imaging (Perkel, 2025a; Vilhekar & Rawekar, 2024a). These capabilities have played a crucial role in accelerating the drug discovery pipelines, evident during pandemics such as Coronavirus (COVID-19), where AI facilitated rapid genome analysis and treatment identification (Ayesiga et al., 2025).

However, while AI applications are progressing suitably in their use for microbiology, they continue to struggle with implementation in resource-constrained settings, especially throughout Africa. Poor digital infrastructure, unstable electricity, and lack of trained personnel impede the effective adoption of AI (Lawanson et al., 2025a; Musa et al., 2023a). Data quality, interoperability, privacy, and data ownership are among technological challenges to the revolution (Barrett et al., 2025a; Chitungo et al., 2020) while issues of data sovereignty and fair benefit sharing are ethical hurdles to be overcome in due course. Challenges to be met over the coming years will be important for managing the transition to an AI-infused microbiology and ensuring that it is beneficial in microbiology-related research.

DOI: 10.1201/9781003727514-5

This chapter explores the intersection of AI and microbiology, emphasizing its applications in genomic analysis, diagnostics, and drug discovery, while critically examining the infrastructural, ethical, and technical challenges in the African context. Strategies for overcoming these barriers, including low-cost hardware solutions, standardized electronic data capture, and explainable AI techniques, were also discussed.

4.2 MICROBIOLOGICAL DATA RESOURCES

4.2.1 Antimicrobial Resistance (AMR) Dataset

AMR is a major threat to global healthcare and causes significant morbidity and mortality, with approximately 5 million deaths (Murray et al., 2022). AMR develops when parasites such as bacteria, viruses, and fungi develop resistance to antimicrobial agents and are made worse by antibiotic misuse (McManus, 1997; Nateghi Haredasht et al., 2025). Addressing this threat requires robust and comprehensive data resources. Nevertheless, integrated databases combining clinical and microbiological information are limited.

The antimicrobial resistance microbiology dataset (ARMD) represents a significant advancement, offering an integrated platform that supports epidemiological analysis and clinical decision-making by consolidating patient histories, antibiotic susceptibility profiles, and microbiological culture data (Nateghi Haredasht et al., 2025). ARMD facilitates trend analysis, risk factor identification, and predictive modeling, making it an important tool in addressing AMR. AI reinforces such efforts by facilitating genetic analysis (Lee & Lee, 2000), accelerating innovative drug discovery (Gershell & Atkins, 2003), improving treatment techniques (Hammond et al., 2025), and enabling tailored therapeutic regimens (Christenson et al., 2025).

The ARMD dataset was developed using de-identified electronic health record (EHR) from Stanford Health Care, encompassing a broad range of microbiological, clinical, and demographic data collected from 1999 to 2024. The dataset integrates microbiology laboratory results, demographic data, clinical encounters, antibiotic exposures, and socioeconomic indicators to enable comprehensive analyses of AMR patterns.

4.2.2 Diagnostics and Pathology Dataset

AI enhances the speed, accuracy, and automation of microbial diagnostics in several ways, such as image analysis, rapid testing (D'Urso & Broccolo, 2024; Hannah, 2025). By gathering pathogen sequencing and phenotypic information, public databases are essential to clinical microbiology. The National Center for Biotechnology Information (NCBI)'s Sequence Read Archive and GenBank are important open-access archives (Panzenhagen et al., 2021). Research on bacterial and viral illnesses is supported by the Bacterial and Viral Bioinformatics Resource Center (BV-BRC) (Olson et al., 2023), while the NCBI pathogen detection system helps with genome analysis for outbreak surveillance. While the global initiative on sharing avian flu data (GISAID) (Shu & McCauley, 2017) and FungiDB focus on viruses and fungi (Basenko et al., 2018), respectively. Particularly during the COVID-19 pandemic under the World Health Organization (WHO)'s Global Initiative on Sharing All Influenza Data (GI-SAID) project, GISAID stands out for its prompt exchange of viral genomic data.

4.2.3 Microbiome Analysis Dataset

The analysis of the human microbiome is a complex, data-rich field where AI is instrumental in uncovering hidden patterns, such as understanding disease, personalized intervention, and functional genomics. The primary sources and resources for microbiome sequencing from animal, environmental, and human-associated systems are summarized. The NIH's Human Microbiome Project (HMP) (USA) (Peterson et al., 2009), Europe's MetaHIT (human gut) (Qin et al., 2010), the citizen-science American Gut Project (McDonald et al., n.d.), and extensive environmental surveys like the Earth Microbiome Project (EMP) (Gilbert et al., 2018) and Tara Oceans (Zhang & Ning, 2015) are important international initiatives. 16S rRNA amplicons and shotgun metagenomes are frequently accessible through portals.

4.2.4 Environmental and Ecological Microbiology Dataset

AI provides new insights into microbial life and its role in global ecosystems, such as Ecosystem analysis and Outbreak prediction. A range of publicly available databases enables global environmental and ecological microbiome research. The EMP used metagenomics and standardized 16S/18S/ITS sequencing to gather more than 27,000 samples from soil, water, plants, animals, and constructed habitats (Gilbert et al., 2018). Tara Oceans has conducted considerable research on marine microbiomes, generating over 250 billion multi-omics data and sampling over 4,300 plankton groups globally (Zhang & Ning, 2015). The first extensive metagenomic atlas of surface-ocean microorganisms was created by the previous Global Ocean Sampling (GOS) Expedition (Williamson et al., 2008). Ocean Sample Day offers a unique one-day overview of marine microbial diversity by combining globally synchronized coastal samples from 191 locations (Kopf et al., 2015).

4.3 AI TECHNIQUES IN MICROBIOLOGY

AI brings the power of automation and the ability to predict, where traditional methods struggle to excel. For example, genomic sequences are the codes encapsulated in the DNA of every organism. Scientists need to study genomic sequences to understand the kind of microorganism they are dealing with and also mutations. When a microbe mutates, there is a chance that it might develop resistance to antibiotics, for example. As a result, scientists must revisit the genomic sequence to understand how to develop drugs to combat the microbe. This process of manually studying microbes might take time. This is where AI can be useful. A trained AI model would read the new genomic sequence and determine where changes are made. It can read and interpret hidden patterns found in the genomic sequence – patterns such as what kind of diseases such uncovered patterns can generate. AI can also be used to detect the existence of mutations that can potentially cause diseases. Furthermore, AI is useful in drug discovery spaces (Vilhekar & Rawekar, 2024b).

An analogy of how AI encapsulates all we have mentioned is given here – AI is able to read billions of DNA codes, spot mutations, predict the kind of disease the mutation would likely lead to, and then find potential drugs to combat it. This approach was replicated during the COVID-19 era. The genomic sequence was established, and then a potential drug to combat it was developed. And as the virus mutates, scientists use AI to research potent drugs to combat the virus (Ayesiga et al., 2025).

AI can carry out image classification, potentially saving time. For example, AI models on image classification can differentiate between *E. coli* and *Staphylococcus aureus*. AI can carry out an automatic diagnosis. For example, a blood sample collected from a patient can be analyzed immediately for tuberculosis infection or malaria (Perkel, 2025b).

4.4 INTERSECTION OF AI AND MICROBIOLOGY CHALLENGES OF AFRICA

4.4.1 Data Availability and Quality

Many laboratories in Africa still rely on inconsistent datasets. For example, laboratories still rely on copying observations in notebooks, instead of relying on computers to automate the collections of observed features. The manual process will potentially result in missing data and an outdated storage format, which will make it impossible to find patients' history. Also, it becomes difficult for hospitals across Africa to collaborate because data is not available in the digital context. To give more context, a Nigerian rural hospital might store tuberculosis samples in a notebook. When the time comes to digitize this data for training an AI model, there will be challenges. The handwritten data might not be recorded professionally; data might be missing, making the development of an AI tuberculosis detection challenging. Another difficulty may result from using old equipment. For example, using old equipment to capture microscopy images will result in generating non-uniform illumination and poor resolution, which makes it difficult to leverage such images for AI training.

The above makes it challenging to develop a robust AI system for pathogen detection (Garba et al., 2022; Musa et al., 2023b). The problem of missing data can be eliminated by ensuring data is captured on a standardized electronic form. In this instance, a record will not be complete until all information regarding a patient's condition is captured. In addition, a database that stores this information will be required. Periodically, the information stored in the database will be backed up to ensure information is not wiped from the database when it crashes. Digital templates for capturing information must also be standardized across all platforms – hospitals and laboratories (Lawanson et al., 2025b; Musa et al., 2023b). Also, inexpensive equipment that can give quality image output should be explored to ensure images captured are useful for AI training. All of these ensure captured data are used effectively for the training of AI models.

4.4.2 Model Interpretability

Another challenge with the AI models is the ability to understand and explain how they carry out predictions. This is very important in microbiology because decisions made by AI need to be explained. For example, experts need to know which genetic markers are responsible for predictions. Without this information, it becomes difficult for medical experts to rely totally on AI prediction. In addition, an AI model built to classify microscopy images might focus on irrelevant parts of the image classification. This is because AI model reasoning is hidden, making experts doubt its output.

In summary, the following should be adhered to – deploy electronic forms to capture data, train staff on how to capture data effectively, and then move to buying low-cost, effective equipment for capturing image data. To solve the challenge of explainable AI, a Grad-CAM heatmap can be used to show areas where the AI model focused attention on making predictions. This gives a sense of traceability – the system decides based on the highlighted portion of the image. This traceability assures medical experts that the AI model understands what it's doing. In the aspect of classification in the genomic sequence space, SHAP can be used to identify which genomic marker dominates to give the classification output (Degoot et al., 2025).

4.4.3 Infrastructure and Cost

AI models require electricity and computing resources for training, such as GPU. This infrastructure could cost thousands of dollars; however, it might be beyond the budget of some African universities. While much of the equipment needed might be donated by developed countries, the electricity and stable Internet to run it might not be available. While cloud computing facilities exist, there might not be enough power to push the AI model together with the dataset to the cloud for AI training. Away from the cost of cloud facilities, the infrastructure needs technical expertise, which might not be available (Musa et al., 2023b; Oktaviana et al., 2025).

To mitigate these challenges, developers could leverage low-cost embedded systems like the Raspberry Pi and the Jetson series, and AI models could be deployed on them to make predictions. There are benefits to this. First, you do not inquire about monthly subscription charges; no Internet connection is needed because it will be deployed locally. This approach will cut down the price of hosting AI models on the cloud. In addition, deployment of solar-based electricity is needed so that reliability is guaranteed to be 100%.

4.4.4 Ethical Considerations

AI leveraged in microbiology has its own challenges around fairness and ownership. Genomic data collected from African countries might be used to train AI models in developed countries with no clear-cut arrangement as to how Africa will benefit from it. In addition, an AI model built on datasets from another continent might perform poorly when used for prediction or AI classification (Barrett et al., 2025b; Dzinamarira et al., 2025).

One way of mitigating this challenge is to develop a data sovereignty agreement that clearly specifies how such data will be used to benefit both parties. Such models should also be tested properly before they are deployed in Africa. Models should be retrained with local data to ensure fairness. The revenue from such collaboration should be shared with African partners. Africa should also build capacity so that they are able to match their foreign counterparts.

4.5 INTERSECTION OF MICROBIOLOGY IN AFRICAN PROSPECTS

African pathogens are not well represented in international collections since genomic sequencing is mostly restricted to a few geographic centers. Large-scale sequencing initiatives are hampered by infrastructure challenges, including unreliable supply chains and reagent shortages. Furthermore, interoperability with international AMR systems is diminished by fragmented data governance, inadequate sectoral integration, and uneven metadata standards (Dall, 2024). Gaps in continuity are caused by funding constraints and reliance on other programs, and data exchange is delayed by sociopolitical issues, such as reporting reluctance. To effectively incorporate Africa into global AMR monitoring, it is imperative to improve local sequencing capabilities, health surveillance, workforce training, digital infrastructure development, and equitable global cooperation.

Only 33% of evaluated African public health facilities have appropriate data infrastructure for contemporary sequencing (Mboowa et al., 2024). A lack of financial resources and restricted or insufficient data access and processing capacity further hamper governance concerns and unwillingness to disclose information. Furthermore, there is a lack of competent bioinformaticians, data scientists, molecular epidemiologists, and modelers (Mboowa et al., 2024). The Africa Center for Disease Control and Prevention (CDC) is collaborating with partners and Member States to improve data infrastructure and expand its workforce as necessary (Tamirat et al., 2024).

In global microbiome data, Africa is noticeably underrepresented, with only 168 published human microbiome studies for Africa, which covers 61% of the continent (Allali et al., 2021). These investigations were mostly by non-African academia and with support from institutions such as the US NIH and the Gates Foundation (Allali et al., 2021). Access to huge datasets is hampered by limited sequencing infrastructure and local finance. Organizations such as H3Africa BioNet and the African Microbiome Project aim to improve regional infrastructure and boost capacity to address these shortcomings (https://africanmicrobiomeproject.org/).

Biodiversity estimates are skewed toward Europe, North America, and Asia due to the relative underrepresentation of African habitats in international environmental microbiology programs. Poor representation in datasets, poor sequencing infrastructure, limited bioinformatics capabilities, a lack of financing and monitoring programs, and concerns about data sovereignty are five major obstacles to advancement (Abdill et al., 2022). Developing sequencing centers and start-up programs with an African emphasis are opportunities for progress (Ayeni et al., 2024). Africa's involvement in microbiology will encourage local research capacities in relation to international consortia and data standards.

4.6 CONCLUSION

This chapter reviews the intersection of AI and microbiology in the light of recent developments globally and considerations for the African continent. The emergence of a new discipline from AI techniques, microbial research, diagnostics, and therapeutic development is reshaping the landscape. AI tools are integral as major instruments for the fast and accurate analysis of genomics, prediction of AMR, and drug discovery, notably in health crises of a global nature, such as COVID-19. These technologies open the way for rapid elucidation of highly complicated data, enhanced imaging, diagnostics, and empower personalized treatment regimens, thus representing a considerable departure from the conventional time-consuming methods. Nevertheless, in Africa, the introduction of AI in microbiology is still emerging due to numerous obstacles. Implementation is impeded by poor digital infrastructure, weak power supply, and lack of skilled human resources. Apart from that, obstacles such as bad data quality, lack of interoperability, and ethical issues, like data sovereignty and fair benefit, sharing, affect the situation. In the absence of precise strategies to address these issues, such restrictions will deepen the technological gap in microbiological research and healthcare worldwide.

REFERENCES

Abdill, R. J., Adamowicz, E. M., & Blekhman, R. (2022). Public human microbiome data are dominated by highly developed countries. *PLoS Biology*, *20*(2), e3001536. https://doi.org/10.1371/journal.pbio.3001536

Allali, I., Abotsi, R. E., Tow, L. A., Thabane, L., Zar, H. J., Mulder, N. M., & Nicol, M. P. (2021). Human microbiota research in Africa: A systematic review reveals gaps and priorities for future research. *Microbiome*, *9*(1), 241. https://doi.org/10.1186/s40168-021-01195-7

Ayeni, K. I., Berry, D., Ezekiel, C. N., & Warth, B. (2024). Enhancing microbiome research in sub-saharan Africa. *Trends in Microbiology*, *32*(2), 111–115. https://doi.org/10.1016/j.tim.2023.11.003

Ayesiga, I., Yeboah, M. O., Okoro, L. N., Edet, E. N., Gmanyami, J. M., Ovye, A., Atimango, L., Gadzama, B. N., Kembabazi, E., & Atwau, P. (2025). Artificial intelligence-enhanced biosurveillance for antimicrobial resistance in sub-Saharan Africa. *International Health*, *17*(5), 795–803. https://doi.org/10.1093/inthealth/ihae081

Barrett, T., Okolo, C. T., Biira, B., Sherif, E., Zhang, A., & Battle, L. (2025a). African Data Ethics: A Discursive Framework for Black Decolonial AI. *Proceedings of the 2025 ACM Conference on Fairness, Accountability, and Transparency*, 334–349. https://doi.org/10.1145/3715275.3732023

Barrett, T., Okolo, C. T., Biira, B., Sherif, E., Zhang, A., & Battle, L. (2025b). African Data Ethics: A Discursive Framework for Black Decolonial AI. *Proceedings of the 2025 ACM Conference on Fairness, Accountability, and Transparency*, 334–349. https://doi.org/10.1145/3715275.3732023

Basenko, E. Y., Pulman, J. A., Shanmugasundram, A., Harb, O. S., Crouch, K., Starns, D., Warrenfeltz, S., Aurrecoechea, C., Stoeckert, C. J., Kissinger, J. C., Roos, D. S., & Hertz-Fowler, C. (2018). FungiDB: An integrated bioinformatic resource for fungi and oomycetes. *Journal of Fungi*, *4*(1), 39. https://doi.org/10.3390/jof4010039

Chitungo, I., Dzobo, M., Hlongwa, M., & Dzinamarira, T. (2020). COVID-19: Unpacking the low number of cases in Africa. *Public Health in Practice*, *1*(July), 100038. https://doi.org/10.1016/j.puhip.2020.100038

Christenson, C., Wu, C., Hormuth, D. A., Ma, J., Yam, C., Rauch, G. M., & Yankeelov, T. E. (2025). Personalizing neoadjuvant chemotherapy regimens for triple-negative breast cancer using a biology-based digital twin. *NPJ Systems Biology and Applications*, *11*(1), 53. https://doi.org/10.1038/s41540-025-00531-z

D'Urso, F., & Broccolo, F. (2024). Applications of artificial intelligence in microbiome analysis and probiotic interventions—An overview and perspective based on the current state of the art. *Applied Sciences*, *14*(19), 8627. https://doi.org/10.3390/app14198627

Dall, C.. (2024). *Report highlights challenges of tackling antimicrobial resistance in Africa.*

Degoot, A., Koné, I., Baichoo, S., Ngungu, M., Liku, N., Kumuthini, J., Nakatumba-Nabende, J., Khomh, F., & Bah, B. (2025). Health data issues in Africa: Time for digitization, standardization and harmonization. *Nature Communications*, *16*(1), 5694. https://doi.org/10.1038/s41467-025-61104-6

Dzinamarira, T., Mbunge, E., Steiner, C., Moyo, E., Akinjeji, A., Yamba, K., Mwila, L., & Muvunyi, C. M. (2025). Practical recommendations for artificial intelligence and machine learning in antimicrobial stewardship for Africa. *Applied AI Letters*, *6*(2). https://doi.org/10.1002/ail2.123

Garba, F. M., Usman, R., Umeokonkwo, C., Okolocha, E. C., Yahaya, M., AbdulQadir, I., Oyeladun, O., Dada, A. O., & Balogun, M. S. (2022). Descriptive epidemiology of measles cases in zamfara state—Nigeria, 2012–2018. *Journal of Interventional Epidemiology and Public Health*, *5*(4). https://doi.org/10.37432/jieph.2022.5.4.69

Gershell, L. J., & Atkins, J. H. (2003). A brief history of novel drug discovery technologies. *Nature Reviews Drug Discovery*, *2*(4), 321–327. https://doi.org/10.1038/nrd1064

Gilbert, J. A., Jansson, J. K., & Knight, R. (2018). Earth microbiome project and global systems biology. *MSystems*, *3*(3). https://doi.org/10.1128/mSystems.00217-17

Hammond, C. V., Williams, S. T., Vega, M. M., Ao, D., Li, G., Salati, R. M., Pariser, K. M., Shourijeh, M. S., Habib, A. W., Patten, C., & Fregly, B. J. (2025). The neuromusculoskeletal modeling pipeline: MATLAB-based model personalization and treatment optimization functionality for OpenSim. *Journal of NeuroEngineering and Rehabilitation*, *22*(1), 112. https://doi.org/10.1186/s12984-025-01629-5

Hannah, A. (2025, March 31). *Microbiome scientist uses AI to redefine the gut-brain axis and deep-sea ecosystems.* Microbiome scientist uses AI to redefine the gut-brain axis and deep-sea ecosystems.

Kopf, A., Bicak, M., Kottmann, R., Schnetzer, J., Kostadinov, I., Lehmann, K., Fernandez-Guerra, A., Jeanthon, C., Rahav, E., Ullrich, M., Wichels, A., Gerdts, G., Polymenakou, P., Kotoulas, G., Siam, R., Abdallah, R. Z., Sonnenschein, E. C., Cariou, T., O'Gara, F., & Glöckner, F. O. (2015). The ocean sampling day consortium. *GigaScience*, *4*(1), 27. https://doi.org/10.1186/s13742-015-0066-5

Lawanson, O., Berleant, D., & Ajiferuke, O. (2025a). Review and recommendations for health informatics in sub-Saharan African countries: Between opportunities and challenges. *Medical Research Archives*, *13*(5). https://doi.org/10.18103/mra.v13i5.6554

Lawanson, O., Berleant, D., & Ajiferuke, O. (2025b). Review and recommendations for health informatics in sub-Saharan African countries: Between opportunities and challenges. *Medical Research Archives, 13*(5). https://doi.org/10.18103/mra.v13i5.6554

Lee, P. S., & Lee, K. H. (2000). Genomic analysis. *Current Opinion in Biotechnology, 11*(2), 171–175. https://doi.org/10.1016/S0958-1669(00)00077-X

Mboowa, G., Tessema, S. K., Christoffels, A., Ndembi, N., Kebede Tebeje, Y., & Kaseya, J. (2024). Africa in the era of pathogen genomics: Unlocking data barriers. *Cell, 187*(19), 5146–5150. https://doi.org/10.1016/j.cell.2024.08.032

McDonald, D., Hyde, E., Debelius, J. W., Morton, J. T., Gonzalez, A., Ackermann, G., Aksenov, A. A., Behsaz, B., Brennan, C., Chen, Y., DeRight Goldasich, L., Dorrestein, P. C., Dunn, R. R., Fahimipour, A. K., Gaffney, J., Gilbert, J. A., Gogul, G., Green, J. L., Hugenholtz, P., & Goldasich, D. L. (n.d.). *American Gut: an Open Platform for Citizen Science Microbiome Research.* https://doi.org/10.1128/msystems.00031-18

McManus, M. C. (1997). Mechanisms of bacterial resistance to antimicrobial agents. *American Journal of Health-System Pharmacy, 54*(12), 1420–1433. https://doi.org/10.1093/ajhp/54.12.1420

Murray, C. J. L., Ikuta, K. S., Sharara, F., Swetschinski, L., Robles Aguilar, G., Gray, A., Han, C., Bisignano, C., Rao, P., Wool, E., Johnson, S. C., Browne, A. J., Chipeta, M. G., Fell, F., Hackett, S., Haines-Woodhouse, G., Kashef Hamadani, B. H., Kumaran, E. A. P., McManigal, B., & Naghavi, M. (2022). Global burden of bacterial antimicrobial resistance in 2019: A systematic analysis. *The Lancet, 399*(10325), 629–655. https://doi.org/10.1016/S0140-6736(21)02724-0

Musa, S. M., Haruna, U. A., Manirambona, E., Eshun, G., Ahmad, D. M., Dada, D. A., Gololo, A. A., Musa, S. S., Abdulkadir, A. K., & Lucero-Prisno, D. E. III (2023a). Paucity of health data in Africa: An obstacle to digital health implementation and evidence-based practice. *Public Health Reviews, 44.* https://doi.org/10.3389/phrs.2023.1605821

Musa, S. M., Haruna, U. A., Manirambona, E., Eshun, G., Ahmad, D. M., Dada, D. A., Gololo, A. A., Musa, S. S., Abdulkadir, A. K., & Lucero-Prisno, D. E. III (2023b). Paucity of health data in Africa: An obstacle to digital health implementation and evidence-based practice. *Public Health Reviews, 44.* https://doi.org/10.3389/phrs.2023.1605821

Nateghi Haredasht, F., Amrollahi, F., Maddali, M. V., Marshall, N., Ma, S. P., Cooper, L. N., Johnson, A. O., Wei, Z., Medford, R. J., Kanjilal, S., Banaei, N., Deresinski, S., Goldstein, M. K., Asch, S. M., Chang, A., & Chen, J. H. (2025). Antibiotic resistance microbiology dataset (ARMD): A resource for antimicrobial resistance from EHRs. *Scientific Data, 12*(1), 1299. https://doi.org/10.1038/s41597-025-05649-7

Oktaviana R, S., Handayani, P. W., Hidayanto, A. N., & Siswanto, B. B. (2025). Healthcare data governance assessment based on hospital management perspectives. *International Journal of Information Management Data Insights, 5*(1), 100342. https://doi.org/10.1016/j.jjimei.2025.100342

Olson, R. D., Assaf, R., Brettin, T., Conrad, N., Cucinell, C., Davis, J. J., Dempsey, D. M., Dickerman, A., Dietrich, E. M., Kenyon, R. W., Kuscuoglu, M., Lefkowitz, E. J., Lu, J., Machi, D., Macken, C., Mao, C., Niewiadomska, A., Nguyen, M., Olsen, G. J., & Stevens, R. L. (2023). Introducing the bacterial and viral bioinformatics resource center (BV-BRC): A resource combining PATRIC, IRD and ViPR. *Nucleic Acids Research, 51*(D1), D678–D689. https://doi.org/10.1093/nar/gkac1003

Panzenhagen, P., Portes, A. B., dos Santos, A. M. P., Duque, S. S., & Conte Junior, C. A. (2021). The distribution of *Campylobacter jejuni* virulence genes in genomes worldwide derived from the NCBI pathogen detection database. *Genes, 12*(10), 1538. https://doi.org/10.3390/genes12101538

Perkel, J. M. (2025a). Beyond AlphaFold: How AI is decoding the grammar of the genome. *Nature, 644*(8077), 829–832. https://doi.org/10.1038/d41586-025-02621-8

Perkel, J. M. (2025b). Beyond AlphaFold: How AI is decoding the grammar of the genome. *Nature, 644*(8077), 829–832. https://doi.org/10.1038/d41586-025-02621-8

Peterson, J., Garges, S., Giovanni, M., McInnes, P., Wang, L., Schloss, J. A., Bonazzi, V., McEwen, J. E., Wetterstrand, K. A., Deal, C., Baker, C. C., Di Francesco, V., Howcroft, T. K., Karp, R. W., Lunsford, R. D., Wellington, C. R., Belachew, T., Wright, M., Giblin, C., & Guyer, M. (2009). The NIH human microbiome project. *Genome Research, 19*(12), 2317–2323. https://doi.org/10.1101/gr.096651.109

Qin, J., Li, R., Raes, J., Arumugam, M., Burgdorf, K. S., Manichanh, C., Nielsen, T., Pons, N., Levenez, F., Yamada, T., Mende, D. R., Li, J., Xu, J., Li, S., Li, D., Cao, J., Wang, B., Liang, H., Zheng, H., & Wang, J. (2010). A human gut microbial gene catalogue established by metagenomic sequencing. *Nature, 464*(7285), 59–65. https://doi.org/10.1038/nature08821

Shu, Y., & McCauley, J. (2017). GISAID: Global initiative on sharing all influenza data – From vision to reality. *Eurosurveillance*, *22*(13). https://doi.org/10.2807/1560-7917.ES.2017.22.13.30494

Tamirat, B., Mazuguni, F., Bamutura, M., Mercy, K., Nyarko, K. M., Tilahun, B., Alinon, K. N., & Tebeje, Y. K. (2024). Africa CDC spearheading the strengthening of health information exchange in Africa. *The Lancet Digital Health*, *6*(6), e382–e384. https://doi.org/10.1016/S2589-7500(24)00068-2

Vilhekar, R. S., & Rawekar, A. (2024a). Artificial intelligence in genetics. *Cureus*. https://doi.org/10.7759/cureus.52035

Vilhekar, R. S., & Rawekar, A. (2024b). Artificial intelligence in genetics. *Cureus*. https://doi.org/10.7759/cureus.52035

Williamson, S. J., Rusch, D. B., Yooseph, S., Halpern, A. L., Heidelberg, K. B., Glass, J. I., Andrews-Pfannkoch, C., Fadrosh, D., Miller, C. S., Sutton, G., Frazier, M., & Venter, J. C. (2008). The sorcerer II global ocean sampling expedition: Metagenomic characterization of viruses within aquatic microbial samples. *PLoS ONE*, *3*(1), e1456. https://doi.org/10.1371/journal.pone.0001456

Zhang, H., & Ning, K. (2015). The Tara oceans project: New opportunities and greater challenges ahead. *Genomics, Proteomics & Bioinformatics*, *13*(5), 275–277. https://doi.org/10.1016/j.gpb.2015.08.003

Integrating Microbial AI with Traditional Microbiology Approaches

5

Ajimah Nnabueze Edmund and Esenogho Ebenezer

5.1 INTRODUCTION

Microbiology has been changed in a number of ways by the use of artificial intelligence (AI) and machine learning (ML). AI can improve our productivity and precision in clinical microbiology, as well as enable us to derive novel insights from data produced in our labs. This is fantastic since patient care can be enhanced by utilizing these talents. AI is widely employed in image analysis, including digital plate readings of bacterial cultures, Gram stains, and ova and parasite tests (Avin et al., 2024). In medical microbiology, computerized measurement and differentiation of bacterial, epithelial, and inflammatory cell morphologies could be used to interpret respiratory Gram stains. AI may, for instance, interpret a picture as having a 90% chance of Gram-positive cocci in clusters and a 10% chance of Gram-positive cocci in chains (Ince et al., 2024; Mukherjee et al., 2024; Washington, 1996). The connection between sequencing data and an organism's or organisms' biological functioning is created by bioinformatics. Through the use of sophisticated algorithms, AI plays a significant role in bioinformatics by facilitating the far more effective analysis of enormous volumes of biological data. utilizing ML algorithms, AI can forecast the best results based on historical data records, offering important insights into DNA sequencing, disease causation, evolutionary patterns, medicine customization, and protein structure prediction. Rapid drug discovery and the resolution of challenging clinical issues may benefit from AI and ML's capacity to mimic human intellect (Ajimah et al., 2023).

Microorganisms can be categorized by AI models according to their phenotypic traits or genetic sequences. Mycobacterium tuberculosis is the root cause of tuberculosis. Since tuberculosis can be deadly if treatment is not received early enough, it is critical to identify patients as soon as possible. The most popular AI technique for analyzing chest X-rays and determining whether a patient has tuberculosis is computer-aided detection (CAD). The radiologists' workload is lessened by this procedure, which eventually expedites the screening process. The first stage in AI-based drug enhancement is ML of a main image, which is followed by processing and sorting of the targets and druggable compounds.

DOI: 10.1201/9781003727514-6

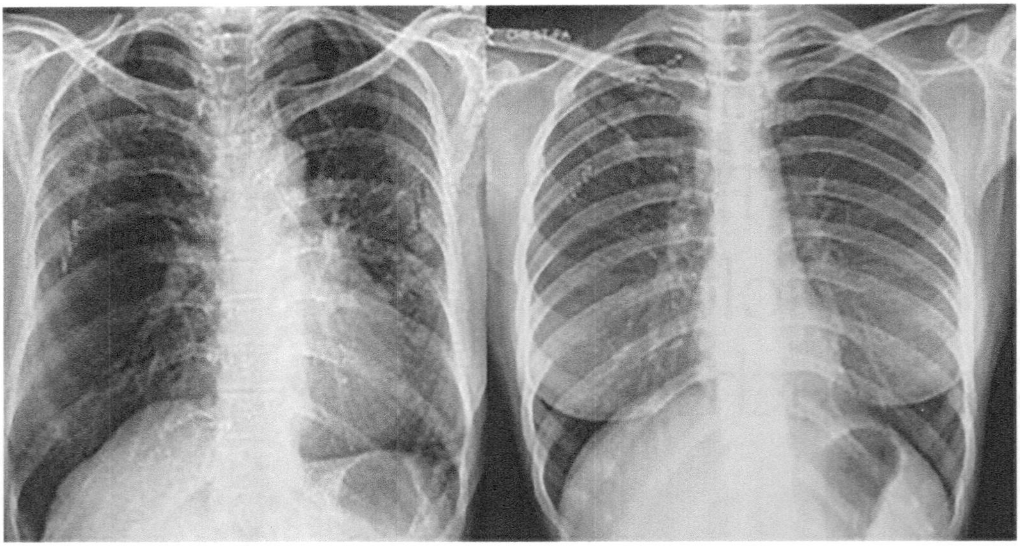

FIGURE 5.1 Chest X-ray for screening for tuberculosis (Bhavin, 2023).

Both were asymptomatic (having no symptoms or indicators), but asymptomatic individuals with active tuberculosis. Perhaps these findings would have been discovered sooner with a routine screening using Mantoux tests or a chest X-ray. Figure 5.1 shows an X-ray screening for a tuberculosis patient, with arrows indicating some degree of damages (Ajimah et al., 2023).

Modern medicine is seriously threatened by the worldwide health epidemic known as antimicrobial resistance (AMR). Therefore, quick and precise AMR diagnostics are required (Imran et al., 2024). But conventional antimicrobial susceptibility testing (AST) takes a lot of time and works only with germs that can be grown. By predicting the resistance of various antibiotics in microorganisms based on gene content and genome composition, AI, particularly ML, is being employed more and more to solve the AMR challenge. By analyzing hundreds of thousands of gene content and genome composition datasets, ML can inform treatment approaches and offer prevention techniques. The study of microbial populations on and within the human body is often referred to as human microbiome analysis. One interesting method for estimating the postmortem interval (PMI) is microbiome analysis (Avin et al., 2024; Bhavin, 2023; Hossainpour & Mahmoudi, 2024). Researchers are now able to make far more accurate predictions thanks to advances in AI that have improved their knowledge of post-mortem microbial ecosystems. AI makes it easier to analyze big datasets and create models from them (Imran et al., 2024).

5.1.1 Traditional Microbial Diagnosis

Microscopy, antigen detection, serology, and culture have typically been incorporated into traditional or conventional microbiological testing. Parasites, fungi, viruses, and bacteria can all cause infections. The pathogen might be endogenous (found in the normal flora) or exogenous (obtained from surroundings, animals, or human beings). Specimens should be representative of the disease process, chosen based on signs and symptoms, and obtained prior to the administration of antimicrobial drugs (Ajimah & Iloanusi, 2024). Test findings are influenced by the quantity of specimens and how quickly they are transported to the lab. Clinical identification is possible for certain infectious diseases due to their unique characteristics. However, a broad range of clinical symptoms in humans can be brought

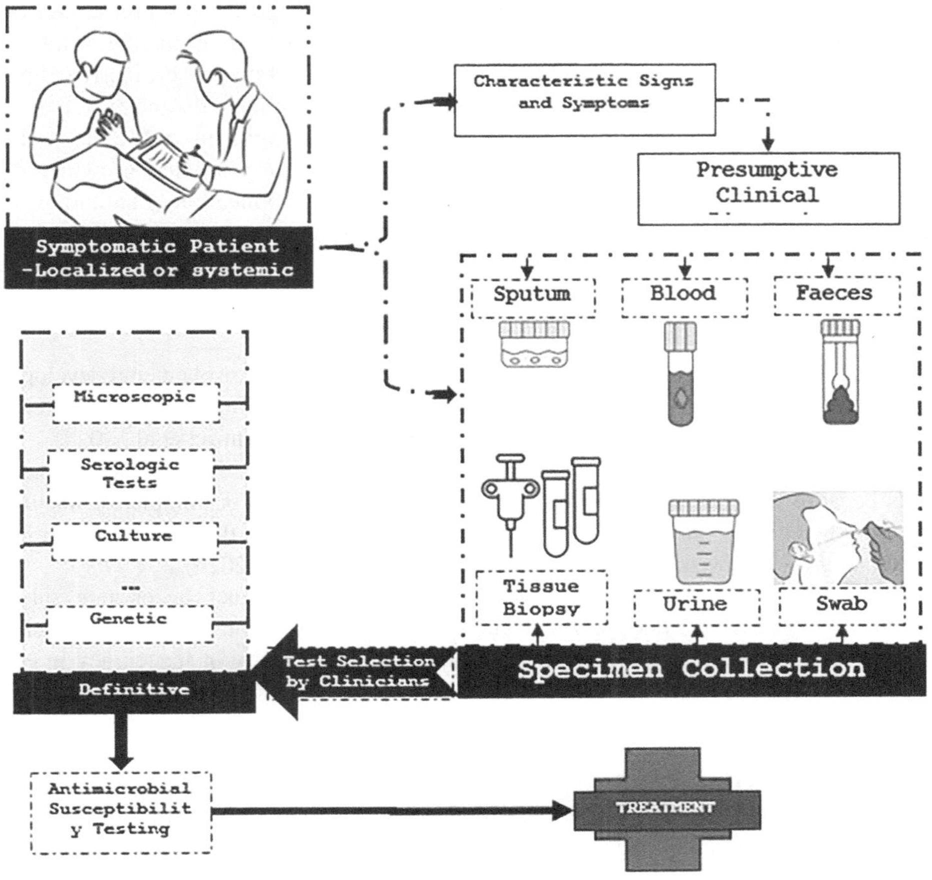

FIGURE 5.2 Laboratory techniques used to validate a clinical diagnosis of a bacterially based infectious disease – TRADITIONAL METHOD.

on by pathogens. On the other hand, infection with any one of several microorganisms can cause a single clinical condition. Numerous respiratory symptoms generated by organisms are clinically indistinguishable from those caused by numerous other viruses. Testing patient specimens for bacteria that cause sickness and providing information on the in vitro activity of antimicrobial medications against the discovered pathogens are the responsibilities of the clinical microbiology laboratory. Microscopy, culture, biochemical testing, and AST are all part of traditional microbial diagnosis (Washington, 1996). These techniques visualize microorganisms using stained samples, cultivate them on media for identification based on colony appearance and biochemical reactions, and then assess antibiotic susceptibility. Traditional methods are seen to be the best, although they can be slow, and some picky organisms might not thrive in culture. See Figure 5.2 for laboratory techniques used to validate a clinical diagnosis of a bacterially based infectious disease (Ajimah et al., 2023).

5.1.2 Traditional Microbial Diagnosis's Drawbacks

Conventional microbiological diagnostic approaches, which are mostly dependent on culture, are sluggish to produce results, necessitate a large number of viable organisms, and cannot identify diseases that cannot be grown. These techniques also need specific knowledge and resources, have limitations in terms of sensitivity and specificity, and are prone to mistakes like contamination or false positives. Particularly with new or picky infections, this can result in incorrect diagnoses, increased patient suffering, and delayed treatment. Consequently, one of the main conventional diagnostic techniques is the culture and isolation of microorganisms. Culture results from this lengthy process. Inaccurate negative outcomes are more likely when microorganisms are cultivated improperly. The growth and isolation of particular organisms may be further impacted by the excessive application of antibiotics in clinical practice. The major disadvantages of conventional microbial diagnostics are as follows (Avin et al., 2024; Bhavin, 2023; Hossainpour & Mahmoudi, 2024; Mohankumar Dongre & M Patel, 2025; Pace et al., 2025):

a. *Slow and Delayed Results:* It can take days or weeks to identify, isolate, and develop pathogens using culture procedures. Rapid, efficient clinical decision-making is hampered by the long wait for results, particularly in cases of severe infections (Imran et al., 2024).
b. *Limited Scope and Sensitivity:* Many pathogens cannot be detected using conventional methods because they are either fastidious (are difficult to culture) or completely uncultivable. Culture techniques may overlook the true causal agent because they only detect organisms that can grow on the selected medium for culture (Imran et al., 2024).
c. *Resource-Intensive and Expertise-Dependent:* These techniques necessitate substantial laboratory resources, trained workers, and specialized equipment. Reading and interpreting cultures requires expert knowledge, which raises the possibility of inaccuracy in environments with few resources (Imran et al., 2024).
d. *Dependence on Viable Organisms:* Only living, viable bacteria can be found using culture-based techniques; pathogens or resistance genes in non-viable cells cannot be found using these techniques (Imran et al., 2024).
e. *Limitations in AST:* Additionally, phenotypic AST techniques such as broth microdilution and disk diffusion are sluggish, yielding results in 18–24 hours or more. Although these techniques can find resistance, they frequently fall short of pinpointing the precise resistance genes or mechanisms at play.
f. *Risk of Errors and Contamination:* Traditional methods' manual nature leaves them open to human error in managing and interpreting samples. Inaccurate results could arise from contamination from other microbes during the culturing procedure itself.

Summarily, an extended duration of empirical antibiotic use (therapy without a conclusive diagnosis) due to delayed diagnosis can contribute to the development of AMR. Healthcare expenses may rise as a result of longer hospital stays, more testing, and possible treatment failures brought on by misdiagnosis.

5.2 ARTIFICIAL INTELLIGENCE (AI)

The experimental science of AI is being developed to better understand the nature of intelligent thought and activity. It is often used for devices or computer applications. Nowadays, AI is mostly, though not entirely, associated with computers. Thus, various academic fields, like psychology, philosophy, science, etc., are also involved in the study of AI. Although the idea and gradual development of AI began in the 1940s, the term "artificial intelligence" was initially used by Stanford University researcher John

McCarthy. Although there isn't a comprehensive definition for AI, which is a developmental science, Dr. McCarthy's statement is still widely accepted (Imran et al., 2024).

> 'There will be an effort to discover how to enable machines to use language, create abstractions and concepts, tackle certain kinds of problems that only humans currently address, and enhance themselves. We believe that if a carefully chosen group of scientists collaborates on one or more of these issues over the summer, it can lead to a significant progress'
>
> – *McCarthy, John*

5.2.1 Applications of AI

AI's capacity to automate, analyze, and make decisions and improve productivity during tasks has transformed a number of industries. Given its reputation for adaptability, some of its uses include the following (Hossainpour & Mahmoudi, 2024; Imran et al., 2024; Li et al., 2024):

a. *Healthcare:* Through data-driven insights and automation, AI is reshaping healthcare by enhancing diagnosis, treatment, and patient care. It aids in the early detection of diseases through medical imaging and predictive analytics, assists physicians in making precise diagnoses, and facilitates personalized treatment strategies tailored to individual health information (Imran et al., 2024). AI finds applications in drug discovery, robotic surgery, and virtual health assistants that offer remote patient support. In the realm of hospital management, AI enhances scheduling, record maintenance, and resource distribution. Moreover, AI-driven wearable gadgets keep track of vital parameters and forecast health threats. On the whole, AI improves efficiency, precision, and accessibility within contemporary healthcare systems (Imran et al., 2024).
b. *Finance:* The finance industry is undergoing a transformation due to AI, which is improving efficiency, accuracy, and decision-making. Its applications are numerous, including fraud detection, risk management, automated trading, credit scoring, and financial forecasting (Imran et al., 2024). Algorithms driven by AI examine extensive datasets to recognize patterns, uncover fraudulent actions, forecast market developments, and evaluate creditworthiness with greater speed than conventional approaches. Moreover, AI chatbots and robo-advisors offer tailored financial guidance and customer service, enhancing the user experience. While AI boosts automation and profit margins, it simultaneously generates worries regarding data privacy, algorithmic bias, and adherence to regulations. This necessitates meticulous supervision to guarantee fairness and security within financial systems (Imran et al., 2024).
c. *Manufacturing and Industries:* Manufacturing and industrial operations are being transformed by AI, which facilitates production processes that are more intelligent, rapid, and efficient. It serves predictive maintenance purposes by identifying equipment failures prior to their occurrence, thus enhancing uptime and lowering costs. Production lines benefit from AI-driven automation and robotics, which improve precision, speed, and quality. Additionally, computer vision systems facilitate real-time quality inspections (Imran et al., 2024). In supply chain management, AI optimizes inventory, logistics, and demand forecasting, ensuring smoother operations. Moreover, AI contributes to product design, energy management, and workplace safety through data analysis that enhances efficiency and decision-making. All in all, AI assists sectors in attaining increased productivity, reduced operational expenses, and enhanced innovation (Imran et al., 2024).

d. *Agriculture:* AI is revolutionizing the agricultural sector by facilitating more intelligent, data-informed farming methods that enhance productivity and sustainability. Precision farming employs it to oversee soil health, weather conditions, and crop growth via sensors and satellite imagery. Farmers can optimize irrigation, fertilizer application, and pest control with the help of AI-powered systems, which reduce waste and enhance yields. Drones and computer vision enable early detection of diseases, weeds, and nutrient deficiencies, while predictive analytics provide forecasts of crop performance and market demand. Moreover, AI aids automated machinery in planting and harvesting, thereby reducing labor expenses. In summary, AI boosts efficiency, sustainability, and decision-making processes in contemporary farming (Imran et al., 2024).
e. *Transportation*: AI is changing the transportation industry by enhancing safety, efficiency, and mobility. It enables self-driving cars to find their way and decide how to drive in the moment based on input from sensors and algorithms. AI systems enhance traffic flow via smart traffic management, leading to decreased congestion and travel duration. In logistics, AI improves route planning, fleet management, and predictive maintenance, resulting in reduced fuel consumption and lower operational costs. AI enhances public transportation with intelligent scheduling and predictions of passenger demand. Moreover, AI contributes to accident prevention, driver assistance, and infrastructure monitoring. In general, AI increases the efficiency and safety of transportation systems while making them more environmentally sustainable (Eldin et al., 2019).
f. *Chatbots and Virtual Assistants:* Chatbots and virtual assistants, driven by AI, simulate human-like dialogues to offer immediate, tailored assistance on multiple platforms. These systems leverage natural language processing (NLP) and ML to comprehend user queries, provide pertinent responses, and enhance their performance through ongoing interaction. AI chatbots in customer service manage standard questions, grievances, and transactions around the clock, alleviating workload and shortening response times. Users can efficiently manage their schedules, control smart devices, and access information with the help of virtual assistants such as Siri, Alexa, and Google Assistant. In general, AI-powered chatbots and virtual assistants improve user experience, accessibility, and operational efficiency in various sectors (Eldin et al., 2019).
g. *Media and Entertainment:* AI is transforming the media and entertainment sector through personalized content creation, recommendations, and distribution. Services such as Netflix and Spotify utilize AI algorithms to examine user preferences and recommend films or music customized to personal tastes. In the realm of content creation, AI aids in editing, scriptwriting, animation, and special effects, thereby accelerating creative workflows. It allows for deepfake technology, automated journalism, and targeted advertising by examining audience behavior. Moreover, AI boosts audience involvement via chatbots, virtual influencers, and interactive media. AI transforms content production and delivery, making media consumption more customized, efficient, and immersive (Imran et al., 2024).
h. *Safety and Security*: AI is crucial for boosting safety and security through capabilities such as real-time surveillance, identification of threats, and swift action against possible dangers. Surveillance systems that utilize AI technology employ facial recognition and behavioral assessment to detect dubious actions and avert criminal acts. In the realm of cybersecurity, AI identifies and addresses hacking attempts, malware, and data breaches through the analysis of network patterns and anomalies (Imran et al., 2024). In the realm of disaster management, it serves to forecast and address natural threats, thereby facilitating a quicker emergency response. Moreover, AI aids in access management, fraud deterrence, and public safety activities. In general, AI enhances security systems by delivering protection that is more intelligent, faster, and more precise in both digital and physical contexts (Ajimah et al., 2017).

5.2.2 AI in Microbiology and Medicine

The field of computer science known as AI is responsible for building intelligent machines that behave and think like human beings. Because of its relevance to everyday life, it is one of the revolutionary technologies that people find fascinating. AI gives robots the ability to understand, learn, and adapt, which improves and automates work in a variety of industries. There are numerous subfields within AI that concentrate on various facets of human mimicry. Among the popular subsets are robotics, deep learning (DL), and NLP. ML is also one of them. From improving consumer experiences to diagnosing diseases, AI has many uses in a variety of industries. Some of its salient characteristics include (Al-Huseini et al., 2025; Alsulimani et al., 2024; Imran et al., 2024; Li et al., 2024):

a. *Capacity to Learn:* In the context of AI microbial systems, learning capacity denotes the potential of AI models, which are based on microbial behavior, to enhance their performance as they engage with data and their surroundings over time. AI microbial systems leverage ML algorithms to mimic adaptation, cooperation, and competition in problem-solving, much like microbes evolve and respond to changing conditions. By continuously learning from feedback, these systems can enhance processes like pattern recognition, resource allocation, and environmental modeling. With this adaptive learning, AI microbial models can enhance their efficiency and intelligence, thus serving valuable roles in biotechnology, environmental science, and computational research (Eldin et al., 2019).
b. *Logical Decision-Making:* In microbial AI, logical decision-making consists of employing AI models that draw inspiration from microbial behavior to make data-driven and adaptive choices in response to environmental conditions. In a way analogous to the way microbes respond to stimuli and refine survival strategies, microbial AI systems utilize logic-driven algorithms to scrutinize data, assess potential consequences, and identify the most efficient solution. This method allows the system to imitate biological reasoning – balancing exploration, cooperation, and competition – to tackle complex problems effectively. In summary, the capacity of microbial AI to adapt, self-organize, and make intelligent decisions that take context into account in changing environments is improved by logical decision-making.
c. *Flexibility:* In the context of microbiology, flexibility in AI pertains to the capacity of AI systems to adjust to a variety of biological data, experimental conditions, and research requirements. AI is capable of analyzing intricate microbial genomes, forecasting behaviors, and adapting to new discoveries or data changes without the need for extensive reprogramming. Its flexibility enables it to manage various microbiological applications, including pathogen detection, AMR prediction, and microbial community analysis (Eldin et al., 2019). AI systems in microbiology can enhance their models and improve accuracy over time by continuously learning from new data. Overall, the ability to adapt ensures that AI can respond promptly, scale appropriately, and prove useful in furthering microbiological research and applications.
d. *Repetitiveness:* In the context of microbiology, the term repetitiveness in AI denotes the ability of AI systems to carry out repetitive tasks with consistency and precision, without experiencing fatigue or making mistakes. In the fields of microbiological research and diagnostics, AI takes over standard procedures that include analyzing images of microbial cultures, entering data, interpreting genome sequences, and identifying patterns in extensive datasets (Eldin et al., 2019). This diminishes the burden on people, reduces mistakes, and accelerates workflows related to experimentation. AI enables scientists to concentrate on intricate analysis and innovation by managing repetitive tasks effectively. In summary, AI's repetitiveness boosts productivity, accuracy, and dependability in microbiological research and lab work (Eldin et al., 2019).

e. *Multipurpose:* AI's multipurpose nature in microbiology stems from its diverse applications in research, diagnostics, and biotechnology. AI aids in the identification and classification of microorganisms, forecasting AMR, examining genomic data, and observing microbial interactions within ecosystems. Its applications also extend to drug discovery (Eldin et al., 2019), vaccine development, and environmental microbiology, where it aids in investigating the roles of microbes in pollution management and bioengineering. AI expedites discoveries and improves precision in lab processes via automation, data analysis, and predictive modeling. AI can be a potent instrument for promoting scientific comprehension and innovation in microbiology, thanks to its versatile capabilities.

5.2.3 AI in Healthcare

The use of ML algorithms, NLP, and DL technologies to improve healthcare services to make life easier for both patients and physicians is known as AI in healthcare (Eldin et al., 2019). Health practitioners can better manage their resources and address different facets of healthcare by utilizing AI's data processing and forecasting capabilities. AI is utilized in various aspects of healthcare, such as identifying novel connections between genetic codes, powering robots that assist with surgery, automating administrative duties, customizing treatment plans, and many more, as listed below (Mukherjee et al., 2024):

a. *Patient Care:* AI is instrumental in patient care, enhancing diagnosis, treatment, and the overall provision of healthcare services. It aids physicians in the early detection of illnesses via medical imaging analysis, forecasts patient risks based on health data, and contributes to the creation of individualized treatment strategies. Virtual assistants and chatbots powered by AI offer round-the-clock support, medication reminders, and health monitoring, thereby improving patient engagement and convenience. In healthcare facilities, AI enhances scheduling, oversees patient records, and tracks vital signs via smart devices. In general, AI improves precision, efficiency, and accessibility in patient care, resulting in enhanced outcomes and a more proactive approach to healthcare management (Mukherjee et al., 2024).
b. *Robotic Surgery:* AI is vital to robotic surgery, as it improves the accuracy, safety, and effectiveness of surgical operations. Surgical robots powered by AI analyze imaging and sensor data to guide instruments with remarkable precision, minimizing human error and tissue damage. These systems help surgeons with operation planning, navigating intricate anatomical structures, and performing real-time modifications during surgery (Eldin et al., 2019). AI makes it possible to perform minimally invasive procedures, which results in quicker recoveries, reduced pain, and shorter hospital stays for patients. Moreover, ML enhances robotic performance by enabling continuous learning from previous surgeries. In summary, AI in robotic surgery merges human skill with technological know-how to provide surgical outcomes that are safer and more effective (Ajimah & Iloanusi, 2024).
c. *Management of Health Data:* AI is essential for health data management, as it effectively organizes, analyzes, and secures large volumes of medical data. It automates the processes of entering, retrieving, and processing data from electronic health records (EHRs), which help to minimize human errors and lessen the administrative burden. AI algorithms can detect patterns and trends in patient data, aiding clinical decision-making and public health research. Moreover, AI improves data security by means of anomaly detection and predictive analytics, which aid in breach prevention and privacy maintenance. AI enhances the accuracy, accessibility, and utility of health data, resulting in improved patient care and more informed healthcare management (Mukherjee et al., 2024).

d. *AI in the Diagnosis of Medicine:* AI revolutionizes the process of identifying illnesses and medical issues through rapid and precise examination of extensive clinical datasets. With the help of image recognition and ML, AI is able to analyze medical images (e.g., MRIs, CT scans, and X-rays) to identify irregularities such as infections, fractures, or tumors. It aids in recognizing patterns across laboratory results, genetic data, and patient histories to facilitate early and accurate diagnosis. AI diagnostic tools assist physicians in making decisions based on data, minimizing mistakes, and providing quicker results. In summary, AI improves diagnostic precision and efficiency in addition to individualized treatment in contemporary medicine (Mukherjee et al., 2024).
e. *The Use of AI in Drug Discovery:* AI has a transformative and complex role in drug discovery, revolutionizing the processes of identifying, designing, and developing new medicines. The process of discovering drugs has conventionally required a large investment of time and money. However, through the analysis of extensive biological, chemical, and clinical datasets to identify patterns and connections that would be hard for people to find, AI greatly speeds this process up. AI can utilize sophisticated ML and DL algorithms to forecast drug-biological target interactions, evaluate the effectiveness and toxicity of compounds, and pinpoint the most promising drug candidates at an early stage of research. It allows for in silico virtual screening of millions of molecules, which significantly cuts down on the necessity for physical testing and lab experiments. AI aids in drug repurposing by identifying new therapeutic uses for existing medications, which saves time and cuts development costs. During the preclinical and clinical phases, AI models mimic human biology and disease advancement, aiding in the refinement of trial design, identification of appropriate patient cohorts, and forecasting possible adverse effects. Furthermore, the use of AI for automating data gathering and analysis improves precision and decision-making while reducing mistakes made by humans. Pharmaceutical companies are progressively incorporating AI at all phases of drug development – from identifying targets and molecular modeling to clinical validation – leading to quicker innovation cycles and more personalized medications. AI's capability to analyze intricate biomedical data, produce insights, and forecast successful results renders it a crucial resource in contemporary drug discovery, facilitating the development of therapeutics that are safer, more effective, and quicker to reach the market (Mukherjee et al., 2024).

AI algorithms such as convolutional neural networks (CNNs), R-CNN, and NLP are used in the classifications of medical images from X-rays, MRIs, and CT scans, as they help in the diagnosis of varying diseases. Other important algorithms encompass support vector machines (SVMs) for classification, Naïve Bayes for diagnostics, and K-Means clustering for patient stratification. These find application in domains such as drug discovery, personalized medicine, robotic surgery, and administrative tasks like billing and coding.

5.3 BIOMEDICAL DIAGNOSTIC IMAGING EQUIPMENT

Diagnostic imaging tools or methods utilize various types of energy (like X-rays, sound waves, and magnetic fields) to generate detailed representations of the body's internal structures for the purposes of diagnosis and treatment planning. Mentioned here are some common types of diagnostic imaging tools (Fonseca et al., 2025; Gurajala, 2024; Rahman et al., 2024; Shelke et al., 2023):

a. Computed Tomography Scanners (CT Scanners)
b. X-ray Machines

c. Magnetic Resonance Imaging (MRI) Scanner
d. Ultrasound Machines
e. Echocardiography Machine
f. Bone Densitometer (DEXA Scanner)
g. Mammography Machine
h. Single-Photon Emission Computed Tomography (SPECT) Scanner
i. Hybrid Scanners (e.g., PET/CT, PET/MRI)

5.3.1 Computed Tomography Scanners (CT Scanners)

Computed tomography (CT) scanners are advanced medical imaging systems that use a combination of X-rays and computer processing to create detailed cross-sectional images of the body. Unlike traditional X-ray machines that produce flat images, CT scanners rotate around the patient, capturing multiple images from different angles. These images are then processed by a computer to form detailed 3D representations of bones, organs, blood vessels, and soft tissues (Ajimah et al., 2023). CT scans are widely used for diagnosing internal injuries, tumors, infections, and vascular diseases, as well as for guiding surgical or biopsy procedures. They provide more precise information than standard X-rays, making them invaluable in emergency medicine and oncology. Modern CT scanners operate quickly and can capture high-resolution images in seconds, minimizing patient discomfort and exposure to radiation while improving diagnostic accuracy. See Figure 5.3 (i).

5.3.2 X-ray Machines (Radiography)

X-ray machines are sophisticated imaging tools that allow for the examination of the internal structures of objects, particularly the human body, without surgical intervention. They operate by directing regulated quantities of X-rays – a type of high-energy electromagnetic radiation – through the body. Various tissues absorb these rays to different extents: denser substances such as bones take in more X-rays and reflect white on the image, whereas softer tissues permit more rays to pass through, resulting in a darker appearance (Avin et al., 2024; Bhavin, 2023). In medical diagnostics, X-ray machines serve as vital instruments for identifying fractures, infections, tumors, and dental problems. In addition to healthcare, they are utilized in security screening, industrial inspection, and scientific research. X-ray machines of modern design frequently come equipped with digital imaging systems, which yield clearer results while using lower radiation doses – a development that enhances safety and accuracy. See Figures 5.3 (ii) (A) and (B), which display X-ray machine and X-ray image, respectively.

5.3.3 Magnetic Resonance Imaging (MRI) Scanner

MRI is a medical imaging method that does not require any surgical procedures and that serves to generate detailed images of the body's internal structures, especially soft tissues such as the brain, muscles, and organs. MRI does not employ ionizing radiation, in contrast to X-rays or CT scans. Rather, it employs strong magnets and radio waves to produce detailed images (Alsulimani et al., 2024; Fonseca et al., 2025). A strong magnetic field is applied around the patient, causing the hydrogen atoms in their body to align. Then, radiofrequency pulses are used to disrupt this alignment. As the atoms revert to their normal state, they emit signals that a computer captures and converts into images. MRI is very effective for diagnosing conditions that impact the brain, spinal cord, joints, and internal organs. It offers superb contrast among various tissue types, rendering it an essential instrument for precisely

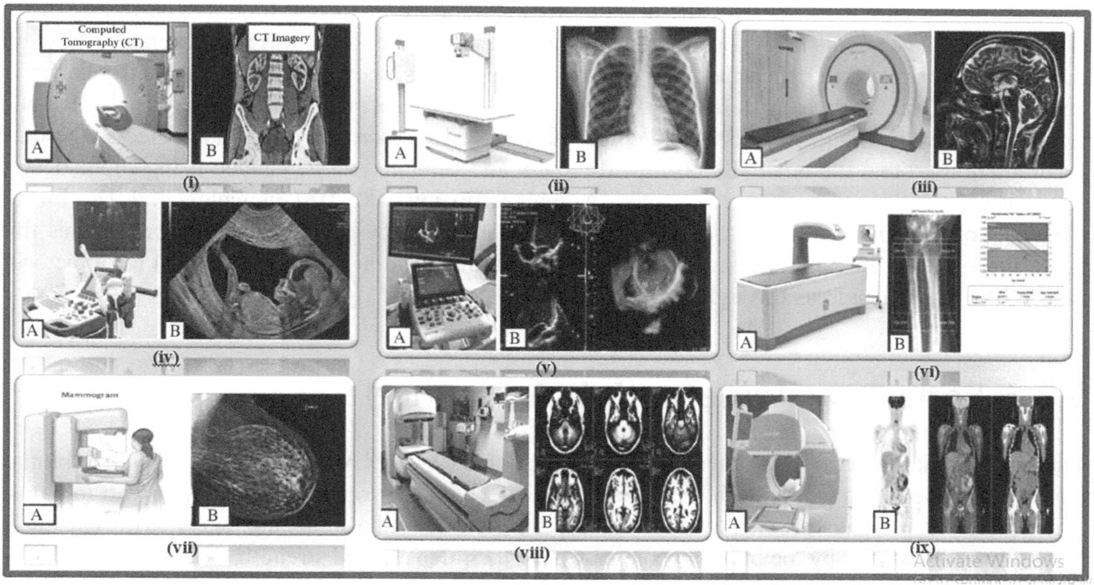

FIGURE 5.3 Images of (i) (A) computed tomography and (B) the imagery; (ii) (A) an X-ray machine and (B) an X-ray image; (iii) (A) an MRI machine and (B) an MRI image of a human head (*Side view*); (iv) (A) an ultrasound machine and (B) ultrasound image of fetus in the womb; (v) (A) echocardiography machine and (B) echocardiography output; (vi) (A) bone densitometer (DEXA scanner) and (B) image of bone as measured through the machine; (vii) (A) mammography machine and (B) image of mammary organ of a female as displayed by the mammography machine; (viii) (A) SPECT scanner and (B) SPECT image of a human brain; and (ix) (A) hybrid scanners and (B) hybrid scanning image.

identifying tumors, brain injuries, and other irregularities. See Figures 5.3 (iii) (A) and (B), which display the MRI machine and MRI image of the human head, respectively.

5.3.4 Ultrasound Machines

Ultrasound machines are medical imaging devices that utilize high-frequency sound waves to create real-time images of the body's interior. Ultrasound, which does not use ionizing radiation like X-ray or CT scans, is a safe and commonly used diagnostic tool. The way the machine operates is by transmitting sound waves via a transducer (probe) into the body. These waves reflect off internal tissues and organs, and a computer captures the returning echoes and converts them into images (Alsulimani et al., 2024; Hossainpour & Mahmoudi, 2024). Ultrasound is widely employed for pregnancy monitoring, examining organs like the heart, liver, and kidneys, as well as identifying irregularities in soft tissues and blood flow via Doppler imaging. It is portable and cost-effective, and it can provide dynamic images that enable physicians to track movements like a heartbeat or a developing fetus in real time. Figures 5.3 (iv) (A) and (B) show an ultrasound machine and ultrasound image of fetus in the womb, respectively (Ajimah et al., 2023).

5.3.5 Echocardiography Machine

An echocardiography machine, commonly known as an echo machine, is a specialized ultrasound device that produces detailed images of the heart. It employs sound waves of high frequency to visualize the chambers, valves, and walls of the heart as well as blood flow in real time. In the course of the procedure, a transducer is positioned on the patient's chest, and the sound waves it generates reflect off the heart structures. After the machine processes these echoes, they become moving images shown on a monitor. Doctors use echocardiography to evaluate heart size, pumping efficiency, and valve performance, as well as to identify issues like heart failure, valve disorders, congenital defects, and fluid accumulation around the heart. Blood flow direction and velocity can be gauged using sophisticated methods such as Doppler echocardiography (Ajimah & Iloanusi, 2024). This yields vital data for the diagnosis of cardiovascular diseases. This imaging technique does not involve any procedures that break the skin, is safe, and is essential for assessing cardiac health. See Figure 5.3 (v) for a pictorial view of the machine and its throughput.

5.3.6 Bone Densitometer (DEXA Scanner)

The bone densitometer, or dual-energy X-ray absorptiometry (DEXA) scanner, is a specialized imaging tool designed for measuring bone mineral density (BMD). It operates by sending two low-dose X-ray beams of varying energy levels through the bones, usually the spine, hip, or forearm. The quantity of X-rays absorbed by the bones aids in assessing their density and strength. The most precise and commonly employed technique for diagnosing osteoporosis, evaluating fracture risk, and tracking bone loss over time is DEXA scanning (Washington, 1996). The method is swift, noninvasive, and entails minimal exposure to radiation. In addition to assessing bone health, DEXA scanners can measure body composition, including fat and muscle mass, which offers valuable insights for overall health evaluation and treatment planning. Figures 5.3 (vi) (A) and (B) show bone densitometer (DEXA Scanner) and image of bone as measured through the machine, respectively.

5.3.7 Mammography Machine

A mammography machine is a specialized X-ray imaging tool used to examine breast tissue for the early detection of breast cancer and other irregularities. It operates by employing low-dose X-rays to obtain detailed images of the breasts' internal structure. The breast is gently compressed between two plates during the procedure to evenly distribute the tissue, which results in clearer images and lessens the required radiation amount. Mammography machines can create images in either 2D or 3D (tomosynthesis) formats, with 3D mammography providing more detailed views and enhanced accuracy for identifying small tumors. This imaging method is vital for regular breast cancer screenings, diagnosing lumps, and conducting follow-up assessments posttreatment (Wang et al., 2020). It is a quick, safe, and highly effective device that is essential for saving lives via prompt detection as well as timely medical intervention. Figures 5.3 (vii) (A) and (B) show mammography machine and image of mammary organ of a female as displayed by the mammography machine, respectively.

5.3.8 Single-Photon Emission Computed Tomography (SPECT) Scanner

A SPECT scanner is a nuclear medicine imaging tool that offers intricate, three-dimensional representations of the internal functions of the body. SPECT, in contrast to traditional imaging applications that display only structure, depicts physiological processes like blood flow, metabolism, and organ function.

A small quantity of a radioactive tracer is injected into the patient's blood circulation as part of the process. The SPECT scanner's rotating gamma cameras detect the emissions from the tracer, which emits gamma rays, from various angles. The data is then reconstructed into cross-sectional or 3D images by a computer. SPECT is frequently utilized for the diagnosis and monitoring of heart diseases, brain illnesses, bone abnormalities, and specific types of cancer (Bhavin, 2023; Hossainpour & Mahmoudi, 2024). It offers important insights into the functioning of organs beyond their appearance. This renders SPECT scanning an essential resource in functional imaging and personalized medicine, providing diagnostic as well as prognostic insights. Figures 5.3 (viii) (A) and (B) are SPECT scanner and SPECT image of a human brain, respectively.

5.3.9 Hybrid Scanners (e.g., PET/CT, PET/MRI)

Hybrid scanners like PET/CT and PET/MRI are sophisticated imaging systems that merge two distinct imaging technologies into a single device, offering both structural and functional data in one scan. In PET/CT, positron emission tomography (PET) uses a radioactive tracer to show metabolic and biochemical activity, while CT offers detailed anatomical images. Likewise, PET/MRI merges the functional imaging capabilities of PET with the excellent soft-tissue contrast provided by MRI. Hybrid scanners enable the accurate detection of metabolic anomalies within anatomical structures by integrating these modalities. They are frequently used in oncology to monitor recurrence, assess therapy response, and identify and stage malignancies (Alsulimani et al., 2024; Fonseca et al., 2025). They are also useful in neurology and cardiology for evaluating brain abnormalities and the activity of the heart. Hybrid imaging is a potent tool in contemporary medical diagnostics and research because it improves diagnostic accuracy, minimizes the need for multiple scans, and provides a complete understanding of the way things work. Figures 5.3 (ix) (A) and (B) are hybrid scanners and hybrid scanning images, respectively.

5.4 DEEP LEARNING (DL) AND MACHINE LEARNING (ML) ALGORITHMS FOR MEDICAL DIAGNOSIS

Medical diagnosis is being revolutionized by DL and ML algorithms, which allow computers to quickly and accurately assess complicated medical data. Combining microbiology and AI methods will improve the comprehension, identification, and control of microbes (Imran et al., 2024). Microbial AI makes it possible to quickly analyze complicated biological data, including genetic sequences, metabolic pathways, and microbial interactions, by fusing data-driven algorithms with microbiological research. By examining antibiotic resistance trends, ML models can detect new species, forecast microbial behavior, and aid in the identification of antibiotics. Additionally, DL algorithms are employed for disease detection, fermentation or biotechnological process optimization, and microscopy picture interpretation. Microbial AI is used in environmental and clinical microbiology to follow disease outbreaks, comprehend microbiome dynamics, and create targeted microbial treatments. This combination of AI with microbiology speeds up research, lessens laboratory work, and improves the precision of microbial diagnostics and bioengineering, opening the door to more intelligent, data-driven approaches to environmental sustainability, agriculture, and health.

5.4.1 Deep Learning (DL) Algorithms in Microbial AI

a. *Convolutional Neural Networks (CNNs):* used to examine microbial colony morphology, genetic patterns, and microscopy images.
b. *Recurrent Neural Networks (RNNs):* used in sequential DNA/RNA data analysis and time-series microbial growth prediction.
c. *Long Short-Term Memory (LSTM) Networks:* for sequence-based forecasts and microbial gene expression modelling.
d. *Autoencoders:* employed in unsupervised representation learning, microbiological feature extraction, and dimensionality reduction.
e. *Generative Adversarial Networks (GANs):* for creating new bioactive compounds, enhancing databases, or producing artificial microbiological data.
f. *Graph Neural Networks (GNNs):* used to research community structures and networks of microbial interactions (e.g., microbiome ecology).
g. *Transformer Models (e.g., BERT, GPT-like architectures):* for functional annotation as well as the understanding of DNA and protein sequences.
h. *Deep Belief Networks (DBNs):* used to learn features from proteomic or genomic data.
i. *Deep Reinforcement Learning (DRL):* for improving synthetic biology experiment design or microbial process control.

Table 5.1 summarizes some of the functionalities of the deep learning algorithms and their applications in microbials.

TABLE 5.1 Deep learning: description and application as it pertains to microbiology

ALGORITHM	*DESCRIPTION/FUNCTION*	*APPLICATIONS IN MICROBIOLOGY*
Convolutional Neural Networks (CNNs)	Extract spatial features from images and genomic patterns	Microbial image analysis, colony morphology classification, pathogen identification
Recurrent Neural Networks (RNNs)	Process sequential or time-series data	Modeling microbial growth, analyzing gene expression sequences
Long Short-Term Memory (LSTM)	Advanced RNNs for long-term dependencies	DNA/RNA sequence prediction, microbial evolution modeling
Autoencoders	Unsupervised feature extraction and data compression	Microbial feature learning, dimensionality reduction in genomic data
Generative Adversarial Networks (GANs)	Generate synthetic data through adversarial learning	Synthetic microbial data generation, drug molecule design
Graph Neural Networks (GNNs)	Learn from graph-structured data	Microbial community network analysis, microbiome interaction prediction
Transformer Models (e.g., BERT, GPT)	Attention-based sequence modeling	Protein function prediction, metagenomic sequence annotation
Deep Belief Networks (DBNs)	Layered probabilistic networks for feature learning	Protein classification, microbial gene function inference
Deep Reinforcement Learning (DRL)	Optimizes decision-making through feedback	Adaptive control of bioprocesses, experimental optimization in synthetic biology

5.4.2 Machine Learning Algorithms in Microbial AI

a. *Support Vector Machines (SVMs)*: frequently employed for genomic data processing, AMR prediction, and microbiological classification.
b. *Random Forest (RF):* for microbiome composition prediction, microbial taxonomy classification, and feature selection.
c. *Decision Trees*: used in challenges involving pattern recognition and microbiological phenotypic prediction.
d. *k-Nearest Neighbors (k-NN):* to predict similarities in function or group related microbial species together.
e. *Naïve Bayes Classifier:* utilized in models for illness prediction and microbial taxonomic identification.
f. *Logistic Regression:* used for tasks involving binary classification, such as antibiotic resistance or the existence of an infection.
g. *Gradient Boosting Machines (GBM):* for increasing the precision of predictions in metagenomic and microbial genomics research.
h. *Extreme Gradient Boosting (XGBoost):* extensively utilized for the classification of large-scale genomic and microbiome data.
i. *Artificial Neural Networks (ANNs):* basic ML algorithms for microbial dataset pattern recognition and prediction.
j. *Clustering Algorithms (e.g., K-Means and Hierarchical Clustering):* for classifying microbial species or groups according to genetic similarities.

Table 5.2 summarizes some of the functionalities of the ML algorithms and their applications in microbials.

TABLE 5.2 Machine learning: description and application as it pertains to microbiology

ALGORITHM	*DESCRIPTION/FUNCTION*	*APPLICATIONS IN MICROBIOLOGY*
Support Vector Machines (SVMs)	Separates data using hyperplanes	Pathogen classification, antibiotic resistance prediction
Random Forest (RF)	Ensemble of decision trees	Microbial taxonomy classification, feature selection in microbiome data
Decision Trees	Hierarchical decision modeling	Microbial phenotype prediction, infection diagnosis
k-Nearest Neighbors (k-NN)	Classifies data based on proximity	Microbial strain clustering, functional similarity detection
Naïve Bayes Classifier	Probabilistic model for categorical data	Microbial taxonomy identification, disease prediction
Logistic Regression	Linear model for binary outcomes	Infection detection, antibiotic susceptibility classification
Gradient Boosting Machines (GBM)	Boosting algorithm for predictive accuracy	Genomic feature ranking, microbial diversity prediction
Extreme Gradient Boosting (XGBoost)	Optimized version of GBM	High-performance microbial data classification
Artificial Neural Networks (ANNs)	Basic neural models for learning complex patterns	Protein-protein interaction prediction, microbial gene function analysis
Clustering Algorithms (K-Means, Hierarchical)	Groups similar data points	Microbiome composition analysis, microbial community profiling

In the concluding section, an experiment to classify a respiratory disorder is carried out by employing a CNN algorithm of a DL algorithm in a MATLAB environment.

5.5 EXPERIMENT ON CLASSIFYING RESPIRATORY DISORDER USING CNN

The aim of this study is to create and assess a DL model for the automatic classification of respiratory illnesses from medical imaging or acoustic data (such as lung sound recordings or chest X-rays) using CNNs.

5.5.1 Materials and Methods

The chest X-ray dataset (such as the NIH ChestX-ray14, COVIDx, or Kaggle Pneumonia Dataset) was utilized. It contains over 10,000 X-ray pictures with the labels "COVID-19", "Viral Pneumonia", and "Normal". The dataset is divided into training, validation, and test sets, with corresponding percentage ratios of 70%, 15%, and 15%. Figure 5.4 displays chest X-ray images A, B, and C, which show normal COVID and pneumonia, respectively.

The following are some of the steps involved in the training of the datasets:

a. Image resizing to 224 × 224 pixels
b. Normalization of pixel values (0–1 range)
c. Data augmentation: rotation, zoom, flip, and shift
d. Label encoding (one-hot encoding for multiclass classification)

5.5.2 Data Preprocessing in MATLAB

```
imds = imageDatastore('Dataset', ...
   'IncludeSubfolders', true, ...
   'LabelSource', 'foldernames');

   imds.ReadFcn = @(x)imresize(imread(x), [224 224]);
[imdsTrain, imdsVal, imdsTest] = splitEachLabel(imds, 0.7, 0.15, 0.15,
'randomized');
```

- Images were resized to **224×224×3 pixels**.
- Data augmentation (rotation, flipping, and translation) was applied using imageDataAugmenter to prevent overfitting.

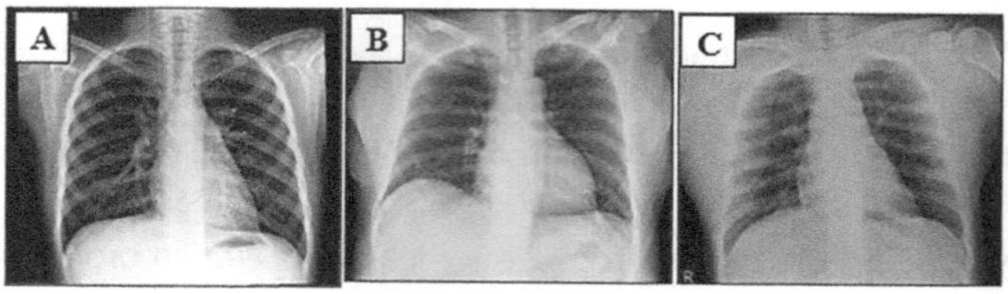

FIGURE 5.4 X-ray images: (A) normal chest, (B) COVID chest, and (C) pnemonia chest.

TABLE 5.3 CNN architecture summary

LAYER TYPE	*KEY PARAMETERS*	*PURPOSE*
Input Layer	224x224x3	Accepts image input
Conv Layer 1	16 filters, 3x3 kernel	Detects basic edges
Max Pooling	2x2	Reduces spatial size
Conv Layer 2	32 filters, 3x3 kernel	Learns mid-level features
Conv Layer 3	64 filters, 3x3 kernel	Learns complex lung features
Flatten	—	Converts 2D to 1D
Dense Layer	128 neurons (ReLU)	Learns class patterns
Output Layer	4 neurons (Softmax)	Classifies disease type

Table 5.3 is the summary of the CNN architecture.

5.5.3 Results

During the testing, validation, and training of the model, the following results were obtained as represented in Table 5.4.

5.5.4 Discussion

Respiratory diseases were accurately and successfully classified by the CNN model. Batch normalization and data augmentation enhanced the model's generalization and decreased overfitting. However, class imbalance and dataset size may have an impact on performance.

Using MATLAB's deepNetworkDesigner program, transfer learning with pretrained networks like ResNet-50, VGG-16, or EfficientNet could be used to enhance outcomes. These models frequently produce better results with fewer training samples by utilizing learned characteristics from sizable datasets like ImageNet.

In MATLAB, a CNN-based model was successfully used to categorize respiratory conditions from chest X-ray pictures. The findings show that the Deep Learning Toolbox in MATLAB offers a reliable platform for creating and assessing medical picture categorization models. CNNs have the potential to be a useful diagnostic assistance tool in respiratory healthcare, as demonstrated by the excellent accuracy attained.

The precision and accuracy for the training set are 98.1% and 94.5%, respectively. For the validation set, they are 95.6% and 93.8% for the accuracy and precision, respectively. Test set accuracy is 94.3% whereas the precision is 93.2%.

TABLE 5.4 Accuracy and precision of the training set, validation set, and test set

METRIC	*TRAINING SET (%)*	*VALIDATION SET (%)*	*TEST SET (%)*
Accuracy	98.1	95.6	94.3
Precision	94.5	93.8	93.2

5.5.5 Conclusion

This chapter makes the case that a more potent, effective, and reliable research and diagnostic pipeline can be produced by fusing traditional microbiology techniques with AI specifically designed for microbial data, or "Microbial AI." It begins by outlining complementary strengths: AI's capacity to find patterns in high-dimensional data (metagenomes, metabolomes, and imaging) and to speed up idea generation, while classical microbiology's experimental rigor, cultivability assays, and mechanistic insight. Practical integration points are described in the middle sections, including standards for data collection and curation, hybrid experimental–computational workflows (e.g., model-driven target selection followed by bench validation), and case studies where AI proposed novel taxa, metabolic pathways, or antimicrobial candidates that were later verified in the laboratory. Important issues are covered, including sample preparation variability, comprehension, reproducibility, dataset bias, and ethical and data-sharing issues. A roadmap for adoption (best-practice standards, interdisciplinary teams, and iterative validation loops) and an outlook on how integrated techniques will change environmental microbiology, drug discovery, and monitoring are presented at the end of the chapter.

REFERENCES

Ajimah, E. N., Iloanusi, O. N., Eze, M. C., & Olisa, S. (2017). Adaptive Encryption Technique to Protect Biometric Template in Biometric Database Using a Modified Gaussian Function. *IJSER*, *8*(12), 722–734.

Ajimah, N. E., & Iloanusi, O. N. (2024). Biometric Voice Recognition System in the Context of Multiple Languages : Using Traditional Means of Identification of Individuals in Nigeria Languages and English Language. *Research in Statistics*, *2*(i), 1–16.

Ajimah, N. E., Christopher, A. A., Oluwaseun, O. T., Agbotiname, L. I., & Tanko, D. S. (2023). Artififcial Intelligence-Assisted Internet of Medical Things Enabling Medical Image Processing. In *Handbook of Security and Privacy of AI-Enabled Healthcare Systems and Internet of Medical Things* (Vol. 1, pp. 309–334). Routledge Publishers: Taylor & Francis Group.

Al-Huseini, L. M. A., Kadhim, N. J., Mahdi, M. S., Ogaili, R. H., & Al-Hammood, O. (2025). Microbial Infection Disease Diagnosis and Treatment by Artificial Intelligence. *Wiadomosci Lekarskie, (Warsaw, Poland)*, *78*(2), 442–447. https://doi.org/10.36740/WLek/200511https://doi.org/10.36740/WLek/200511

Alsulimani, A., Akhter, N., Jameela, F., Ashgar, R. I., Jawed, A., Hassani, M. A., & Dar, S. A. (2024). The Impact of Artificial Intelligence on Microbial Diagnosis. *Microorganisms*, *12*(6), 1–20. https://doi.org/10.3390/microorganisms12061051

Avin, J. B., Adinath, P., Christy, S., & Noel, J. A. (2024). AI & ML Applications in Microbiology: A Comprehensive Review. *International Journal of Research Publication and Reviews*, *5*(3), 3823–3826. https://ijrpr.com/uploads/V5ISSUE3/IJRPR23844

Bhavin, J. (2023). *Chest X-ray for Screening for Tuberculosis*. https://www.atmasvasth.com/chestxray/

Eldin, C., Parola, P., & Raoult, D. (2019). Limitations of Diagnostic Tests for Bacterial Infections. *Medecine et Maladies Infectieuses*, *49*(2), 98–101. https://doi.org/10.1016/j.medmal.2018.12.004

Fonseca, D. C., da Rocha Fernandes, G., & Waitzberg, D. L. (2025). Artificial Intelligence and Human Microbiome: A Brief Narrative Review. *Clinical Nutrition Open Science*, *59*, 134–142. https://doi.org/10.1016/j.nutos.2024.12.009

Gurajala, S. (2024). Artificial Intelligence (AI) and Medical Microbiology: A Narrative Review. *Indian Journal of Microbiology Research*, *11*(3), 156–162. https://doi.org/10.18231/j.ijmr.2024.029

Hossainpour, H., & Mahmoudi, H. (2024). Revolutionizing Microbial Infection Diagnosis: The Role of Artificial Intelligence. *Iranian Journal of Medical Microbiology*, *18*(2), 66–79. https://doi.org/10.30699/ijmm.18.2.66

Imran, K., N, R., & V V, N. (2024). Role of Artificial Intelligence in Microbiology: Implementations and Obstacles. *Futuristic Trends in Biotechnology Volume 3 Book 17, 3*, 189–198. https://doi.org/10.58532/v3bjbt17p3ch1

Ince, V., Bader-El-Den, M., Arabikhan, F., & Sari, O. F. (2024). Machine Learning for Bacterial Growth Prediction and Examination of Bacterial Impact on Growth. *2024 IEEE 12th International Conference on Intelligent Systems (IS)*, 1–6. https://doi.org/10.1109/IS61756.2024.10705237

Li, Y., Cui, X., Yang, X., Liu, G., & Zhang, J. (2024). Artificial Intelligence in Predicting Pathogenic microorganisms' Antimicrobial Resistance: Challenges, Progress, and Prospects. *Frontiers in Cellular and Infection Microbiology*, *14*(November), 1–13. https://doi.org/10.3389/fcimb.2024.1482186

Mohankumar Dongre, S., & M Patel, A. (2025). Sustainable Food Systems: The Role of AI and Microorganisms in Korean Fermented Foods. *International Journal of Scientific Engineering and Research*, *13*(2), 17–23. https://doi.org/10.70729/se25216134145

Mukherjee, R., Vidic, J., Auger, S., Wen, H. C., Pandey, R. P., & Chang, C. M. (2024). Exploring Disease Management and Control through Pathogen Diagnostics and One Health Initiative: A Concise Review. *Antibiotics*, *13*(1), 1–16. https://doi.org/10.3390/antibiotics13010017

Pace, R., Schiano Di Cola, V., Monti, M. M., Affinito, A., Cuomo, S., Loreto, F., & Ruocco, M. (2025). Artificial Intelligence in Soil Microbiome Analysis: A Potential Application in Predicting and Enhancing Soil Health – A Review. *Discover Applied Sciences*, *7*(2). https://doi.org/10.1007/s42452-024-06381-4

Rahman, H., Anjuman, S., Aunni, A., Ahmed, B., Rahman, M., & Shabuj, M. H. (2024). Artificial Intelligence for Improved Diagnosis and Treatment of Bacterial Infections. *Microbial Bioactives*, *7*(1), 1–18. https://doi.org/10.25163/microbbioacts.7110036

Shelke, Y. P., Badge, A. K., & Bankar, N. J. (2023). Applications of Artificial Intelligence in Microbial Diagnosis. *Cureus*, *15*(11), 1–8. https://doi.org/10.7759/cureus.49366

Wang, Y., Lei, B., Elazab, A., Tan, E. L., Wang, W., Huang, F., Gong, X., & Wang, T. (2020). Breast Cancer Image Classification via Multi-Network Features and Dual-Network Orthogonal Low-Rank Learning. *IEEE Access*, *8*, 27779–27792. https://doi.org/10.1109/ACCESS.2020.2964276

Washington, J. A. (1996). Principles of Diagnosis – Medical Microbiology – NCBI Bookshelf. In S. Baron (Ed.), *Medical Microbiology* (4th edition). https://www.ncbi.nlm.nih.gov/books/NBK8014/

AI Modelling in Microbiology

6

Jogipeta Harihara, B. S. R. S. Sai Kaustubh, Yedla Akanksha, P. Gnana Deepu, and Kondakindi Venkateswar Reddy

6.1 INTRODUCTION

Artificial intelligence (AI) could be a cutting-edge technique in microbiology research for new discoveries in this increasingly digital world (Wang and Gao, 2025a). AI has the potential to revolutionise microbiology research by streamlining processes. In addition to simplifying our understanding of microbial dynamics, AI, particularly through machine learning (ML) and deep learning (DL), gives us the accelerating tools to create a variety of models for drug discovery and diagnostics. The intricacy of data produced by multiomics and next-generation sequencing (NGS) may be too much for conventional data analytics techniques to handle. The creation of strong tools for categorisation, pattern recognition, and prediction may be aided by AI models.

Learning non-linear representations, reducing dimensionality, imputing missing data, and integrating various data modalities are all made possible by ML and DL techniques (Ballard et al., 2024). In particular, deep generative models like variational autoencoders (VAEs) and generative adversarial networks (GANs) have been utilised to generate missing modalities, correct batch effects, and learn the latent shared embeddings between modalities (Baião et al., 2025). In microbiology, supervised/unsupervised ML techniques and deep neural architectures—such as CNNs, autoencoders, graph neural networks, and other generative models—come to mind when the term "AI modelling" is used. These techniques have already been used in a number of microbiological applications, including host–microbe interaction modelling, functional annotation, taxonomic profiling, and more (X.-W. Wang et al., 2024). For example, convolutional neural nets have been used to improve the interpretability and classification accuracy of microbiome data portrayed as graphs or images (Oshrit et al., 2022).

DL techniques offer adaptable data integration methodologies for multi-omics integration. Depending on the objective and data structure, integration can be carried out in three ways: early integration, which involves concatenating features; intermediate integration, which involves learning a shared latent space; or late integration, which involves an ensemble of modality-specific models. Additionally, DL techniques enable generative architectures to handle missing modalities (e.g., when not all omics layers are present on every sample) (Ballard et al., 2024). In this chapter, we will discuss the overview of the role of AI and models in the growing field of microbiological research.

DOI: 10.1201/9781003727514-7

6.2 FUNDAMENTALS OF AI FOR MICROBIOLOGY

6.2.1 What Is Machine Learning?

ML is the branch of AI that grasps knowledge from the raw data available using special algorithms and models in order to improve the understanding of the variations present in data (X.-W. Wang et al., 2024). ML is broadly categorised into three models of learning, which include supervised learning, unsupervised learning, and reinforcement learning. The detailed explanation of these models is further discussed in the next sub-sections. Four major steps lead to the development of ML-based algorithms. The initial step involves gathering data from a source using a suitable technique, and that is followed by the utilisation of clustering in order to prepare the operational taxonomic units (OTU) table. After obtaining the OTU table, the selection process is initiated to increase the performance and accuracy. The ultimate step involves building the model using the training data and then checking how well it summarises using a test set (Qu et al., 2019).

6.2.1.1 Supervised learning

The supervised learning works on the principles of regression analysis and statistical classification. In supervised learning, a model is built from categorised data to identify the unpredictable data (Knights et al., 2010). For example, clinical microbiologists, to mark the presence of pathogenic organisms, may annotate stained slide images. These expert annotations are then used to build ML systems capable of rapidly detecting pathogens in diagnostic images (Burns et al., 2023). The most widely used algorithms of supervised learning include decision trees (DT), ordinary least squares regression (OLS), naive bayesian classification (NB), logistic regression (LR), support vector machine (SVM), ensemble methods (EM), random forests linear regression (RF), linear discriminant analysis (LDA), k-nearest neighbour (k-NN) algorithm, multi-layer perceptron (MLP), and convolutional neural networks (CNN) (C. Xu et al., 2025). Each algorithm has its own positive and negative elements. The NB model of supervised learning is used for the sequential learning of dealing with multiple tasks involved in classification for limited-scale data (Jiang et al., 2022). RF model is quite advantageous in the growing world full of data for faster training of large samples, but it can still overfit if the data has too much noise or if some features are split into too many tiny value groups (C. Xu et al., 2025). k-NN algorithm uses the principle of matching the template to solve the problems, but the low-paced prediction speed of this model allows one to opt for alternatives, though this is highly accurate and efficient. With a clear vision of well-defined coefficients, a simple and efficient model named the LR model was developed, which has the capacity to identify the probability of classifying a sample as positive. The major limitation of this model is low accuracy and its vulnerability to unfitting (Jiang et al., 2022). SVMs work well in complex, high-dimensional problems, even when features exceed samples. But they are not ideal for big data and struggle with missing information (C. Xu et al., 2025).

The following are the steps involved in supervised learning (Goodswen et al., 2021)

1. Data collection with labels: High-quality, diverse examples drive accuracy and generalisation; flawed input yields flawed predictions.
2. Extract features: Organise data in tables and consider noise; various algorithms are more resilient against spurious values.

3. Prepare data: Clean and preprocess raw data so models can train reliably and produce useful outputs.
4. Split data: Into training and evaluation sets, such as 80/20; or into train/test/validation for tuning parameters.
5. Choose a model: Problem, data, hardware, accuracy needs, ease, speed, and interpretability; compare several models.
6. Training the Model: Improve the model's performance by continuously learning predictive patterns over updates.
7. Model Evaluation: Test on unseen data to estimate real-world performance and reliability.
8. Refine parameters: If necessary, adjust data, steps, or hyperparameters; optimisation is often experimental.
9. Make predictions: Use the trained model to make predictions on real-world data and check results whenever possible.

6.2.1.2 Unsupervised learning

A category of ML independent of human insights. For example, by using an unaided model, an image can be grouped based on similar and dissimilar features (Burns et al., 2023). It includes methods like clustering, dimensionality reduction, and association rule extraction. k-means clustering, principal component analysis (PCA), principal coordinate analysis (PCoA), t-distributed stochastic neighbour embedding (t-SNE), and the Apriori algorithm are the common methods in unsupervised learning (X. Wang et al., 2024). Specifically, PCA is a linear method that offers high interpretability at low information loss; t-SNE, which does not have these features, seems to be used mostly in two-dimensional visualisations in single-cell transcriptomics. However, dimensionality reduction may imply some loss of information, decreased interpretability in certain transformations, and higher technical expertise. In contrast, clustering algorithms group similar data points based on measures of their similarity to predict patterns or subgroups within a dataset. Applications in microbiology include gene expression analysis, protein sequence-structure-function mapping, classification of microbial communities, and detection of infectious disease outbreaks. Accordingly, K-means allows the identification of diverse clinical phenotypes and estimation of bacterial community composition, while agglomerative clustering discloses distinct community types. The major strengths of clustering are simplicity, flexibility, and tolerance against noise, whereas limitations encompass high computational complexity, sensitivity to parameter choices, and the need to presuppose the number of clusters in many applications (C. Xu et al., 2025).

The steps in unsupervised learning include the following (Goodswen et al., 2021)

1. Collect labelled examples: Labelled data is only useful for downstream evaluation of clustering performance and not for training.
2. Extract features: Compute and tabulate features from the data as you would in supervised ML.
3. Data preparation: Data cleaning, selection of distance/dissimilarity metric appropriate for the application, and computation of pairwise distance matrix as input to clustering.
4. Model selection: Run several clustering algorithms on the distance matrix and choose a suitable one for your purpose.
5. Choose k (cluster number): Choose an appropriate k (e.g., using existing methods) such that a constraint of assignment into k labels will be enforced; note hierarchical clustering forms nested subsets.
6. Cluster evaluation: If labels are available, check whether same-label samples co-cluster and calculate relevant quality measures.

6.2.1.3 *Reinforcement learning*

Reinforcement learning is all about teaching agents to learn through trial and error in changing environments. In view of labelled data, the agent tries different actions, sees the results, and adjusts to maximise rewards. That way, it explores on its own and acquires knowledge. This is especially useful for problems that are much too complex to program directly (X. Wang et al., 2024).

6.2.2 Deep Learning

DL is a branch of ML algorithms that relies on artificial neural nodes or units, which convert the input to output via nodes. These nodes are arranged differently to form layers in a network with connectivity (Medina et al., 2022). DL shows strong performance on large datasets, often surpassing traditional ML methods that are limited as the data size increases (Roy et al., 2024). Within DL, CNNs and recurrent neural networks (RNNs) are specialised tools designed for different types of data.

6.2.2.1 *Convolutional neural networks (CNNs)*

The development of CNNs was based on the functioning of cells in the cortex that aid in visual mechanisms, and hence, they are frequently used in image processing and classification. CNNs consist of convolutional layers, which apply shared filters to extract features, and pooling layers, which summarise feature maps to reduce variance and preserve key information (Cao et al., 2018). They are used in the structural characterisation of the microbiome and also in metagenomics in the process of sequencing classification using common patterns (Roy et al., 2024).

6.2.2.2 *Recurrent neural networks (RNNs)*

RNNs are suitable for the processing of temporal or sequential data using the cycles they have in connections. The most important models of RNNs in use today are the long short-term memory (LSTM) networks, which capture long-term dependencies, and GRUs, which introduce selective input or forgetting. These networks can be used to analyse DNA sequences for predicting the presence of specific elements, such as phages, or to track the abundance of microbial species, enabling predictions about the dynamics and evolution of microbial ecosystems in the future. (Roy et al., 2024).

6.3 HANDLING AND PREPROCESSING MICROBIOME-SPECIFIC DATA: FROM SEQUENCES TO FEATURES

6.3.1 Genomics & Metagenomics Data: From Raw Reads to Taxonomic/Functional Features

The analysis of microbiome genomic and metagenomic datasets starts with the conversion of raw sequencing reads into biologically relevant features. Current NGS platforms, including Illumina for short reads and Nanopore/PacBio for long reads, produce billions of reads that undergo an exhaustive preprocessing step. Standard pipelines begin with quality filtering using software such as FastQC and

Trimmomatic, which remove low-quality bases and adapter contamination to guarantee reliability in downstream analyses (Bolger et al., 2014).

The cleaning of the reads must then be followed by the removal of host reads, usually done by alignment to the human genome using Bowtie2 to ensure that microbial diversity estimates are not inflated by host contamination. For taxonomic profiling, Kraken2 and MetaPhlAn3 are among the current tools used to classify reads based on their k-mer signatures or clade-specific markers, respectively, enabling species-level or even strain-level identification with high accuracy (Truong et al., 2015; Wood et al., 2019). Because microbiome data are inherently compositional—each sample sums to a constant—classical statistical methods produce misleading correlations. The solution to compositionality comes through transformations such as the centred log-ratio (CLR) described by Gloor et al. (2017).

6.3.2 Transcriptomics (Bulk & Single-Cell)

Transcriptomic profiling in microbiome research observes the pattern of gene expression and microbial responses to environmental alterations. In the case of bulk microbial RNA-seq, the workflow would include rRNA depletion, quality control, and mapping either to reference genomes or metagenomic assemblies. With lightweight pseudo-alignment implemented, Salmon and Kallisto can achieve highly accurate quantification (Patro et al., 2017b). The preprocessing comprises barcode correction, UMI collapsing, and normalisation by using scVI and Scanpy, which enables representation learning in noisy microbial single-cell datasets (Lopez et al., 2018). Single-cell transcriptomics for microbiomes, though challenging due to bacterial cell walls and low mRNA content, has continued to evolve with methods such as in situ combinatorial indexing known as PETRI-seq (Blattman et al., 2020), and smRandom-seq, droplet-microfluidics combined with in situ reverse transcription (Z. Xu et al., 2023). To reduce the batch effects, algorithms like Combat-seq or Harmony are used; this way, it is ensured that the biological variation surpasses the technical noise (Korsunsky et al., 2019). It enables metatranscriptomics when combined with metagenomics and thus allows one to correlate the abundance of genes with real-time functional activity essential to understand microbial metabolism in disease and/or environmental systems.

6.3.3 Proteomics & Metabolomics

Metaproteomics refers to the identification and quantification of proteins expressed by microbial communities. Sample preparation generally involves protein extraction, digestion (e.g., trypsinisation), and separation by liquid chromatography, followed by tandem mass spectrometry (MS/MS). The complexity of microbial proteomes, which are often comprised of thousands of taxa-derived proteins, presents a formidable computational challenge. MS-based proteomics of microbial systems often necessitates advanced MS instruments and bioinformatics pipelines suited to mixed-community samples. In metaproteomics, database search represents one major bottleneck. The standard way to improve peptide identifications by researchers is by building custom protein databases derived from metagenomic assemblies. FDR control, especially over large sequence spaces, is critical but non-trivial in the microbial context (Dumas et al., 2024). More recently, newer computational methods, including DL algorithms for peptide-spectrum matching, have been developed to improve identifications. For example, methods such as DeepFilter have been proposed to enhance confident peptide identification from complex metaproteomic datasets (X. Guo and Feng, 2020). Normalisation and quantitation in metaproteomics are also a challenge: sample-to-sample variation in extraction efficiency, digestion, and instrument response needs to be corrected, usually by means of internal standards or label-free quantitation methods. Though there are classical proteomics methods, community-wide workflows are not yet fully standardised. The Metaproteomics Initiative, a collective effort, aims at the standardisation of protocols, data sharing, and bioinformatic pipelines in metaproteomics across studies (Van Den Bossche et al., 2021). Metaproteomic data analysis commonly involves filtering low-quality peptide identifications, normalising protein abundances, and assigning proteins to microbial taxa based on taxonomic and functional annotations from metagenomic or reference databases. Community complexity and homologous proteins across species make the disentanglement of protein origins a challenge (Heyer et al., 2025).

6.3.4 High-Throughput Imaging in Microbiome Research

High-throughput imaging of microbial communities is generally carried out by fluorescence microscopy, phase-contrast, or time-lapse live-cell imaging in order to capture highly resolved morphological, spatial, and dynamic phenotypes in individual microbes or microbial consortia. Raw microscopy data often suffer from non-uniform illumination, background noise, and optical distortions, which have to be corrected by preprocessing steps including flat-field correction and deconvolution to improve the signal-to-noise ratio for reliable quantitative analysis (Carpenter et al., 2006). Specific cell segmentation is one of the crucial steps in high-throughput imaging. Bacteria pose particular difficulties because they are small, often densely packed, and may take radical shapes under variable conditions. Such algorithms very often cannot define sharp boundaries between cells where the latter are elongated, branched, or in morphological transition. In turn, deep-learning approaches include Omnipose, which is a neural network algorithm utilising a distance field gradient for precision, morphology-independent segmentation of bacterial cells, outperforming classical and earlier deep-learning approaches across mixed microbial cultures, antibiotic-treated populations, and highly elongated or branched cells (Cutler et al., 2022). Synthetic micrographs, as generated by approaches like SyMBac, can therefore be used for training or retraining segmentation networks in cases where annotated bacterial training data are limited. SyMBac simulates realistic bacterial images under different imaging conditions, and retraining with SyMBac data significantly improves the segmentation accuracy of Omnipose, especially in microfluidic devices and across different morphologies (Hardo et al., 2022). After segmentation, single-cell quantitative features are extracted. Image analysis platforms such as CellProfiler enable high-throughput cellular metric extraction, including cell size, fluorescence intensity, shape descriptors (e.g., elongation and compactness), and spatial relationships between cells, to enable downstream statistical or machine-learning analyses of microbial phenotypes (Carpenter et al., 2006). In addition, sophisticated tracking pipelines incorporate segmentation with time-lapse microscopy to follow cellular behaviours in their timescales of interest, including growth, division, motility, and interaction dynamics. For example, the RABiTPy pipeline incorporates Omnipose for segmentation and then extracts trajectories, speed distributions, and density heatmaps, thereby enabling high-content motility analysis in bacterial populations (Sen et al., 2025).

In summary, high-throughput imaging in microbiome research creates a pipeline for the translation of rich microscopy data into quantitative phenotypic readouts by combining rigorous image preprocessing, DL-based segmentation, synthetic image augmentation, and automated feature extraction/tracking (Figure 6.1). Such phenotypic readouts can be integrated with genomic, transcriptomic, or other "omics" layers to achieve a multi-modal understanding of microbial function and behaviour.

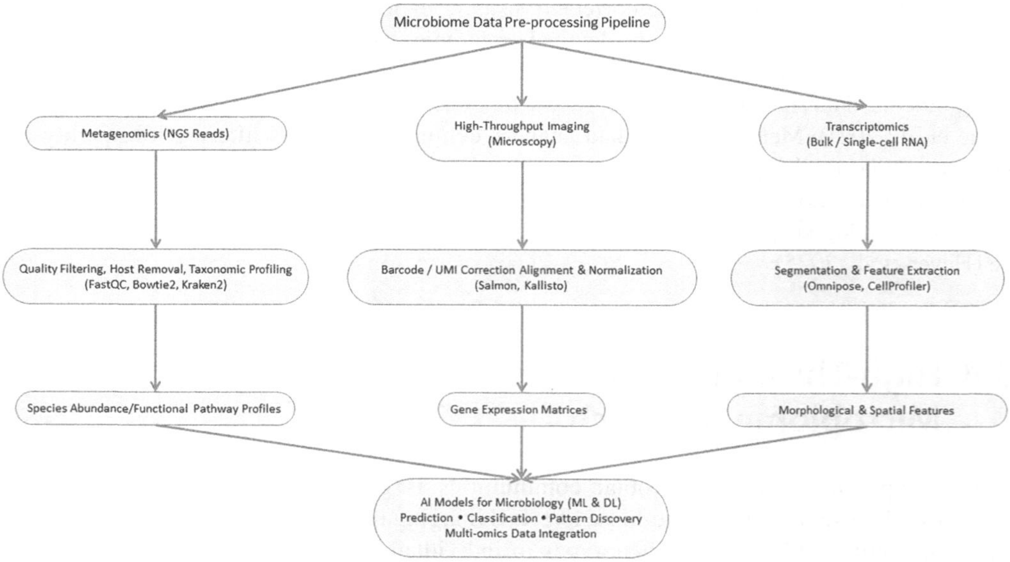

FIGURE 6.1 Microbiome data preprocessing pipeline.

6.4 INTEGRATIVE MODELLING TECHNIQUES FOR MICROBIOLOGY

6.4.1 Multi-omics Data Integration

Multi-omics modelling is a systematic computational technique that combines and analyses data from different "omics" technologies using the same set of biological samples to gain a more comprehensive and in-depth understanding of a biological system (Subramanian et al., 2020).

This technique can be used to determine the host–microbe interactions and microbiome function, understand the mechanisms of antibiotic resistance and pathogenicity, and find new natural products and biosynthetic pathways (Blin et al., 2021; Lin et al., 2021; Zhou et al., 2019).

While the data integration tools are used to obtain biological solutions, each tool is developed in a different model to perform integration using distinct methods.

6.4.1.1 Dimensionality reduction and clustering methods

To visualise high-dimensional omics data and find overarching structures, dimensionality reduction is an essential initial step. Techniques such as multi-omics factor analysis (MOFA) have been developed to analyse many data modalities at the same time. A limited number of latent variables that reflect the common causes of variation across all input data types (such as genotypic, microbiome, and metabolomic) are identified by MOFA, an unsupervised linear model. MOFA identifies the essential characteristics from each omics layer that cause these variances by enforcing sparsity. For example, it has been effectively used to uncover integrated genotypic, microbiome, and metabolomic variables linked to the risk of colorectal cancer and adenoma (Chetty and Blekhman, 2024).

In order to find physiologically significant patterns, clustering techniques are used in conjunction with dimensionality reduction. Clustering samples to identify "enterotypes", or different conditions of the human gut microbiota, is a common use in microbiology. Methods such as iCluster use matrix factorisation to accomplish simultaneous clustering across various omics layers for integrative analysis. iCluster finds reliable, system-wide classifications and emphasises the most instructive biomarkers from each data type by maximising variance explained while enforcing sparsity requirements (Chetty and Blekhman, 2024).

6.4.1.2 Correlation-based methods

Correlation and association approaches establish unique pairwise correlations between characteristics at various omics levels, resulting in hypotheses regarding microbial interactions with host chemicals. Pairwise similarity metrics between features from various datasets are computed using basic methods. While Spearman correlation and Kendall's tau discover monotonic non-linear correlations but still struggle with compositionality, Pearson correlation finds linear relationships but is sensitive to compositional data. By converting data to Euclidean space or assuming sparse correlation networks, specialised techniques like the CLR transformation and algorithms like SparCC handle compositionality (Chetty and Blekhman, 2024).

Multivariate techniques find coordinated multi-omic programs that go beyond pairwise correlations. Canonical correlation analysis (CCA) identifies optimally correlated linear feature combinations between two datasets. Through L1 regularisation, its extension, sparse CCA (sCCA), enhances interpretability by generating components that emphasise just the most important microorganisms and chemicals. Additional generalisations, such as sparse generalised CCA (SGCCA), make it possible to integrate three or more omics tables at once, exposing intricate networks of interactions between host plasma metabolites, faecal metabolites, and gut microbiota (Chetty and Blekhman, 2024).

6.4.1.3 Regression and classification methods

Models of regression and classification are effective supervised learning techniques for forecasting certain results from multi-omics data. Disease severity and other continuous responses are predicted using regression models, and different data distributions can be accommodated by generalised linear models (GLMs). Regularised regression techniques (LASSO, Ridge, and Elastic Net) are crucial for variable selection and preventing overfitting in high-dimensional microbiological data when features outnumber samples. In order to identify critical regulatory interactions, such as those between steroid hormones and immune cells during pregnancy, advanced designs like the nested elastic net use two-stage modelling, first creating individual models for each data modality and then combining their predictions in a second-level model (Chetty and Blekhman, 2024; Ghaemi et al., 2019).

Samples are categorised into predetermined groups using classification models. While random forests produce reliable classifiers from ensembles of DT, LDA and its sparse variation (sPLS-DA) identify feature combinations that optimally divide classes. These techniques have found biomarkers for diseases, including major depressive disorder and hepatic steatosis, as well as keystone species in the microbiomes of cystic fibrosis (Chetty and Blekhman, 2024).

6.4.1.4 Network methods

Reconstructing a thorough, system-wide map of interactions is the ultimate objective of many multi-omics investigations, which go beyond pairwise relationships and prediction models. A strong framework for this is provided by network (or graph) analysis, which simultaneously visualises and measures the intricate web of connections between microorganisms, host molecules, and clinical phenotypes. This method enables researchers to view the system's "forest" rather than simply certain "trees". The network consists of a set of nodes (such as a particular microbe, metabolite, or gene) and edges (the connections between them, which indicate a strong statistical link like a correlation or mutual information).

Network construction workflows include:

- Computing pairwise correlations (such as SparCC and Pearson correlation) between every feature across omics layers.
- Filtering out weak correlations by using a significance threshold.
- Forming the network by joining nodes with edges that surpass this threshold.

After a network is built, network topology may be examined to determine modules and cliques, which are groupings of nodes with a high degree of connectivity that might constitute functional units (e.g., a group of co-occurring microorganisms and their linked metabolites) and to locate hubs, a highly linked nodes that might be important regulatory chemicals or keystone species (Chetty and Blekhman, 2024).

Bayesian networks: Bayesian Networks (BNs) provide a more sophisticated framework for determining possible directed and causal linkages, whereas correlation networks detect associations. An arrow from Node A to Node B indicates that Node A may have an impact on Node B, and a BN depicts the joint probability distribution of all variables as a directed acyclic graph. Approximation methods such as hill-climbing, which rapidly add edges to optimise a model fit criterion, are frequently used to learn BN structure from data (Chetty and Blekhman, 2024).

6.4.1.5 Spatial and temporal multi-omics integration methods

Moving beyond the composition of samples to comprehend their environment in geography and time is the last frontier in multi-omics integration. These methods acknowledge that interactions between microbial communities are dynamic processes and that these populations are not uniform mixes. The best insight into the causal dynamics and functional organisation of host–microbe interactions is obtained by integrating spatial and temporal dimensions (Chetty and Blekhman, 2024).

Spatial integration approaches provide information about the physical position of molecular and microbial interactions. High-resolution spatial maps of microbial species directly within a tissue, such as the gut, may be created using techniques like HiPR-FISH, and gathering multi-omics data from designated, mapped areas and spatially correlating them constitutes the main integration approach.

By gathering multi-omics data from the same individual over several time periods, time-series integration makes it possible to examine when and how interactions change (Chetty and Blekhman, 2024).

Studies of Longitudinal Cohorts reveal stability and dynamics and produce ideas regarding interactions (Mihindukulasuriya et al., 2021). DBNs, or Dynamic Bayesian networks (DBNs), predict the likelihood of a future state (such as microbe abundance) based on the system's historical histories. Temporal integration can be used to monitor functional responses in environmental microbiology. In order to discover unique functional niches, methods like multidimensional scaling (MDS) and correlation of metatranscriptomic and metaproteomic data over time can be used (Chetty and Blekhman, 2024).

6.4.2 Generative AI for Microbiology

Generative AI is a specialised paradigm that comes after integrative modelling and produces new data instead of only analysing it. While generative models such as generative adversarial networks (GANs) learn underlying data distributions to generate realistic, synthetic outputs, classical AI is excellent at identifying patterns in existing datasets. This makes it possible to create synthetic microbial genomic data to supplement small datasets in microbiology, create novel protein sequences or metabolic pathways in silico, and simulate the dynamics of microbial communities, all of which speed up discovery and hypothesis generation in ways that were previously limited by data scarcity (Roy et al., 2024).

6.4.2.1 Variational autoencoders (VAEs) & generative adversarial networks (GANs)

Generative AI presents a framework for producing new biological data by building on analytical models. De novo molecular design is made possible by two main architectures: GANs and VAEs. A compressed latent representation of biological sequences, such as antimicrobial peptides (AMPs), is learnt by VAEs. Researchers can produce new, theoretically active peptide candidates for medicinal discovery by carefully sampling and interpolating within this latent space, for example, between areas of high and low bioactivity (Dean and Walper, 2020). On the other hand, GANs use a competitive framework in which a discriminator separates sequences from actual proteins while a generator generates sequences from random noise. This combative process teaches the generator the detailed rules governing protein sequences, making it able to generate a variety of viable enzyme variants, as shown by "ProteinGAN", which was trained on bacterial malate dehydrogenase. When combined, these generative models significantly increase the explorable sequence space, surpassing natural variation to computationally create novel functional biomolecules (Repeka et al., 2021).

6.4.2.2 Large language models (LLMs) for sequences

LLMs pre-trained on evolutionary-scale sequence data offer a potent, all-purpose framework for comprehending protein function that goes beyond creating particular molecules. By anticipating masked amino acids in sequences, the evolutionary scale model (ESM), a transformer-based architecture trained on 250 million protein sequences, learns the basic "syntax" and "semantics" of proteins. The 3D structures of proteins are predicted with the help of massive biological information captured by internal representations (embeddings), which ESM learns. The embeddings acquire the data without direct structural supervision. Additionally, these representations may be refined to predict protein function and mutation effects. This redefinition provides a universal tool to comprehend the functional landscape of metagenomic data and microbial proteomes, changing the researchers' ability to annotate unidentified proteins from a variety of microbiomes (Rives et al., 2021).

6.4.3 Causal Inference and Mechanistic Modelling

A paradigm shift towards detecting causality and biochemical mechanisms responsible for it in host–microbe interactions is represented by causal inference and mechanistic modelling, which build upon pattern-discovery techniques. In order to be certain that microbial activity actually causes phenotypic changes and to clarify the precise metabolic pathways involved, this methodology goes beyond simply finding correlations (Ruiz-Perez et al., 2021).

DBNs are an important technique for determining causal links; they use longitudinal multi-omics data to gather directional interactions across variables over time. DBNs identify sequences of biological events through modelling temporal relationships and infer causal processes from observational data. For example, they can determine that a metabolite fluctuation precedes and may cause a shift in microbial abundance (Ruiz-Perez et al., 2021).

In addition, by reconstructing an organism's whole metabolic network, genome-scale metabolic models (GEMs) offer a mechanistic framework. These models function as biochemical "wiring diagrams" that may be limited by omic data to simulate metabolic flow rates under various circumstances. By imitating the flow of metabolites through interconnected pathways, GEMs allow researchers to predict how microbial communities react to deviations, such as dietary changes or the introduction of new species. This gives researchers the ability to generate testable hypotheses about the biochemical mechanisms governing microbial ecology and host interactions (Magnusdottir et al., 2017).

A combination of DBNs and GEMs offers a powerful toolset for moving beyond statistical correlations to causal, mechanistic knowledge of microbial systems.

6.5 THE CRITICAL ROLE OF MODEL INTERPRETATION AND EXPLAINABLE AI (XAI)

6.5.1 The "Black Box" Problem

The "black box" character of AI models in microbiology raises a major obstacle to biological discovery and clinical use as they get more complicated. Knowing why a model predicts a certain disease state or microbial interaction is just as important as the prediction's accuracy to obtain reliable and useful results (Arrieta et al., 2020; Kelly et al., 2019).

Although advanced models like ensemble techniques and deep neural networks can obtain great prediction accuracy, they provide little fundamental information about how they make decisions (Castelvecchi, 2016). In clinical microbiology, where mechanistic knowledge is the ultimate aim and diagnostic judgements need to be justified, the lack of transparency is a significant obstacle. To unlock these mysteries, XAI offers a set of tools (Antoniadi et al., 2021; Riberio et al., 2016.

6.5.2 Tools for Model-Agnostic Interpretation

These techniques can be used post-hoc to explain the predictions of any model.

6.5.2.1 SHapley additive exPlanations (SHAP)

SHAP measures the marginal contribution of each input characteristic (such as the quantity of a particular microorganism) to a single prediction. It is based on cooperative game theory. By displaying the degree to which each feature influenced the model's output for a given sample towards a certain class (such as "disease" or "healthy"), it offers a consistent measure of feature relevance (Lundberg and Lee, 2017).

6.5.2.2 Local interpretable model-agnostic explanations (LIME)

LIME uses a modified version of the original data to create a smaller, interpretable model (like linear regression) that approximates the complicated model locally around a particular prediction. The characteristics that were most important in that specific case are highlighted by this local surrogate model (Riberio et al., 2016).

6.5.3 Feature Importance

A lot of models, like Random Forests, include built-in techniques for determining global feature importance. Usually, this involves figuring out how much a feature reduces impurity across all of the forest's trees (Breiman, 2001).

6.5.4 Linking Model Outputs to Biological Insight

The ability to convert XAI's results into provable biological hypotheses is what gives it its power. For example, a consistently high SHAP value for a particular, unidentified microbe in a model predicting treatment indicates that the microbe plays a major role. This type of computational finding immediately makes the microbe eligible for additional in vitro confirmation, including isolating it and seeing how it interacts with the medication. Therefore, XAI ensures that AI functions as a tool for discovery rather than just a guide by connecting the gap between causal microbiology and predictive statistics (Riberio et al., 2016).

6.6 CONCLUSION OF INTEGRATING AI INTO THE MICROBIAL SCIENCES: FROM PATTERN RECOGNITION TO MECHANISTIC INSIGHT

Inculcating AI, specifically ML and DL, into microbiology represents a paradigm shift in researchers' ability to understand the microbial world. From the foundational ideas of ML to the uses of DL, this chapter has established a clear narrative: AI is a new fundamental pillar of the scientific method in microbiology, not just a statistical instrument.

The large amounts of data from multi-omics technologies made it impossible to extract deep, nonlinear insights using typical analytical techniques. The basic ML paradigms of supervised, unsupervised, and self-supervised learning offer the foundations required to convert the unprocessed data from multi-omics into knowledge that can be used for all sorts of analysis, from disease classification to the identification of new microbial community structures.

In order to create reliable and physiologically significant models, the distinctive features of microbiological data, such as sparsity and compositionality, must be handled using methods like CLR transformation. The model selection is equally important. Researchers observed that:

- For many classification and ranking problems, traditional models such as random forests and SVMs provide a strong, comprehensible foundation.
- DL architectures like CNNs for pictures and sequences, RNNs for time series, and autoencoders for dimensionality reduction offer the adaptability and ability to imitate the large complexity present in biological systems.
- New paradigms like Generative AI and graph neural networks allow the creation of synthetic data to get beyond the constraints of limited sample sizes and represent intricate ecological relationships.

However, the potential of these models to develop provable biological theories is ultimately more valuable than their predicted accuracy alone. At this point, XAI becomes a scientific and ethical need rather than just a technological tool. By converting a model's output into a particular, mechanistic inquiry, tools like SHAP and LIME serve as a link between the "black box" and the bench scientist.

Therefore, developing a deep, integrative partnership between computational models and experimental validation is more important for the future of AI in microbiology. To identify the most promising leads, such as the essential species, genes, or interactions causing a phenotype, AI must sort through the vast complexity of microbial data. In order to close the circle and establish a positive cycle of discovery, the microbiologist's job is to create detailed tests to confirm these computer predictions.

In brief, AI modelling has permanently changed the field of microbiology from observation to prediction and mechanistic comprehension. It enables going beyond just cataloguing correlations to understand the causal laws governing microbial life. As this border is approached, the most exciting discoveries will come from multidisciplinary teams combining pipettes and algorithms, ultimately revealing the significant questions that lie within the vast, invisible majority of life on Earth.

REFERENCES

Antoniadi, A. M., Du, Y., Guendouz, Y., Wei, L., Mazo, C., Becker, B. A., & Mooney, C. (2021). Current challenges and future opportunities for XAI in machine learning-based clinical decision support systems: A systematic review. *Applied Sciences*, *11*(11), 5088. https://doi.org/10.3390/app11115088

Arrieta, A. B., Diaz-Rodriguez, N., Del Ser, J., Bennetot, A., Tabik, S., Barbado, A., Garcia, S., Gil-Lopez, S., Molina, D., Benjamins, R., Chatila, R., & Herrera, F. (2020). Explainable artificial intelligence (XAI): Concepts, taxonomies, opportunities and challenges toward responsible AI. *Information Fusion*, *58*, 82–115. https://doi.org/10.1016/j.inffus.2019.12.012

Baião, A. R., Cai, Z., Poulos, R. C., Robinson, P. J., Reddel, R. R., Zhong, Q., Vinga, S., & Gonçalves, E.. 2025. "A Technical Review of Multi-omics Data Integration Methods: From Classical Statistical to Deep Generative Approaches." *arXiv (Cornell University)*, January. https://doi.org/10.48550/arxiv.2501.17729

Ballard, J. L., Wang, Z., Li, W., Shen, L., & Long, Q. (2024). Deep learning-based approaches for multi-omics data integration and analysis. *BioData Mining*, *17*(1), 38. https://doi.org/10.1186/s13040-024-00391-z

Blattman, S. B., Jiang, W., Oikonomou, P., & Tavazoie, S. (2020). Prokaryotic single-cell RNA sequencing by in situ combinatorial indexing. *Nature Microbiology*, *5*(10), 1192–1201. https://doi.org/10.1038/s41564-020-0729-6

Blin, K., Shaw, S., Kloosterman, A. M., Charlop-Powers, Z., van Wezel, G. P., Medema, M. H., & Weber, T. (2021). antiSMASH 6.0: Improving cluster detection and comparison capabilities. *Nucleic Acids Research*, *49*(W1), W29–W35. Oxford Academic. https://doi.org/10.1093/nar/gkab335

Bolger, A. M., Lohse, M., & Usadel, B. (2014). Trimmomatic: A flexible trimmer for illumina sequence data. *Bioinformatics*, *30*(15), 2114–2120. https://doi.org/10.1093/bioinformatics/btu170

Breiman, L. (2001). Random forests. *Machine Learning*, *45*(1), 5–32. https://doi.org/10.1023/A:1010933404324

Burns, B. L., Rhoads, D. D., & Misra, A. (2023). The use of machine learning for image analysis artificial intelligence in clinical microbiology. *Journal of Clinical Microbiology*, *61*(9), e0233621. https://doi.org/10.1128/jcm.02336-21

Cao, C., Liu, F., Tan, H., Song, D., Shu, W., Li, W., Zhou, Y., Bo, X., & Xie, Z. (2018). Deep learning and its applications in biomedicine. *Genomics Proteomics & Bioinformatics*, *16*(1), 17–32. https://doi.org/10.1016/j.gpb.2017.07.003

Carpenter, A. E., Jones, T. R., Lamprecht, M. R., Clarke, C., Kang, I. H., Friman, O., Guertin, D. A., Chang, J. H., Lindquist, R. A., Moffat, J., Golland, P., & Sabatini, D. M. (2006). CellProfiler: Image analysis software for identifying and quantifying cell phenotypes. *Genome Biology*, *7*(10), R100. https://doi.org/10.1186/gb-2006-7-10-r100

Castelvecchi, D. (2016). Can we open the black box of AI? *Nature*, *538*(7623), 20–23. https://doi.org/10.1038/538020a

Chetty, A., & Blekhman, R. (2024). Multi-omic approaches for host-microbiome data integration. *Gut Microbes*, *16*(1), 2297860. PubMed Central National Center for Biotechnology Information. https://doi.org/10.1080/19490976.2023.2297860

Cutler, K. J., Stringer, C., Lo, T. W., Rappez, L., Stroustrup, N., Peterson, S. B., Wiggins, P. A., & Mougous, J. D. (2022). Omnipose: A high-precision morphology-independent solution for bacterial cell segmentation. *Nature Methods*, *19*(11), 1438–1448. https://doi.org/10.1038/s41592-022-01639-4

Dean, S. N., & Walper, S. A. (2020). Variational autoencoder for generation of antimicrobial peptides. *ACS Omega*, *5*(33), 20746–20754. https://doi.org/10.1021/acsomega.0c00442

Dumas, T., Pinna, R. M., Lozano, C., Radau, S., Pible, O., Grenga, L., & Armengaud, J. (2024). The astounding exhaustiveness and speed of the Astral mass analyzer for highly complex samples is a quantum leap in the functional analysis of microbiomes. *Microbiome*, *12*(1), 46. https://doi.org/10.1186/s40168-024-01766-4

Ghaemi, M. S., DiGiulio, D. B., Contrepois, K., Callahan, B., Ngo, T. T. M., Lee-McMullen, B., & Lehallier, B.. (2019). Multiomics modeling of the immunome, transcriptome, microbiome, proteome and metabolome adaptations during human pregnancy. *Bioinformatics*, *35*(1), 95–103. https://doi.org/10.1093/bioinformatics/bty537

Gloor, G. B., Macklaim, J. M., Pawlowsky-Glahn, V., & Egozcue, J. J. (2017). Microbiome datasets are compositional: And this is not optional. *Frontiers in Microbiology*, *8*, 2224. https://doi.org/10.3389/fmicb.2017.02224

Goodswen, S. J., Barratt, J. L. N., Kennedy, P. J., Kaufer, A., Calarco, L., & Ellis, J. T. (2021). Machine learning and applications in microbiology. *FEMS Microbiology Reviews*, *45*(5). https://doi.org/10.1093/femsre/fuab015

Guo, X., & Feng, S. (2020). Deep Learning for Peptide Identification from Metaproteomics Datasets. *arXiv (Cornell University)*. https://doi.org/10.48550/arxiv.2009.11241

Hardo, G., Noka, M., & Bakshi, S. (2022). Synthetic Micrographs of Bacteria (SyMBac) allows accurate segmentation of bacterial cells using deep neural networks. *BMC Biology*, *20*(1), 263. https://doi.org/10.1186/s12915-022-01453-6

Heyer, R., Wolf, M., Benndorf, D., Uzzau, S., Seifert, J., Grenga, L., Pabst, M., Schmitt, H., Mesuere, B., Van Den Bossche, T., Haange, S., Jehmlich, N., Di Luca, M., Ferrer, M., Serrano-Villar, S., Armengaud, J., Bode, H. B., Hellwig, P., Masselot, C. R., & Wilmes, P. (2025). Metaproteomics in the one health framework for unraveling microbial effectors in microbiomes. *Microbiome*, *13*(1), 134. https://doi.org/10.1186/s40168-025-02119-5

Jiang, Y., Luo, J., Huang, D., Liu, Y., & Li, D. (2022). Machine Learning Advances in Microbiology: A Review of Methods and applications. *Frontiers in Microbiology*, *13*, 925454. https://doi.org/10.3389/fmicb.2022.925454

Kelly, C. J., Karthikesalingam, A., Suleyman, M., Corrado, G., & King, D. (2019). Key challenges for delivering clinical impact with artificial intelligence. *BMC Medicine*, *17*(195). https://doi.org/10.1186/s12916-019-1426-2

Knights, D., Costello, E. K., & Knight, R. (2010). Supervised classification of human microbiota. *FEMS Microbiology Reviews*, *35*(2), 343–359. https://doi.org/10.1111/j.1574-6976.2010.00251.x

Korsunsky, I., Millard, N., Fan, J., Slowikowski, K., Zhang, F., Wei, K., Baglaenko, Y., Brenner, M., Loh, P., & Raychaudhuri, S. (2019). Fast, sensitive and accurate integration of single-cell data with Harmony. *Nature Methods*, *16*(12), 1289–1296. https://doi.org/10.1038/s41592-019-0619-0

Lin, Y., Huang, L., Zhang, X., Yang, J., Chen, X., Li, F., Li, J., & Huang, R. (2021). Multi-omics analysis reveals anti-Staphylococcus aureus activity of Actinomycin D originating from Streptomyces parvulus. *International Journal of Molecular Sciences*, *22*(22), 12231. Multidisciplinary Digital Publishing Institute. https://doi.org/10.3390/ijms222212231

Lopez, R., Regier, J., Cole, M. B., Jordan, M. I., & Yosef, N. (2018). Deep generative modeling for single-cell transcriptomics. *Nature Methods*, *15*(12), 1053–1058. https://doi.org/10.1038/s41592-018-0229-2

Lundberg, S., & Lee, S.-I. (2017). A Unified Approach to Interpreting Model Predictions. *arXiv (Cornell University)*, *2*. https://doi.org/10.48550/arXiv.1705.07874

Magnusdottir, S., Heinken, A., Kutt, L., Ravcheev, D. A., Baur, E., Noronha, A., Greenhalgh, K., Baginska, J., Wilmes, P., Fleming, R. M. T., & Thiele, I. (2017). Generation of genome-scale metabolic reconstructions for 773 members of the human gut microbiota. *Nature Biotechnology*, *35*, 81–89. https://doi.org/10.1038/nbt.3703

Medina, R. H., Kutuzova, S., Nielsen, K. N., Johansen, J., Hansen, L. H., Nielsen, M., & Rasmussen, S. (2022). Machine learning and deep learning applications in microbiome research. *ISME Communications*, *2*(1), 98. https://doi.org/10.1038/s43705-022-00182-9

Mihindukulasuriya, K. A., Mars, R. A. T., Johnson, A. J., Ward, T., Priya, S., Lekatz, H. R., Kalari, K. R., Droit, L., Zheng, T., Blekhman, R., D'Amato, M., Farrugia, G., Knights, D., Handley, S. A., & Kashyap, P. C.. (2021). Multi-omics analyses show disease, diet, and transcriptome interactions with the Virome. *Gastroenterology*, *161*(4), 1194–1207.e8. https://doi.org/10.1053/j.gastro.2021.06.077

Oshrit, S., Haim, I., Sondra, T., Omry, K., & Yoram, L.. 2022. "Image and Graph Convolution Networks Improve Microbiome-based Machine Learning Accuracy." *arXiv (Cornell University)*, May. https://doi.org/10.48550/arxiv.2205.06525

Patro, R., Duggal, G., Love, M. I., Irizarry, R. A., & Kingsford, C. (2017b). Salmon provides fast and bias-aware quantification of transcript expression. *Nature Methods*, *14*(4), 417–419. https://doi.org/10.1038/nmeth.4197

Qu, K., Guo, F., Liu, X., Lin, Y., & Zou, Q. (2019). Application of machine learning in microbiology. *Frontiers in Microbiology*, *10*, 827. https://doi.org/10.3389/fmicb.2019.00827

Repeka, D., Jauniskis, V., Karpus, L., Rembeza, E., Rokaitis, I., Zrimec, J., Poviloniene, S., Laurynenas, A., Viknander, S., Abuajwa, W., Savolainen, O., Meskys, R., Engqvist, M. K. M., & Zelezniak, A.. (2021). Expanding functional protein sequence spaces using generative adversarial networks. *Nature Machine Intelligence*, *3*, 324–333. https://doi.org/10.1038/s42256-021-00310-5

Riberio, M. T., Singh, S., & Guestrin, C. (2016). "Why Should I Trust You?": Explaining the Predictions of Any Classifier. In B. Krishnapuram (Ed.), *Proceedings of the 22nd ACM SIGKDD International Conference on Knowledge Discovery and Data Mining* (pp. 1135–1144). Association for Computing Machinery. https://doi.org/10.1145/2939672.2939778

Rives, A., Meier, J., Sercu, T., Goyal, S., Lin, Z., Liu, J., Guo, D., Ott, M., Zitnick, C. L., Ma, J., & Fergus, R. (2021). Biological structure and function emerge from scaling unsupervised learning to 250 million protein sequences. *Proceedings of the National Academy of Sciences*, *118*(5), e2016239118. https://doi.org/10.1073/pnas.2016239118

Roy, A. D., Dasgupta, S., Roy, R. D., & Das, D. (2024). The use of generative artificial intelligence (AI) in teaching and assessment of postgraduate students in pathology and microbiology. *Indian Journal of Microbiology Research*, *11*(3), 140–146. https://doi.org/10.18231/j.ijmr.2024.027

Roy, G., Prifti, E., Belda, E., & Zucker, J. (2024). Deep learning methods in metagenomics: A review. *Microbial Genomics*, *10*(4). https://doi.org/10.1099/mgen.0.001231

Ruiz-Perez, D., Lugo-Martinez, J., Bourguignon, N., Mathee, K., Lerner, B., Bar-Joseph, Z., & Narasimhan, G. (2021). Dynamic Bayesian networks for integrating multi-omics time series microbiome data. *mSystems*, *6*(2). https://doi.org/10.1128/msystems.01105-20

Sen, S., Vairagare, I., Gosai, J., & Shrivastava, A. (2025). RABiTPy: an open-source Python software for rapid, AI-powered bacterial tracking and analysis. *BMC Bioinformatics*, *26*(1), 127. https://doi.org/10.1186/s12859-025-06145-w

Subramanian, I., Verma, S., Kumar, S., Jere, A., & Krishnanpal, A. (2020). Multi-omics data integration, interpretation, and its application. *Bioinformatics and Biology Insights*, *14*(1177932219899051). https://doi.org/10.1177/1177932219899051

Truong, D. T., Franzosa, E. A., Tickle, T. L., Scholz, M., Weingart, G., Pasolli, E., Tett, A., Huttenhower, C., & Segata, N. (2015). MetaPhlAn2 for enhanced metagenomic taxonomic profiling. *Nature Methods*, *12*(10), 902–903. https://doi.org/10.1038/nmeth.3589

Van Den Bossche, T., Arntzen, M., Becher, D., Benndorf, D., Eijsink, V. G. H., Henry, C., Jagtap, P. D., Jehmlich, N., Juste, C., Kunath, B. J., Mesuere, B., Muth, T., Pope, P. B., Seifert, J., Tanca, A., Uzzau, S., Wilmes, P., Hettich, R. L., & Armengaud, J. (2021). The Metaproteomics Initiative: A coordinated approach for propelling the functional characterization of microbiomes. *Microbiome*, *9*(1), 243. https://doi.org/10.1186/s40168-021-01176-w

Wang, J., & Gao, G. F. (2025a). AIM: AI for microbiology. *National Science Review*, *12*(6). https://doi.org/10.1093/nsr/nwaf225

Wang, X.-W., Wang, T., & Liu, Y.-Y.. 2024. "Artificial Intelligence for Microbiology and Microbiome Research." *arXiv (Cornell University)*, November. https://doi.org/10.48550/arxiv.2411.01098.

Wood, D. E., Lu, J., & Langmead, B. (2019). Improved metagenomic analysis with Kraken 2. *Genome Biology*, *20*(1), 257. https://doi.org/10.1186/s13059-019-1891-0

Xu, C., Zhao, L., Ye, C., Xu, K., & Xu, K. (2025). The application of machine learning in clinical microbiology and infectious diseases. *Frontiers in Cellular and Infection Microbiology*, *15*, 1545646. https://doi.org/10.3389/fcimb.2025.1545646

Xu, Z., Wang, Y., Sheng, K., Rosenthal, R., Liu, N., Hua, X., Zhang, T., Chen, J., Song, M., Lv, Y., Zhang, S., Huang, Y., Wang, Z., Cao, T., Shen, Y., Jiang, Y., Yu, Y., Chen, Y., Guo, G., & Wang, Y. (2023). Droplet-based high-throughput single microbe RNA sequencing by smRandom-seq. *Nature Communications*, *14*(1), 5130. https://doi.org/10.1038/s41467-023-40137-9

Zhou, W., M, R. S., Contrepois, K., Zhou, Y., Ahadi, S., Leopold, S. R., Zhang, M. J., Rao, V., Avina, M., Mishra, T., Johnson, J., Lee-McMullen, B., Chen, S., Metwally, A. A., Tran, T. D. B., Nguyen, H., Zhou, X., Albroght, B., Hong, B.-Y., … Snyder, M. (2019). Longitudinal multi-omics of host-microbe dynamics in prediabetes. *Nature*, *569*, 663–667. https://doi.org/10.1038/s41586-019-1236-x

SECTION TWO

Applications of MAI

The Place of Artificial Intelligence in Gnotobiotic Life

7

Ganiyu Ayodele Ajibade, Gafari Lukumon, and Ebenezer Esenogho

7.1 INTRODUCTION

7.1.1 Defining Gnotobiotic Life in the Modern Context

Gnotobiotic systems are optimal for establishing causality, ensuring reproducibility, and dissecting mechanistic pathways in various fields, including physiology, immunology, metabolism, and development. Gnotobiotic life refers to biological systems in which microbial inhabitants are fully known and controlled; these include germ-free (axenic) organisms, mono-, oligo-, and defined community-colonized hosts, as well as systems in which microbial populations are monitored and characterized taxonomically, genomically, and functionally. Studies have shown that the usage of gnotobiotic models is not limited to animals. It is also used in plants, and even in cell or organoid systems with defined microbiota or synthetic microbial consortia to examine host–microbe interactions. Other environmental factors are kept constant at minimal variation (Basic et al., 2019; Jans & Vereecke, 2024).

They are also used for breaking down mechanistic pathways in research areas like physiology, immunology, metabolism, and development. A schematic depiction of modern gnotobiotic research frameworks is displayed in Figure 7.1.

Traditionally, gnotobiotic organisms consist of germ-free mice. Meanwhile, the scope has recently been broadened to include humanized gnotobiotic mice (with human microbial communities transplanted), fish, insects, and plants under gnotobiotic or axenic circumstances (Gómez de la Torre Canny et al., 2023; Wu et al., 2023). The modern context also incorporates high-throughput sequencing and other -omic (metabolomic, transcriptomic, and proteomic) and functional analyses to fully characterize the microbiota, as well as the use of in silico modeling to simulate possible interactions. The use of increasingly precise measurement instruments and computer capacity has allowed for clearer definitions of "known life" in gnotobiotic situations. These include long-term stability, reproducibility across facilities, and clearly specified functional repertoires (Basic et al., 2019; Darnaud et al., 2021).

DOI: 10.1201/9781003727514-9

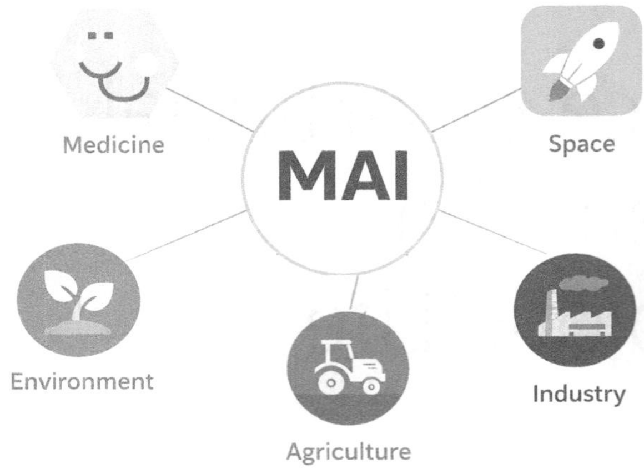

FIGURE 7.1 A schematic representation of modern gnotobiotic research frameworks. Gnotobiotic organisms with fully defined microbial communities serve as controlled biological models for studying host–microbe interactions. A comprehensive characterization of microbial functions and their effects on host phenotypes is enabled by the integration of multi-omic methods such as metabolomics, transcriptomics, and proteomics, together with computational modeling.

7.1.2 The Evolution of Gnotobiotic Research: From Manual Observation to Computational Analysis

The development of gnotobiotic models stemmed from intensive manual laboratory work involving the procedures of derivation of germ-free organisms, sterile rearing, colonization with gnotobiotic cultured isolates, and evaluation of various gross morphological, immunological, metabolic, and developmental characteristics. During these laboratory studies, several differences were noted in gut morphology, pathogen susceptibility, and immune organ size that were present with colonized organisms as opposed to germ-free organisms. Immensely influential initial studies from the mid-20th century laid the groundwork for neutral comparative physiology studies that utilized both germ-free and conventionally colonized subjects, such as those demonstrating cecal hypertrophy and composition of immune cell trafficking alterations. In a basic historical review of gnotobiotic models (Basic et al., 2019), the author observes that, until that time, most of the gnotobiotic model studies were of the descriptive type, and most of the findings were non-predictive. An illustration of gnotobiotic research is presented in Figure 7.2.

The availability of sequencing technologies (16S rRNA and, later, whole metagenomic sequencing); high-resolution taxonomic profiling; metatranscriptomic and metabolomic as well as multi-omic approaches; and technologies for gnotobiotic systems has resulted in an unprecedented expansion of gnotobiotic systems. The availability of cutting-edge and advanced computational approaches has facilitated the integration of several advanced computational and statistical approaches with longitudinal gnotobiotic systems. These approaches include network and ecological modeling, as well as various statistical approaches. Recently, the use of machine learning (ML) has also been introduced (D'Urso & Broccolo, 2024). Gnotobiotic studies now often combine controlled colonization with high temporal resolution, environmental perturbations (diet, infection, and microbiota transfer), and detailed molecular phenotyping. This then enables computational causal inference and prediction.

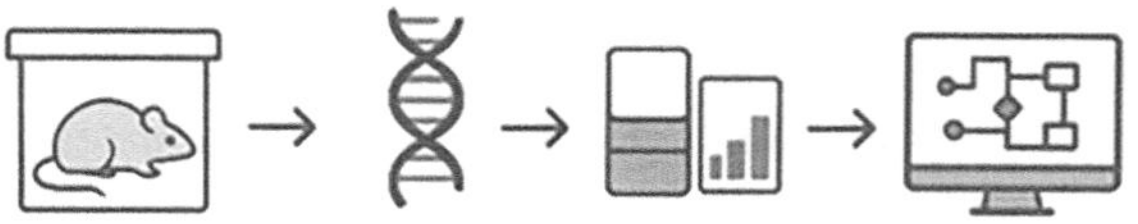

FIGURE 7.2 The evolution of gnotobiotic research, from manual observation to computational analysis. The graphic depicts the progression from early germ-free animal investigations to molecular biology, high-throughput sequencing, and AI-powered computational modeling.

7.1.3 The Convergence of Artificial Intelligence and Gnotobiotic Life: A New Era of Controlled Biological Systems

The combination of artificial intelligence (AI) with gnotobiotic life is one of the most innovative areas of current research in biology. The ability to understand and control complex host–microbe relationships is a game-changer. Gnotobiotic life forms, defined as those with known and controllable relationships with their associated microbes, are becoming important tools in the study of host–microbe relationships (Basic et al., 2019; Fiebiger et al., 2016). The combination of such biological systems with AI created new possibilities in the study of therapeutic discovery and host physiology.

The first work on gnotobiotic animal models dates back to the late 19th and early 20th centuries, where some researchers first succeeded in maintaining germ-free animals (Umesaki, 2014). These early investigations set the groundwork for today's understanding of the distinctions between conventional and germ-free animals. They emphasized the role of microorganisms in host growth, immune system maturation, and metabolism (Rawls et al., 2004). The following growth of gnotobiotic technology, particularly the creation of flexible-film isolators and advanced life support systems, has made it possible to maintain controlled microbial populations for an extended period. These are essential research tools for studying host–microbe interactions (Basic et al., 2019).

Modern gnotobiotic research employs various experimental methodologies. These comprise completely germ-free organisms to basic microbial consortia. Such approaches permit research on distinct host–microbe interactions, unconstrained by confounding variables of classical models (Darnaud et al., 2021; Steimle et al., 2021). As a result, these models have become key to uncovering basic mechanisms of immune system formation and regulation, metabolism, and disease development. For example, studies have shown that using gnotobiotic mouse models has identified specific bacteria, like segmented filamentous bacteria. This is critical to T-helper 17 cell formation and mucosal immune system maturation (Umesaki, 2014). Also, the Oligo-MM12 model and other simplified microbial consortia have shown that a defined microbial community can be reproduced by many metabolic and physiological effects of a complex microbiome while maintaining experimental control (Darnaud et al., 2021).

The rapid expansion of biological data and the need for automation have driven the convergence of AI technology and gnotobiotic research. Automation and AI are now essential to derive valuable conclusions from increasingly complex, multidimensional data. Machine learning algorithms (MLAs) excel at processing high-dimensional microbiome data. They can identify patterns that cannot be detected by traditional statistical methods (Boodaghidizaji et al., 2025; Yang et al., 2025). These computational approaches are effective at integrating diverse data types; this includes genomic, transcriptomic, proteomic, and metabolomic information. They also provide comprehensive views of biological systems that were previously unattainable (Pan et al., 2024).

Utilizing AI with gnotobiotic systems has specific benefits in contrast to standard microbiome exploration. First, gnotobiotic environments are far more controlled and predictable, thereby easing the complexity in variable microbial community systems. This ultimately means AI systems would be more able to discern relevant biological signals and eliminate experimental noise in the analysis (Abavisani et al., 2024). Second, gnotobiotic systems allow for the composition of microbial systems to be manipulated, thereby giving rise to productive, systematic empirical studies that can be used to train ML systems. Finally, longitudinal studies are possible in gnotobiotic systems given the temporal stability and reproducibility to study dynamic host–microbe relationships over long periods (Hoces et al., 2022).

Deep learning models that accurately predict host phenotypes and dependent microbial community structures have far surpassed the 95% accuracy mark and show promise for microbiome-based diagnostic applications (Boodaghidizaji et al., 2025). Advances in AI and the analysis of the microbiome have already started to shift our knowledge of host–microbe interactions in significant ways. New knowledge graph-based frameworks have given rise to the prediction of innovative therapies by clarifying the relationships that exist between microbial species and host functions, enabling the prediction of novel therapeutic targets and microbial interventions (Pan et al., 2024). Furthermore, transformer-based architectures, originally developed for natural language processing, have shown remarkable success in analyzing microbial sequence data and predicting functional relationships within microbial communities (Yang et al., 2025).

Combining AI analysis with multi-omics technologies strengthens gnotobiotic research platforms. As a result of the integration of genomics, transcriptomics, proteomics, and metabolomics data of tightly controlled microbial systems, researchers are now able to develop and refine comprehensive models of host and microbial system interactions with respect to molecular and functional mechanisms (Müller et al., 2024; Yang et al., 2025). These comprehensive models helped answer complex research questions, illustrating the microbial community and host physiological interaction and influence spanning multiple chronic and interlinked pathways, direct metabolite production, the immune system, and epigenetic mechanisms and regulation (Dakal et al., 2025).

AI-boosted gnotobiotic systems have the potential to transform therapeutic development, personalized medicine, and biotechnology well beyond fundamental research. For example, AI algorithms utilizing the gnotobiotic systems, integromics technologies, and research systems pivot to personalized precision medicine by predicting individual outcomes to microbiome therapies (Kashyap et al., 2017; Nazir et al., 2024). Moreover, the gnotobiotic systems offer a unique opportunity to test synthetic biology concepts because we can design, generate, and control complex microbe communities and systems, allowing the integration of AI for design and optimization (Nazir et al., 2024).

Combining the potential of AI with gnotobiotic life opens the opportunity to examine complex biological phenomena more deeply. This includes understanding the biological aging process and age-related neurological and metabolic disorders. As noted in the literature for AI and gnotobiotic life, the microbiome's composition allows prediction of biological age. Gnotobiotic systems can also be used for empirical research to unravel causal relationships between aging and microbial communities (Yang et al., 2025). Research on the gut-brain axis is another pivotal area. Here, gnotobiotic life and AI techniques could help explore neurological and behavioral variables (Luczynski et al., 2016).

Studies have shown that AI can support gnotobiotic life systems. It can also reduce the time needed to translate laboratory research into clinical practice. This confirms the potential of AI and gnotobiotic life in future microbiome research. Given that creating clinical medicines is expensive and time-consuming, the capacity to test thousands of microorganism combinations swiftly and affordably is revolutionary (Romano et al., 2025). In addition, the advancement of digital twins—computational models that replicate real biological systems—may facilitate virtual experimentation. They may also pave the way for the outcome prediction before the commencement of expensive laboratory studies (Probul et al., 2024).

However, the effective combination of AI and gnotobiotic research gives rise to some problems that researchers must tackle. For example, it is often required to have reliable standards for data collection, processing, and interpretation, which are frequently necessary, as variations in experimental approaches can make it difficult to apply AI models across various teams or research settings (Basic et al., 2019). Furthermore, the interpretability of complex AI models is still a big challenge. The understanding of the biological foundations of AI predictions is important for developing actionable insights and assuring safety in therapeutic contexts (Hernández Medina et al., 2022).

Ethical issues related to AI-enhanced gnotobiotic research also need to be carefully thought about. Especially when these technologies start to affect clinical decision-making and the creation of new treatments. To be responsible in this field, AI models must be trained on a wide range of datasets that are indicative of the real world, algorithmic decision-making must not be biased, and computational approaches must be clear (Probul et al., 2024).

We are at a time when the intersection of biological intricacy with computational power is becoming a reality. Introducing AI tools alongside gnotobiotic life systems is about to redefine our comprehension of life. The gnotobiotic systems provide a controlled experimental framework, and AI's pattern recognition and prediction tools provide a framework for transformative discovery and innovation. More importantly, the integrated use of these tools helps us to transcend the study of intricate biological systems from a purely technological perspective. The potential insights from this study will redefine medicine and biotechnology and fundamentally redefine our understanding of the life continuum.

The sections that follow in this book chapter will look at the gnotobiotic life and AI, the specific mechanisms by which AI technologies are being used in gnotobiotic research, case studies demonstrating successful applications, and the future directions and potential impacts of this rapidly evolving field. Through this comprehensive analysis, we hope to give readers a clear grasp of how the merging of AI and gnotobiotic life is transforming the landscape of biological research and opening up new avenues for scientific discovery and therapeutic innovation.

7.2 FOUNDATIONS OF GNOTOBIOTIC LIFE AND ARTIFICIAL INTELLIGENCE

Gnotobiotic life revolves around the creation of organisms whose microbial communities are entirely known and regulated, allowing the most accurate study of host–microbe relationships. The word gnotobiotic comes from the Greek word gnosis ("known") and bios ("life"), referring to germ-free organisms and those that are colonized with defined microbiota (Basic et al., 2019). The history of gnotobiotic research begins in the late 19th century with Nuttall and Thierfelder (1896), who were the first to successfully raise germ-free guinea pigs, setting the framework for future breakthroughs in isolator technology and animal husbandry (Basic et al., 2019). Early breakthroughs included the 1930s production of high-pressure sterilized steel isolators at the University of Notre Dame's Bacteriology Laboratory under the direction of Reyniers as well as the 1950s flexible-film isolators that substantially increased ease of use (Basic et al., 2019).

Breeding gnotobiotic animals, typically rodents like mice or zebrafish, entails techniques such as aseptic hysterectomy, embryo transfer, and isolation breeding. These methods ensure that the microorganisms that surround the organism are fully missing or characterized (Nicklas et al., 2015; Steimle et al., 2021). They are also very important in biomedical research. In addition, they elucidate the impact of specific microbial composition through the study of host development, maturation of the immune system, metabolism, and the progression of diseases (Fiebiger et al., 2016). However, despite the models' initial successes, gnotobiotic models fundamentally lag in progress due to the unique physiologic burdens of microbial absence or abnormal composition, such as immunodeficiency, aberrant metabolism, and the absence of microbiota, necessitating more attention to experimental design and results interpretation (Basic et al., 2019; Umesaki, 2014).

New techniques in gnotobiotic research have broadened the types of questions that can be asked, including those that involve gnotobiotic animals or isobiotic animals with simplified and well-defined microbial consortia (Darnaud et al., 2021). In these systems, it is possible to examine the host response to specific microorganisms in controlled conditions and with less chance of confounding elements essential to traditional models that are colonialized by complex, undefined microbial assemblages (Basic et al., 2019; Steimle et al., 2021). Additionally, the employment of gnotobiotic models serves as a translational bridge that enhances synthetic biology strategies focused on engineering microbial consortia for therapeutic and biotechnological purposes (Nazir et al., 2024).

Simultaneously, with the advancement of gnotobiotics, AI has undergone a significant transformation in recent years. In the past, AI depended on rule-based expert systems. It now employs ML and deep learning. These approaches are powerful because they excel at recognizing patterns. They can also help with predictions and decisions based on complex biological data (Hernández Medina et al., 2022; Müller et al., 2024).

AI refers to computer systems that can do tasks that used to need human intelligence. This includes learning from data, solving problems, and adapting when things change (Russell & Norvig, 2020). ML is a part of AI. It enables computers to find patterns in large datasets, even without being told what to look for. Deep learning is a newer approach. It uses layers of neural networks to mimic how people interpret and organize information (LeCun, Bengio, & Hinton, 2015).

Several basic AI methods are especially useful in biology. For example, supervised learning is used for classification and prediction. Unsupervised learning groups data and simplifies complex information. Reinforcement learning helps computers to make decisions over time (Hernández Medina et al., 2022). These tools are very important for understanding the complicated biological data, such as omics data (Müller et al., 2024). More recently, models called transformers, first used for language tasks, have been adapted to analyze microbiome sequences and to combine various biological data types (Pan et al., 2024; Yang et al., 2025).

Utilization of AI with gnotobiotic research centers on addressing fundamental issues surrounding the host–microbiome interaction's complexity, multidimensionality, and contextual dependency (Abavisani, Rezaei, Kuchaki Rafsanjani, & Heidarzadeh, 2024). Given their controlled microbiome and ecosystem features, gnotobiotic systems become the ideal candidates for testing AI approaches on rich and reproducible datasets (Basic et al., 2019; Steimle et al., 2021). AI advancements allow for non-hypothesis-driven research, identifying concealed microbiome patterns associated with specific host characteristics, forecasting microbial manipulation outcomes, and customizing microbial communities for specific functions (Nazir et al., 2024).

Furthermore, AI combines and correlates data within and across various omics datasets (including genomics, transcriptomics, proteomics, and metabolomics) with phenotype data from gnotobiotic experiments. This then enables the development of systems-level models on the dynamic relationship between the host and microbiota systems (Müller et al., 2024; Yang et al., 2025). This multi-scale methodology corresponds with the advancing framework of precision medicine, wherein AI-generated insights from gnotobiotic models enhance tailored microbiome therapies and synthetic microbiota medicines (Kashyap et al., 2017; Nazir et al., 2024).

Apart from the successful integration of AI in gnotobiotic research by utilizing several sophisticated algorithms, a key component that facilitates these improvements is the availability of quality data, its standardization, and the interpretability of data and models (Basic et al., 2019; Hernández

Medina et al., 2022). There are still challenges to overcome. For example, researchers need to agree on common standards for how data is created, labeled, and shared in gnotobiotic studies. In addition, there is a need for better computational tools that can combine biological knowledge with AI predictions. This way, the results can lead to new ideas that can actually be tested in experiments (Müller et al., 2024; Probul et al., 2024).

Gnotobiotic research and AI each provide something essential. Gnotobiotic systems let scientists control biological variables in ways that aren't possible otherwise. At the same time, AI brings powerful tools for making sense of all the complex information found in host–microbiome interactions. When these two fields work together, they have the potential to completely change our understanding of the microbiome and how it can be applied in real-world settings.

7.3 APPLICATIONS OF ARTIFICIAL INTELLIGENCE IN GNOTOBIOTIC RESEARCH

For the past decades, researchers in gnotobiotic research have traditionally relied heavily on manual observation and sterile techniques to explore various gnotobiotic systems. However, this approach is only feasible for less complicated microbial–host interactions. The advent of AI has made deeper exploration possible because of its capacity to recognize patterns, predict outcomes, and automate experimental routines that were previously done manually. AI uses data and machine-guided experimentation to transform gnotobiotic studies from mere surface descriptions to predictive and well-designed scientific inquiries (Rozera et al., 2025). In what follows, we discuss how this new AI-driven approach is applied, from data integration to digital simulation.

7.3.1 Data Integration and Pattern Recognition

The way scientists continue to advance in how sequencing and metabolomics are done has made the data generated in gnotobiotic studies heavy and voluminous (Wang et al., 2024). These data need to be processed to explore pattern recognition. However, for data integration and pattern recognition, the traditional approaches scientists use make it difficult to get detailed information about the relationship between microbes and host responses. MLAs, often referred to as AI models, make this seamless in the sense that they now allow researchers to integrate multi-omic data, observe key microbial elements, and predict some physiological outcomes that might have been missed because of noise in the data (Oshrit et al., 2022). In this ML approach, supervised learning by its nature can classify communities of microbes linked with specific immune or metabolic states. Unsupervised learning, on the other hand, can reveal novel microbial groupings that can influence development or how diseases are resisted (Wang et al., 2024). In the same realm, convolutional neural networks can process data in image form from tissues or microfluidic chips (Oshrit et al., 2022). This then correlates visual patterns with microbial composition or metabolic shifts. All these tools are helpful for scientists to uncover the dynamics that make metabolites drive host outcomes. This is made possible because AI has the ability to learn from data rather than from predefined hypotheses. But to optimally harness the fruit of AI, one needs to carefully curate datasets and validate them transparently to avoid bringing about misleading correlations. When used ethically and responsibly, pattern recognition models are great for exploring microbial complexity. In the next section, we discuss predictive modeling, which asks why and how patterns emerge.

7.3.2 Predictive Modeling of Host–Microbe Interactions

But why is host–microbe interaction difficult to model manually? This is so because understanding how defined microbial communities function is far beyond just descriptive correlations. For predictive models based on AI, scientists find it easier to simulate immune reactions, metabolic exchanges, and the ecological stability present in gnotobiotic systems (Treloar et al., 2020). To infer how some microbial combinations affect gene expression and nutrient pathways in the host, one can use deep learning architectures that have been trained on transcriptomic and metabolomic data. To optimize microbial compositions dynamically, reinforcement learning can offer a way to assist configurations that sustain stability or health-related outcomes (Sabzevari et al., 2022). For example, to know whether introducing a new bacterium into a germ-free animal will promote or disturb intestinal balance, one can resort to predictive systems. These predictions assist in designing experiments and also reduce the number of trials required to obtain meaningful results. Also, probabilistic models assist researchers in quantifying uncertainty in the data. This can be done by showing when predictions are strong enough and when new data are required. Aside from that, interpretability is a problem because even if complex neural networks perform better, they can still disguise themselves as "black boxes." However, explainable AI helps clarify what actually drives predictions.

7.3.3 Automated Experimentation and Robotics in Gnotobiotic Labs

As discussed above, predictive systems improve theoretical insight, but automation comes into play when bringing these capabilities into everyday lab practice. However, getting a very serene environment with consistent microbial conditions is far from simple. It requires a lot of work. Based on this, AI-driven robotics are now being tasked with these delicate operations (Sabzevari et al., 2022). They are equipped with sensors in their arms and also with computer vision, which can handle sample transfers, feed animals, and clean chambers without human contamination. ML-based systems then check the temperature, humidity, and gas composition to ensure that micro-adjustments maintain optimal growth conditions (Rozera et al., 2025).

But the process differs from one lab to another. In some labs, AI algorithms ensure that camera feeds detect behavioral or physiological changes that occur in gnotobiotic animals. This makes it possible for researchers to be alerted to deviations in real time. So, automation saves time and, in addition, improves reproducibility through the removal of human variability. It is worth stating that robots can also execute complex experiments that might not be possible manually, such as precise nutrient dosing. But to achieve automation, one requires high setup costs as well as calibration to maintain reliability. Also, it cannot be as potent as humans when it comes to interpreting unexpected outcomes. Human beings sometimes see what machines cannot see or interpret.

7.3.4 AI in Microbial Engineering and Synthetic Ecology

Now, as researchers continue to automate experimentation, the focus is shifting toward how designing and engineering microbial systems themselves can be done more effectively. Beyond observation, AI generative algorithms can propose combinations of species that are likely to coexist through the analysis of interaction networks (Treloar et al., 2020). This is made possible through evolutionary computation and Bayesian optimization techniques, which explore many possible configurations and select those predicted to maintain balance and resilience (Sabzevari et al., 2022). These are already being used to construct synthetic consortia for gut health studies as well as for the enhancement of agricultural soil (Rozera et al., 2025).

In the context of gnotobiotics, it makes it possible for researchers to engineer ecosystems that can produce metabolites that modulate immunity. To predict how genetic changes affect metabolic pathways and cross-feeding relationships, deep learning can also guide genome editing (Wang et al., 2024). Doing this saves time and reduces reliance on trial-and-error experimentation. But since real environments may introduce unforeseen variables, predictive design can still suffer from shortcomings. Others include ethical and biosafety concerns when AI designs organisms with novel properties. No matter the case, AI has still become an active collaborator in building living systems.

7.3.5 Digital Simulation and "In Silico Gnotobiotics"

However, how can virtual experiments complement the physical ones through replication? "In silico gnotobiotic" models, that is, simulated systems mirroring real microbial communities, can create virtual organisms and microbial consortia that develop similarly to real experimental data (Sizemore et al., 2024). Agent-based modeling approaches and deep neural networks are used to achieve this. These types of simulations capture metabolic exchanges, population dynamics, and environmental feedback that would be difficult to measure directly (Shrestha Gurung et al., 2025). Thousands of virtual trials can help researchers test hypotheses, identify sensitive parameters, and plan experiments more effectively (Treloar et al., 2020). Through this, dependence on animal models is reduced, in line with the call to minimize in vivo experimentation (Sizemore et al., 2024). Gnotobiotic systems are updated with fresh data to mirror and replicate the real ones. But one of the shortcomings of this approach is ensuring that virtual models accurately replicate the stochastic, context-dependent nature of life. However, as computational power and data quality improve, simulations evolve into predictive environments that allow hypotheses to be tested virtually before proceeding to the bench. AI thus assists and extends gnotobiotic research into a new realm where both biology and computation develop together.

7.4 FUTURE DIRECTIONS, CHALLENGES, AND ETHICAL CONSIDERATIONS

The future phase of gnotobiotic research lies at the intersection between experimental biology and powerful intelligent computation. Currently, AI systems can merge visual, molecular, and behavioral data streams. This development allows for a more complete picture of host–microbe interactions. In order to effectively harness the full potential of microbial influence, future efforts can ensure that multimodal learning models process sequencing data alongside microscopic images and physiological readings. This will enable the prediction of which microbes matter, when, and how their effects manifest over time. Since precision medicine adapts treatment to an individual's genome, researchers could use personalized gnotobiotic models to tailor microbial communities to metabolic profiles. Future efforts also should ensure collaboration between computer scientists, microbiologists, and bioengineers to maintain interpretability and ensure the biological meaning of predictions. Also, a feedback loop should be formed in situations where the boundary between computational modeling and living experimentation becomes increasingly porous.

In terms of the technical and conceptual challenges, we note that despite the progress of AI in gnotobiotic research, full adoption is still limited. One of these is data quality. Many datasets are noisy and not consistent across different laboratories. This makes it difficult for algorithms to learn necessary patterns. Data formats should thus be standardized and metadata included to ensure reproducibility.

Model transparency is another challenge. For example, researchers can achieve impressive accuracy via deep neural networks, but these provide little insight into how decisions are made, thereby risking predictions that lack a mechanistic explanation.

Another issue is sustainability due to computational costs and energy demands with growing models. Also, integrating AI output with biological reasoning might require expertise because algorithms may reveal patterns that biologists may struggle to interpret. To make AI a trusted scientific partner and computational aid, addressing these issues—which will require both technical innovation and cultural change in research communities—becomes necessary.

Conducting gnotobiotic research with AI raises questions about how far scientists should go. Yes, it can yield medical and environmental benefits, but at the same time, it challenges naturalness and biological autonomy. The scientific community should thus have clear consent procedures and data anonymization, and open science should be considered. Also, when AI makes a mistake, who takes responsibility—the developer, researcher, or institution? Ethical frameworks should therefore be proactive and integrate moral reflection into every stage of design and deployment. Courses in computational ethics and microbiological safety should be made compulsory for scientists.

7.5 CONCLUSION

In conclusion, gnotobiotic research presents an avenue to bridge the intelligence of machines with human curiosity as well as moral judgment. AI is doing great in reshaping how we understand and manipulate gnotobiotic research. That is why researchers can now use ML, robotics, and simulations to visualize microbial dynamics, predict host responses, and design ecosystems precisely. These developments make experimentation more reliable and deepen our knowledge of how biological systems change and adapt. But all these are not without shortcomings. Ethical foresight, including data integrity, should be emphasized to avoid the abuse and misrepresentation of living systems. We should harmonize the capabilities between AI and biological meaning.

REFERENCES

Abavisani, M., Khoshrou, A., Foroushan, S. K., Ebadpour, N., & Sahebkar, A. (2024). Deciphering the gut microbiome: The revolution of artificial intelligence in microbiota analysis and intervention. Current Research in Biotechnology, 7, 100211. https://doi.org/10.1016/j.crbiot.2024.100211

Basic, M., Bolsega, S., Smoczek, A., Buettner, M., Eberl, C., Amann, A., Bleich, A., Stecher, B., & Bleich, A. (2019). Gnotobiotics: Past, present and future. Laboratory Animals, 53(2), 124–135. https://doi.org/10.1177/0023677219836715

Boodaghidizaji, M., Jungles, T., Chen, T., Zhang, B., Yao, T., Landay, A., Keshavarzian, A., Hamaker, B., & Ardekani, A. (2025). Machine learning based gut microbiota pattern and response to fiber as a diagnostic tool for chronic inflammatory diseases. BMC Microbiology, 25, 353. https://doi.org/10.1186/s12866-025-04072-7

Carter, P. B., Norin, E., & Swennes, A. G. (2020). Gnotobiotics and the microbiome. In M. Suckow, S. H. Weisbroth, & C. L. Franklin (Eds.), The Laboratory Rat (pp. 827–848). Academic Press. https://doi.org/10.1016/B978-0-12-814338-4.00021-0

D'Urso, F., & Broccolo, F. (2024). Applications of artificial intelligence in microbiome analysis and probiotic interventions—An overview and perspective based on the current state of the art. Applied Sciences, 14(19), 8627. https://doi.org/10.3390/app14198627

Dakal, T. C., Xu, C., & Kumar, A. (2025). Advanced computational tools, artificial intelligence and machine-learning approaches in gut microbiota and biomarker identification. Frontiers in Medical Technology, 6, 1434799. https://doi.org/10.3389/fmedt.2024.1434799

Darnaud, M., Fink, C., Hamzaoui, N., Rouland, L., Serafini, N., Théret, N., Di Santo, J. P., Gerbe, F., & Jay, P. *et al.* (2021). A standardized gnotobiotic mouse model harboring a minimal 15-member mouse gut microbiota recapitulates SOPF/SPF phenotypes. Nature Communications, 12, 6686. https://doi.org/10.1038/s41467-021-26963-9

Fiebiger, U., Bereswill, S., & Heimesaat, M. M. (2016). Dissecting the interplay between intestinal microbiota and host immunity in health and disease: Lessons learned from germfree and gnotobiotic animal models. European Journal of Microbiology and Immunology, 6(4), 253–271. https://doi.org/10.1556/1886.2016.00036

Hernández Medina, R., Kutuzova, S., Nielsen, K. N., Johansen, J., Hansen, L. H., Nielsen, M., & Rasmussen, S. (2022). Machine learning and deep learning applications in microbiome research. ISME Communications, 2(1), 98. https://doi.org/10.1038/s43705-022-00182-9

Hoces, D., Lan, J., Sun, W., Geiser, T., Stäubli, M. L., Cappio Barazzone, E., Arnoldini, M., Challa, T. D., Klug, M., Kellenberger, A., Nowok, S., Faccin, E., Macpherson, A. J., Stecher, B., Sunagawa, S., Zenobi, R., Hardt, W.-D., Wolfrum, C., & Slack, E. (2022, September 20). Metabolic reconstitution of germ-free mice by a gnotobiotic microbiota varies over the circadian cycle. PLoS Biology, 20(9), e3001743. https://doi.org/10.1371/journal.pbio.3001743

Jans, M., & Vereecke, L. (2024, March 24). A guide to germ-free and gnotobiotic mouse technology to study health and disease. FEBS Journal. https://doi.org/10.1111/febs.17124

Kashyap, P. C., Chia, N., Nelson, H., Segal, E., & Elinav, E. (2017). Microbiome at the frontier of personalized medicine. Mayo Clinic Proceedings, 92(12), 1855–1864. https://doi.org/10.1016/j.mayocp.2017.10.004

Luczynski, P., McVey Neufeld, K. A., Oriach, C. S., Clarke, G., Dinan, T. G., & Cryan, J. F. (2016). Using germ-free animals to assess the influence of the gut microbiota on anxiety and cognition. International Journal of Neuropsychopharmacology, 19(8). https://doi.org/10.1093/ijnp/pyw020

Müller, E., Shiryan, I., Borenstein, E., Korem, T., & Geiger, T. (2024). Multi-omic integration of microbiome data for identifying disease-associated modules. Nature Communications, 15, 2920. https://doi.org/10.1038/s41467-024-46888-3

Nazir, A., Nadeem, H. F., Hussain, N., & Raza, A. (2024). Advancing microbiota therapeutics: The role of synthetic biology in engineering microbial communities for precision medicine. Frontiers in Bioengineering and Biotechnology, 12, 1511149. https://doi.org/10.3389/fbioe.2024.1511149

Nicklas, W., Keubler, L., & Bleich, A. (2015). Maintaining and monitoring the defined microbiota status of gnotobiotic rodents. ILAR Journal, 56(2), 241–249. https://doi.org/10.1093/ilar/ilv029

Nuttall, G. H. F., & Thierfelder, H. (1896). Thierisches Leben ohne Bakterien im Verdauungskanal. Zeitschrift für physiologische Chemie, 21(1), 109–121.

Oshrit, S., Haim, I., Sondra, T., Omry, K., & Yoram, L. (2022). Image and Graph Convolution Networks Improve Microbiome-based Machine Learning Accuracy. *arXiv preprint arXiv:2205.06525.*

Pan, J., Zhang, Z., Li, Y., Yu, J., You, Z., Li, C., Wang, S., Zhu, M., Ren, F., Zhang, X., Sun, Y., & Wang, S. (2024). A microbial knowledge graph-based deep learning model for predicting candidate microbes for target hosts. Briefings in Bioinformatics, 25(3). https://doi.org/10.1093/bib/bbae119

Probul, N., Hedberg, N., & Shmueli, G. (2024). AI in microbiome-related healthcare. Engineering Reports, 6(11). https://doi.org/10.1002/eng2.12886

Rawls, J. F., Samuel, B. S., & Gordon, J. I. (2004). Gnotobiotic zebrafish reveal evolutionarily conserved responses to the gut microbiota. Proceedings of the National Academy of Sciences, 101(13), 4596–4601. https://doi.org/10.1073/pnas.0400706101

Romano, S., Savva, G. M., Bedarf, J. R., Charles, I. G., Hildebrand, F., & Narbad, A. (2025). Machine learning-based meta-analysis reveals gut microbiome biomarkers for Parkinson's disease classification. Nature Communications, 16, 1429. https://doi.org/10.1038/s41467-025-56829-3

Rozera, T., Pasolli, E., Segata, N., & Ianiro, G. (2025). Machine learning and artificial intelligence in the multi-omics approach to gut microbiota. Gastroenterology. 169(3), 487–501.

Russell, S., & Norvig, P. (2020). Artificial Intelligence: A Modern Approach (4th ed.). Pearson.

Sabzevari, M., Szedmak, S., Penttilä, M., Jouhten, P., & Rousu, J. (2022). Strain design optimization using reinforcement learning. PLoS computational biology, 18(6), e1010177.

Shrestha Gurung, B. D., Rayamajhi, M., Maharjan, N., Do, T., Bhandari, D., Yadav, R., & Gnimpieba, E. Z. (2025). Forecasting Urban Wastewater Microbiome Dynamics Using a Digital Twin Framework. *bioRxiv*, 2025-07.

Sizemore, N., Oliphant, K., Zheng, R., Martin, C. R., Claud, E. C., & Chattopadhyay, I. (2024). A digital twin of the infant microbiome to predict neurodevelopmental deficits. Science Advances, 10(15), eadj0400. https://doi.org/10.1126/sciadv.adj0400

Steimle, A., De Sciscio, A., Neumann, M., Grant, E. T., Pereira, G. V., Ohno, H., Martens, E. C., & Desai, M. S. (2021). Constructing a gnotobiotic mouse model with a synthetic human gut microbiome to study host–microbe cross talk. STAR Protocols, 2(2), 100607. https://doi.org/10.1016/j.xpro.2021.100607

Treloar, N. J., Fedorec, A. J., Ingalls, B., & Barnes, C. P. (2020). Deep reinforcement learning for the control of microbial co-cultures in bioreactors. PLoS computational biology, 16(4), e1007783.

Umesaki, Y. (2014). Use of gnotobiotic mice to identify and characterize key microbes responsible for the development of the intestinal immune system. Proceedings of the Japan Academy, Series B, 90(8), 313–332. https://doi.org/10.2183/pjab.90.313

Wang, Q., Fan, X., Wu, S., & Su, X. (2024). PM-CNN: Microbiome status recognition and disease detection model based on phylogeny and multi-path neural network. Bioinformatics Advances, 4(1), vbae013.

Wu, J., Wang, Q., Wang, D., Wong, A. C. N., & Wang, G. H. (2023). Axenic and gnotobiotic insect technologies in research on host–microbiota interactions. Trends in Microbiology, 31(8), 858–871. https://doi.org/10.1016/j.tim.2023.02.007

Gómez de la Torre Canny, S., Nordgård, C. T., Mathisen, A. J. H., Lorentsen, E. D., Vadstein, O., & Bakke, I. (2023). A novel gnotobiotic experimental system for Atlantic salmon (Salmo salar L.) reveals a microbial influence on mucosal barrier function and adipose tissue accumulation during the yolk sac stage. Frontiers in Cellular and Infection Microbiology, 12, 1068302. https://doi.org/10.3389/fcimb.2022.1068302

LeCun, Y., Bengio, Y., & Hinton, G. (2015). Deep learning. Nature, 521(7553), 436–444. https://doi.org/10.1038/nature14539

Esteva, A., Robicquet, A., Ramsundar, B., Kuleshov, V., DePristo, M., Chou, K., … Dean, J. (2019). A guide to deep learning in healthcare. Nature Medicine, 25(1), 24–29. https://doi.org/10.1038/s41591-018-0316-z

Topol, E. (2019). High-performance medicine: the convergence of human and artificial intelligence. Nature Medicine, 25, 44–56. https://doi.org/10.1038/s41591-018-0300-7

Libbrecht, M. W., & Noble, W. S. (2015). Machine learning applications in genetics and genomics. Nature Reviews Genetics. https://doi.org/10.1038/nrg3920

Yang, S. Y., Han, S. M., Lee, J. Y., Kim, K. S., Lee, J. E., & Lee, D. W. (2025). Advancing gut microbiome research: The shift from metagenomics to multi-omics and future perspectives. Journal of Microbiology and Biotechnology, 35, e2412001. https://doi.org/10.4014/jmb.2412.12001

AI Use in Microbial Genomics Application

8

Prudence Ejordenore Ehizuenlen
and Pius Uagbae Ejodamen

8.1 INTRODUCTION TO MICROBIAL GENOMICS AND ITS SIGNIFICANCE

Microbial genomics is an interdisciplinary field that combines microbiology and genomics to study the genetic material of microorganisms. The term itself is derived from "microbial," referring to microscopic organisms such as bacteria, viruses, fungi, and archaea, and "genomics," a branch of molecular biology focused on the structure, function, evolution, and mapping of genomes—the complete set of genetic material within an organism (Choffnes et al., 2013). By analyzing microbial genomes, scientists gain critical insights into the biology, behavior, and evolutionary dynamics of microorganisms.

The advent of high-throughput sequencing technologies has revolutionized microbial genomics, enabling researchers to decode microbial DNA with unprecedented speed and accuracy. These advancements have illuminated previously hidden aspects of microbial life, including how bacteria adapt, evolve, and interact with their environments. This deeper understanding has significantly enhanced our ability to investigate microbial roles in health, disease, and ecological systems.

In the realm of human health, microbial genomics has transitioned from a purely research-oriented discipline to a practical tool in clinical diagnostics. Early studies focused on understanding virulence factors and pathogenic mechanisms, offering valuable information about bacterial infections and their evolutionary pathways (Fournier et al., 2007). Today, genomic technologies allow for rapid sequencing of bacterial genomes—once a months-long process, now achievable in under a week at a cost comparable to routine diagnostic procedures such as MRI scans. This capability supports timely identification of pathogens, outbreak tracking, and the development of targeted therapies and vaccines.

Beyond healthcare, microbial genomics plays a pivotal role in addressing environmental challenges. Rapid industrialization and human activity have led to severe environmental degradation, including pollution of air, water, and soil (Medfu Tarekegn et al., 2020). Traditional physical and chemical remediation methods often fall short in effectively eliminating pollutants. In contrast, biological approaches—particularly those leveraging microbial genomics—offer sustainable and cost-effective alternatives. Through bioremediation, genetically characterized microbes equipped with specific enzymes are introduced into contaminated environments, where they degrade pollutants and convert

DOI: 10.1201/9781003727514-10

them into harmless byproducts (Kugarajah et al., 2023; Ojuederie & Babalola, 2017). This approach not only restores ecological balance but also minimizes health risks to humans and animals.

In biotechnology, microorganisms have long been harnessed for the production of essential goods such as bread, wine, and beer. However, the integration of microbial genomics has expanded their utility far beyond traditional fermentation. Today, microbes are engineered to produce bioactive compounds, including antibiotics, antifungals, and anticancer agents, as well as to drive innovations in molecular biology (Santos-Beneit, 2024). Moreover, microbial genomics underpins the development of diagnostic tools, gene expression systems, and biocompatible materials like polyhydroxyalkanoates used in medical devices.

Given its broad applications and transformative potential, microbial genomics continues to be a cornerstone of scientific advancement. As this field evolves, the integration of artificial intelligence (AI) presents new opportunities to accelerate discovery, enhance data interpretation, and optimize microbial applications. This chapter explores the intersection of AI and microbial genomics, examining current methodologies, tools, challenges, and future directions for this rapidly advancing discipline.

8.2 ARTIFICIAL INTELLIGENCE AS A TRANSFORMATIVE TOOL IN GENOMICS

AI is a field of study that allows machines to mimic human reasoning or thought processes. Its innovative features have led to applications in several areas, such as but not limited to education, industry, and health. In its use in academic research, AI has resulted in the development of several large language models that have helped to reshape the way scientists conduct studies as well as how the findings obtained are shared. This has increased the number of data repositories, thus creating a fertile ground for the deployment of AI and aiding researchers to obtain meaningful insight from complex datasets.

In modern times, AI is gradually poised to revolutionize the field of genomics, thus proving to be a transformative tool for its advancement. Genomics, as a groundbreaking field, is dedicated to the exploration of genomes (Smith, 1943). This field of study has remarkably altered the way researchers understand genetic structures, functions, evolutionary patterns, the intricate connections between genes, their impact on health, and diseases of both plants and humans.

Genomics, which is mainly concerned with genes and their roles, usually requires a deeper understanding of the health and diseases of humans. As such, researchers often employ several computational techniques for the purpose of extracting essential biological insights from existing genomic data. However, with the increasing need for a more comprehensive dataset and data analysis, data scientists, through extensive engagement with their community of researchers, explored ways to create innovative solutions. One of which led to the incorporation of AI in genomics with the use of a machine learning (ML) tool that is embodied with the ability to learn from past experiences, after which it applies the knowledge gained to new scenarios. This has resulted in the following transformative roles (Taqi et al., 2025):

a. The analysis of genomic data is a pivotal stage in translational research.
b. The impact analysis of DNA mutation to determine the health and diseases of plants and humans.
c. The application of ML techniques and computational algorithms for the advancement of genome assemblage in genomic annotation.
d. The prediction of genomic mutation in plants.

Significantly speaking, further studies in the application of AI in genomics will help to redefine genomics in its application in the healthcare sector while encouraging researchers and clinicians to enhance existing workflows or embark on new projects for the purpose of developing innovative applications (Taqi et al., 2025).

8.3 HISTORY OF MICROBIAL GENOMICS: FROM CULTURE-BASED METHODS TO NEXT-GENERATION SEQUENCING

The history of microbial genomics can be traced back to the 1940s, with the works of notable researchers such as George Beadle, Edward Tatum, and Joshua Lederberg leading to its foundational discoveries. As time progressed, precisely in the 1960s, key concepts such as, but not limited to, the universality of the genetic code were established, further giving conceptual ground to the study of microbial genomics. Notably, in 1995, microbial genomics had its first-ever complete sequencing of the Haemophilus influenzae and Mycoplasma genitalium, thus bringing about an era of spontaneous growth in genomic database from Sanger sequencing to next-generation sequencing (NGS). Recent research has shown that microbial genomics has grown from a simple study of isolated individual microbial to an in-depth analysis of entire microbial communities through the use of metagenomics and from the usual hypothesis-driven approach to an extensive data-driven science with the increase in genomic data. This advancement has significantly improved the understanding of microbial diversity and provided critical insights into the genomics of pathogens, aiding the understanding, control, and treatment of infectious diseases as well as antimicrobial resistance (AMR). It has also helped scientists identify microbial enzymes and genes that are key players in ecological processes so that the roles of microbes are clearly understood in terrestrial and other ecosystems.

8.4 EVOLUTION OF ARTIFICIAL INTELLIGENCE IN LIFE SCIENCES

Advancements in technology, in general, have brought ease to processes that usually take a longer time to execute. For example, AI in its distinctive forms, such as ML and deep learning, can now be used to automate processes, improve data analysis, and enhance decision-making in various life science sectors, thus assisting in better drug discovery, patient care, and overall industry efficiency. All of these are possible because AI has the potential to aid the understanding of acquired data from any notable electronic health records, which will enhance accurate diagnosis of patients' ailments for appropriate and reliable treatment. With the vast volume of data being generated in life sciences, which are usually unstructured, heterogeneous, non-standard, and incomplete in nature, there is an increased need to employ the use of suitable AI and bioinformatics tools for data analytics to aid efficient pattern discovery. An improved reliability of the acquired data often increases the demand to develop efficient, sensitive, and cost-effective medical solutions for disease prevention, efficient resource planning, and optimal patient-centered care operations.

8.5 CHALLENGES IN MICROBIAL GENOMICS

Like in many other fields of study, the deployment of AI tools in microbial genomics has helped to resolve many operational challenges usually experienced in the use of traditional approaches in its processes. For example, the traditional microbial genomics diagnostic process usually involves manual interpretation of laboratory results, microscopy images, and genomic sequences. This approach usually has issues with sample handling, culture difficulties, misidentification, and delays in determining susceptibility (Alsulimani et al., 2024). The introduction and adoption of AI in microbial genomics have brought about quick analysis of the acquired datasets, thus making it easier to identify subtle patterns and connections that ordinarily would have gone unnoticed by humans, thereby addressing the concerns surrounding data complexity. The deployment of AI has also helped healthcare practitioners achieve accelerated medical condition diagnoses to ensure prompt interventions and enhanced patient outcomes (Alsulimani et al., 2024). Since AI relies heavily on datasets, it utilizes data analysis, pattern recognition, and diagnostic processes to identify disease conditions in their early stages, provide advanced and individualized treatment plans, and conduct epidemic surveillance (Bohr & Memarzadeh, 2020). Notably, the use of AI in microbial diagnosis has shown a significant potential for personalized medicine, thus enabling customized treatment approaches to be developed for individual patients based on their unique profile (Alsulimani et al., 2024). In all, AI is revolutionizing microbial genomics as it enhances the speed, accuracy, scope of diagnosis, research, and disease surveillance activities, thus resulting in accelerated genome analysis, variant detection, and the interpretation of complex hereditary patterns, enabling breakthroughs in personalized medicine (Gupta et al., 2025). Finally, the fusion of AI with microbial genomics will predictably advance human health, agricultural biotechnology, and evolutionary biology.

8.6 ARTIFICIAL INTELLIGENCE TECHNIQUES USED IN MICROBIAL GENOMICS

The implementation of AI in microbial genomics requires the use of several AI techniques, such as ML, deep learning, and natural language processing. These techniques have improved their application in pathogen surveillance, genomic annotation, drug discovery, and microbiome research.

a. Machine learning technique: ML is a subfield of AI and can learn patterns from data and improve its performance using algorithms rather than explicit programming. ML gains insights from analyzing vast and complex datasets for informed decision-making, such as predicting disease progression and detecting antibiotic resistance genes (Dhaarani & Kiranmai Reddy, 2025; Gao & Liu, 2024). ML can be classified into supervised learning, unsupervised learning, and reinforcement learning (Lin et al., 2021).

   In supervised learning, labeled data is available to train the model to learn the underlying patterns. Tasks such as classification and regression require known outputs, unlike unsupervised learning, whose output is unknown. Clustering and dimensionality reduction are common tasks suitable for unsupervised learning. Reinforcement learning is an interactive learning technique that uses the trial-and-error method while maximizing cumulative reward. It is concerned with the reaction of software agents to changes in an environment (Lin et al., 2021). The most commonly used ML algorithms in genomics include support vector machines (SVM), random forests, *k*-nearest neighbor, decision trees, and logistic regression (Dalsaniya et al., 2025; Lin et al., 2021). These classification algorithms have been widely applied to analyze genomic sequences and microbial species.

Unsupervised ML is widely used in microbial genomics due to its effectiveness at extracting insights into microbial diversity and its community structures. K-Means clustering, when applied to complex genomic data, identifies discrete clusters of similar data points by iteratively refining the mean of the clusters (Jia et al., 2024). Due to the high dimensionality of microbial genomic data, there is often a need to concentrate on the most relevant features that influence outcomes. Hence, principal component analysis (PCA) is an unsupervised learning technique that is applied to enhance the detection of patterns in genomic data while maintaining variance (Contreras-Peruyero et al., 2024). DEICODE (robust Aitchison PCA), a variant of PCA optimized for compositional data, was applied to analyze the beta diversity of functional metabolic pathways between healthy and disease groups in samples (Baker et al., 2021).

Reinforcement learning technique is an AI technique used to train software to make appropriate decisions for the purpose of achieving optimal results. It usually requires an agent to learn in a given environment by performing an action and observing the results of the action (Kalaisriram, 2021). Primarily, reinforcement learning consists of two components, namely, the agent and the environment. The agent performs two key activities: exploration and training (Biswas, 2025). During the exploration stage, the agent collects its experience from the environment where it is being deployed by taking actions and determining what happens next. Then, during the training stage, it uses the experiences obtained to improve itself and its actions. Once the agent selects an action, the deployment environment is updated and is expected to return a reward depending on how well the agent is operating. In its application in microbial genomics, it is desired for exploring high-dimensional genomic data due to its adaptive decision-making abilities and reward signaling for process optimization to provide insights into microbial behavior.

b. Deep learning technique: This is also a subset of AI with multiple processing layers of artificial neural networks that learn from hierarchical data representation. It supports the extraction of complex features and patterns from high-dimensional datasets such as microbial genomics, without the need for feature engineering. A deep learning model learns progressively on each processing layer, from the hidden patterns and non-linear relationships and complex interactions in the data. Deep learning techniques have become popular due to their high-performance computing capability (Divya & Aiswarya, 2021). Convolutional neural network, recurrent neural network, generative adversarial network, deep reinforcement learning, and Boltzmann machine are examples of deep learning techniques. These techniques can handle a huge amount of unstructured data with greater functionality and flexibility.
c. Natural language processing: This is a branch of AI that focuses on giving machines the ability to understand human language for the purpose of creating programs that can process and analyze large volumes of text as well as spoken words. Natural language processing integrates computational linguistics, ML, and deep learning in the form of natural language understanding (NLU) and natural language generation (NLG) in the development of its machines (Jayaprada, 2024). The key techniques in natural language processing are tokenization, part-of-speech tagging, named entity recognition, word embedding, and transformer models (Reddy, 2025). The incorporation of these techniques enables machines to

understand and generate human language with remarkable accuracy. In relation to its application in microbial genomics, NLP is used to treat genomic sequence data for the accurate prediction of gene functions, identification of novel antimicrobial peptides, and an understanding of microbial traits.

8.7 APPLICATIONS OF ARTIFICIAL INTELLIGENCE IN MICROBIAL GENOMICS

AI has found several applications in microbial genomics. Some notable applications are as follows:

i. Genome Annotation & Feature Prediction: Genome annotation is described as the process of identifying and characterizing genes and their individual functional elements within microbial genomes (Gray, 2025). Feature prediction involves the use of data in forecasting future events, behaviors, or characteristics. The use of AI-driven tools has revolutionized the way these two processes are now being conducted. It has been observed that their use has commendably improved the accuracy, speed, and efficiency of gene annotation, especially when it involves the analysis of large and complex microbial genomes (Umedum, 2025). The analysis, which relies heavily on a vast dataset, is used to identify code regions and genes that are associated with antibiotic resistance, predict gene locations as well, and classify the functional regions within microbial genomes.
ii. Microbial Identification & Classification: Microbial identification is the process used to determine the genus, species, or strain of a microorganism, like yeast, bacteria, etc. While microbial classification involves categorizing these microorganisms based on their characteristics into their major groups. Traditional methods used in the identification of microbial, such as biochemical tests, molecular and culturing techniques, microscopy, and mass spectrometry, are known to be time-consuming and labor-intensive. To achieve faster, more accurate microbial identification, AI-driven tools that are capable of analyzing large datasets from sequencing, imaging, and laboratory tests are now being employed (Reddy, 2025). For example, AI tools have been developed to automate culture analysis and the interpretation of growth patterns (Umedum, 2025). These AI models can also monitor real-time culture dynamics, thus significantly reducing the process time and improving their efficiency.
iii. Antibiotic Resistance Prediction: There are increasing concerns about the effect of AMR on the implementation of pathogen treatment strategies. However, there are three possible ways to predict antibiotic resistance, namely, genomic analysis, ML, and ecological modeling. With the use of AI-driven tools, there has been a recorded increase in the rate of discovery of new antibiotics as well as the accurate prediction of resistance mechanisms in these pathogens, thus ensuring the implementation of better treatment strategies. This involves the use of AI models for the analysis of data obtained from bacterial genomes and clinical records to predict possible antibiotic resistance patterns (Gray, 2025). The identification of these resistance genes and mutations will aid in the selection of effective treatments.
iv. Metagenomics & Microbiome Analysis: Metagenomics is concerned with sequencing DNA obtained from environmental samples for the purpose of studying and understanding microbial communities. Microbiome analysis is the study of microbial communities and their

collective genomes in a given or specific environment, such as the human gut. AI tools such as Kraken2, MetaPhlAn3, and DeepMicrobes are built with the capacity to process terabytes of sequencing data to profile microbial communities (Gupta et al., 2025). A vast amount of metagenomic data is analyzed to identify new pathogens, antibiotic resistance genes, as well as the nature of interactions existing between the microbiomes. The information obtained is used to appropriately classify the microbial species and aid in the proper understanding of their various roles in disease progression. The deployment of AI tools also aids in the acceleration of genomic analysis through the identification of mutations, predicting pathogenicity, and the analysis of microbial communities (Gray, 2025). AI models are used to analyze genomic sequences to predict the nature of a microorganism as to whether it is pathogenic or not. By comparing the genetic features of harmless microbes and known pathogens, the AI system is able to identify any possible threat.

v. Pathogen Surveillance & Outbreak Tracking: To ensure efficient pathogen surveillance, AI tools are used to analyze huge datasets from existing medical records as well as data obtained from environmental sensors deployed in suspected regions. The result of the analysis can be used to predict and track infectious disease outbreaks. Some AI-driven platforms, such as BlueDot and HealthMap, have been used to identify and detect early warning signs of disease outbreaks such as COVID-19 (Gray, 2025). The deployment of a ML surveillance-based system, which relies on data obtained from climate, travel, and demography, helps to track any possible outbreak, thus aiding in curbing the spread through timely interventions such as vaccination campaigns and quarantine measures. In addition, AI-based tools can now be used to monitor and predict the evolution trend of microbial, such as the emergence of new viral variants. The ability of these tools to analyze mutation rates, structural changes, and host adaptation has helped scientists to stay ahead of any possible emerging threats (Gupta et al., 2025).

vi. Synthetic Biology & Microbial Engineering: Synthetic biology is the field of biological engineering that designs and constructs new biological parts, systems, and devices or redesigns already existing ones. Microbial engineering is an interdisciplinary field that utilizes biological, chemical, and engineering principles for the modification of microorganisms tailored to specific applications. The integration of AI techniques with synthetic biology has led to the acceleration of the design, testing, and optimization of engineered biological constructs across multiple domains (Groff-Vindman et al., 2025). This has resulted in knowledge acquisition and refinement; functionality prediction, increased access, and a reduction in skill threshold; and the acceleration of design cycles as well as the automation of the Design-Build-Test-Learn cycle. The application of AI in microbial engineering includes strain development and metabolic pathway optimization, the acceleration of drug and compound discovery, improving fermentation and bioprocess control, and, finally, the enabling of rapid microbial identification and diagnosis.

8.8 AI-POWERED TOOLS, PLATFORMS, AND DATABASES

i. AI-Powered Tools: Many AI-powered tools based on ML and deep learning approaches have been developed to use advanced algorithms to learn from existing data in order to reveal complex patterns and relationships that are ordinarily not captured by traditional analytical methods (D'Urso & Broccolo, 2024). One of such notable AI-powered models employed for microbial genome analysis is DeepMicro. DeepMicro is a deep learning model used for the analysis of high-dimensional microbiome data to achieve dimensionality reduction, health state classification, and disease prediction. It can undergo training based on the appropriate microbiome sequencing data that is being provided. After the training, the outcome is used to predict the nature of a specific microbial condition since the model has learnt the pattern that links the presence or absence of certain microbial species to possible health outcomes.

Metagenomic Phylogenetic Analysis (MetaPhlAn) is another AI-powered tool that has enhanced the advancement of microbial genomes. MetaPhlAn is a bioinformatics computational tool designed to process, analyze, and interpret complex microbial communities that are obtained from metagenomic sequencing data (Pita-Galeaana et al., 2025). The integration of ML and MetaPhlAn has brought about enhancements in microbial profiling, pathogen and disease prediction, and improved insight into cellular states and disease mechanisms.

Also notable is Kraken, an AI-powered tool that is an ultrafast and highly accurate program for assigning taxonomic labels to metagenomic DNA sequences (Wood & Salzberg, 2014). It is designed primarily to transform complex, high-volume sequencing data into rapid, accurate, and actionable biological insights. It is often known for its speed and accuracy in assigning taxonomic labels to DNA sequences. Kraken is built on a database that can map k-mers to the lowest common ancestor (LCA) in a taxonomic tree for all the organisms that contain that k-mer. Beyond just classification, Kraken is capable of analyzing more complex microbial genome datasets.

ii. Integration of AI-Powered Tools with Databases: An AI-powered database management system is a system designed to aid the processing of high-dimensional vector data, which may be in the form of text, images, or audio, usually presented in a format that it can semantically understand. The databases are usually integrated with AI-powered tools, thus, combining AI techniques with traditional database systems to enhance efficient data processing, storage of large volumes of data, data analysis, and appropriate decision-making capabilities. Generally speaking, data integration involves combining data from different sources into a unified format (Scannapieco, 2006). The integration of AI into the database relates to the harmonization of AI-generated data with existing data obtained from traditional database systems (Unuriode et al., 2023). This integration improves functionality, efficiency, data quality, data cleansing, scalability, and adaptability and speeds up the decision-making process. The synergy aids in the extraction of valuable insight from the data to ultimately drive innovation and competitiveness in the domain of application (Kim, 2020).

The integration of AI-powered tools with a database can be achieved through the following steps (Sarvaiya, 2025):

a. Audit and classify the data landscape: This involves identifying the nature of the data, whether from structured or unstructured sources.
b. Build an AI-ready data layer: This involves preprocessing the data obtained into a usable format.
c. Train the AI for schema mapping and transformation: This involves training the ML models to be able to detect changes and adapt dynamically.
d. Automate AI-powered Data Quality and Cleansing: This involves the cleansing of the data obtained to ensure that anomalies and inconsistencies are detected and removed.
e. Deploy AI-driven integration pipelines: This is to ensure smooth end-to-end integration.
f. AI-enabled governance and compliance: This helps to enforce data security, privacy, and data compliance.
g. Continuous monitoring and optimization: This requires continuous observation of the dashboards and retraining of the integration models to sustain their accuracy level.

Following through with these steps ensures a trusted data integration with AI-powered tools that empower the system to make smarter decisions that trigger innovations. Examples of some databases that are AI-powered tools enabled are the National Center for Biotechnology Information (NCBI), the Emerging Markets Bond Index (EMBL), and Metagenomic Rapid Annotations using Subsystems Technology (MG-RAST). These are open-source web application servers and repositories that contain genetic materials from microbial communities.

iii. Cloud-based platforms and open-source frameworks: Cloud-based platforms and open-source frameworks are used for the analysis of microbial genomes to provide accessibility and scalability during the implementation of research solutions. The cloud-based platforms ensure the use of user-friendly web interfaces with minimal bioinformatics expertise to offer scalable computation and data storage. Microbial data analysis cloud platform (MDACP) is one example of a cloud platform with a friendly user interface that is easy to operate and allows users to upload data and analyze in parallel with multiple tasks (Han et al., 2020). Another example is the IBM functional genomics platform, built on a database that connects genotype to phenotype for bacterial life and provides a powerful programming and development framework that queries data and generates answers to questions of microbial life (Seabolt et al., 2019).

Open-source frameworks are designed as command-line tools that are usually installed on a local server or deployed on cloud instances using containers. Common examples of open-source frameworks are AMRomics, MicroScope, CABGen Galaxy, and many more. These frameworks provide valuable services to individual research groups, giving them a comprehensive workflow of data from its raw form to the final annotated form, serving as a basis for an in-depth analysis and appropriate decision-making.

8.9 CHALLENGES AND LIMITATIONS

Despite the remarkable results that have been achieved in the integration of AI with microbial genomics, some applications have shown that there are several key challenges and limitations that are being experienced. Some of these challenges and limitations are related to data management, reliability of the developed model, and issues surrounding the integration into clinical workflows, as well as ethical considerations. For example, the issue of data quality and standardization occurs due to inconsistency and poor data annotation (Bao et al., 2021). Mistrust, data privacy concerns, and the ethical use of personal data have posed a serious barrier for the effective implementation of developed systems (Fu et al., 2021). Finally, a lack of diversity in the existing dataset can lead to a possible bias in the results generated and an exacerbated health disparity (Becht et al., 2018). Hence, these concerns have limited the full-scale adoption and integration of AI in microbial genomics as well as other areas of AI integration.

8.10 FUTURE DIRECTION

The challenges mentioned in the last subheading have fostered an urgent need to ensure that these concerns are addressed. It is important that the relevant bodies establish standard data practices, address any ethical concerns, and ensure that data integrity is at its topmost level, thus ensuring the full realization of the potential available through the integration of AI in microbial genomics (Mondal et al., 2025). This will result in a more sustainable, effective, reliable, and responsible future in the field of microbial genomics.

8.11 CONCLUSION

This chapter has provided a brief introduction to microbial genomics and its significance in the health sector. It further delved into its integration with AI while considering common AI techniques employed for its adoption and its application areas in relation to microbial genomics, AI-powered tools, platforms, and databases. It concludes by identifying some notable challenges and limitations encountered during the implementation of an integrated system and a future direction to improve the level of adoption.

REFERENCES

Alsulimani, A., Akhter, N., Jameela, F., Ashgar, R. I., Jawed, A., Hassani, M. A., & Dar, S. A. (2024). "The Impact of Artificial Intelligence on Microbial Diagnosis." *Microorganisms*, Vol. 12. No. 6. p. 1051.

Baker, J. L., Morton, J. T., Dinis, M., Alvarez, R., Tran, N. C., Knight, R., & Edlund, A. (2021). "Deep Metagenomics Examines the Oral Microbiome During Dental Caries, Revealing Novel Taxa and Co-Occurrences with Host Molecules." *Genome Research*, Vol. 31. No. 1. pp. 64–74.

Bao, Y., Medland, N. A., Fairley, C. K., Wu, J., Shang, X., & Chow, E. P. F. (2021). "Predicting the Diagnosis of HIV and Sexually Transmitted Infections Among Men Who Have Sex with Men Using Machine Learning Approaches." *Journal Infection*, Vol. 82. pp. 48–59.

Becht, E., Mclnnes, L., healy, J., Dutertre, C. A., Kwok, I. W. H., & Ng, L. G. (2018). "Dimensionality Reduction for Visualizing Single-Cell Data Using UMAP." *National Biotechnology*, Vol. 37. pp. 38–44.

Biswas, A. (2025). *The Reinforcement Learning Handbook: A Guide to Foundational Questions*, Toward Data Science. https://towardsdatascience.com.

Bohr, A., & Memarzadeh, K. (2020). "The Rise of Artificial Intelligence in Healthcare Applications." pp. 25–60. [Google Scholar].

Choffnes, E. R., Olsen, L. A., & Wizemann, T. (2013). "The Science and Applications of Microbial Genomics: Workshop Summary." *National Academy of Sciences*, pp. 1–429. doi: 10.17226/18261

Contreras-Peruyero, H., Nuñez, I., Vazquez-Rosas-Landa, M., Santana-Quinteros, D., Pashkov, A., Carranza-Barragán, M. E., Perez-Estrada, R., Guerrero-Flores, S., Balanzario, E., Muñiz Sánchez, V., Nakamura, M., Ramírez-Ramírez, L. L., & Sélem-Mojica, N. (2024). "CAMDA 2023: Finding Patterns in Urban Microbiomes." *Frontiers in Genetics*, Vol. 15. p. 1449461.

D'Urso, F., & Broccolo, F. (2024). "Applications of Artificial Intelligence in Microbiome Analysis and Probotic Interventions – An Overview and Perspective Based on the Current State of the Art." *Applied Sciences*, Vol. 14. No. 19.

Dalsaniya, A., Beladiya, U., & Kothari, R. K. (2025). "Predictive Modeling of Microbial Functionality from 16S rDNA Sequences Using Machine Learning. World." *World Journal of Biology, Pharmacy, and Health Sciences*, Vol. 21. No. 02. pp. 545–554.

Dhaarani, R., & Kiranmai Reddy, M. (2025). "Progressing Microbial Genomics: Artificial Intelligence and Deep Learning Driven Advances in Genome Analysis and Therapeutics." *Intelligence-Based Medicine*, Vol. 11. p. 100251.

Divya, P., & Aiswarya, V. B. (2021). "Deep Learning: Techniques and Applications." *Journal of Emerging Technologies and Innovative Research*, Vol. 8. No. 7. pp. a125–a129.

Fournier, P. E., Drancourt, M., & Raoult, D. (2007). "Bacterial Genome Sequencing and Its Use in Infectious Diseases." *Lancet Infectious Diseases*, Vol. 7. pp. 711–723.

Fu, C., Zhang, X., Veri, A. O., Iyer, K. R., Lash, E., & Xue, A. (2021). "Leveraging Machine Learning Essentially Predictions, and Chemogenomic Interactions to Identify Antifungal Targets." *National Commun*, Vol. 12. No. 1. p. 6497.

Gao, Y., & Liu, M. (2024). "Application of Machine Learning Based Genome Sequence Analysis in Pathogen Identification." *Frontiers in Microbiology*, Vol. 15. p. 1474078. doi: 10.3389/fmicb.2024.1474078.

Gray, S. (2025). "Artificial Intelligence in Pathogenic Microorganism Research." *Clinical Microbiology*, Vol. 14. No. 1. pp. 1–2.

Groff-Vindman, C. S., Trump, B. D., Cummings, C. L., Smith, M., Titus, A. J., Oye, K., Prado, V., Turmus, E., & Linkov, I. (2025). "The Convergence of AI and Synthetic Biology: The Looming Deluge." *Biomedical Innovations*. Vol. 2, No. 20. pp. 1–11.

Gupta, P., Singh, R. K., Katiyar, S., Azmi, A., Awasthi, A., & Kaur, J. (2025). "Artificial Intelligence in Microbiology and Genetics: Revolutionizing Diagnostics, Research, and Surveillance." *The Academic International Journal of Multidisciplinary Research*, Vol. 3. No. 4. pp. 228–253.

Han, N., Miao, J., Zhang, T., Qiang, Y., Peng, X., Li, X., & Zhang, W. (2020). "MDACP: A Pathogen Genome and Metagenome Analysis Cloud Platform." *Frontiers in Genetics*. Vol. 11. p. 1007. doi. 10.3389/fgene.2020.01007.

Jayaprada, G. L. N. (2024). "Natural Language Processing. Lecture Notes – Department of Computational Intelligence." *Malla Reddy College of Engineering & Technology*, Autonomous Institution – UGC, Government of India, pp. 1–95.

Jia, X., Lin, S., Wang, Y., JiaZhang, Q., Miao, J., Li, M., & Wang, H. (2024). "Recruitment and Aggregation Capacity of Tea Trees to Rhizosphere Soil Characteristic Bacteria Affects the Quality of Tea Leaves." *Plants*, Vol. 13. No. 12. p. 1686.

Kalaisriram, S. (2021). "Reinforcement Learning." Seminar Report Submitted to the Department of Information Technology Manakula Vinayagar Institute of Technology, Kalitheerthalukuppam Pundicherry University.

Kim, S. (2020). "Artificial Intelligence and Database Systems: A Comprehensive Review." *IEEE Transactions on Knowledge and Data Engineering*, Vol. 32. No. 2. pp. 1502–1519.

Kugarajah, V., Nisha, K. N., Jayakumar, R., Sahabudeen, S., Ramakrishnan, P., & Mohamed, S. B. (2023). "Significance of Microbial Genome in Environmental Remediation." *Microbiological Research*, Vol. 271. pp. 1–11.

Lin, Y., Wang, G., Yu, J., & Sung, J. J. Y. (2021). "Artificial Intelligence and Metagenomics in Intestinal Diseases." *Journal of Gastroenterology and Hepatology*, Vol. 36. No. 4. pp. 841–847.

Medfu Tarekegn, M., Zewdu Salilih, F., & Ishetu, A. I. (2020). "Microbes Used as a Tool For Bioremediation of Heavy Metals from the Environment." *Cogent Food Agric*, Vol. 6. No. 1. p. 1783174.

Mondal, R. S., Akter, L., & Bhuiyan, M. N. A. (2025). "Integrating AI and ML Techniques in Modern Microbiology." *Applied IT & Engineering*. pp. 1–10. doi: 10.25163/engineering. 3110323.

Ojuederie, O. B., & Babalola, O. O. (2017). "Microbial and Plant-Assisted Bioremediation of Heavy Metal Polluted Environments: A Review." *International Journal of Environmental Research and Public Health*, Vol. 14. No. 12. p. 1504.

Pita-Galeaana, M. A., Ruhle, M., Lopez-Vazquez, L., de Anda-Jauregui, G., & Hernandez-Lemus, E. (2025). "Computational Metagenomics: State of the Art." *International Journal of Molecular Sciences*. Vol. 26. No. 18. pp. 1–48.

Reddy, P. S. (2025). "Natural Language Processing (NLP) and Understanding." *International Journal of Advanced Research in Electrical, Electronics, and Instrumentation Engineering*, Vol. 14. No. 2. pp. 1–7.

Santos-Beneit, F. (2024). "What is the Role of Microbial Biotechnology and Genetic Engineering in Medicine?" *Microbiology Open*, Published by John Wiley & Sons Ltd., pp. 1–31.

Sarvaiya, S. (2025). "AI Data Integration for Enterprises: How to Implement It Successfully." *Artificial Intelligence – Azilen Technologies.*

Scannapieco, M. (2006). "Data Quality: Concepts, Methodologies and Techniques. Data-Centric Systems and Applications." *Springer.*

Seabolt, E. E., Nayar, G., Krishnareddy, H., Agarwal, A., Beck, K. L., Terrizzano, I., Kandogan, E., Kunitomi, M., Roth, M., Mukherjee, V., & Kaufman, J. H. (2019). "IBM Functional Genomics Platform: A Cloud-Based Platform for Studying Microbial Life at Scale." *IEEE/ACM Transactions on Computational Biology and Bioinformatics*, Vol. 16. No. 7. pp. 1–14.

Smith, H. H. (1943). "Effects of Genome Balance, Polyploidy, and Single Extra Chromosomes on Size in Nicotiana," *Genetics*, Vol. 28, No. 3, pp. 227–236. doi: 10.1093/genetics/28.3.227.

Taqi, M., Younus, S., & Yousaf, A. (2025). "Role of Artificial Intelligence in Genomics." *International Journal of Research Publication and Reviews*, Vol. 6. No. 1. pp. 3149–3156.

Umedum, C. U. (2025). "Artificial Intelligence and Machine Learning in Microbial Diagnostics." *Novel Diagnostic Methods in Clinical Microbiology.* IIP Series, Chapter 9. pp. 216–259.

Unuriode, O. A., Durojaiye, M. O., Yusuf, Y. B., & Okunade, O. L.. (2023). "The Integration of Artificial Intelligence into Database Systems (AI-DB Integration Review)." *International Journal on Cybernetics & Informatics (IJCI).* Vol. 12, No. 2. pp. 1–12.

Wood, D. E., & Salzberg, S. L. (2014). "Kraken: Ultrafast Metagenomic Sequence Classification Using Exact Alignment." *Genome Biology*, Vol. 15. No. 3. p. R46.

AI and Metagenomics in Relation to Human Health

9

Gbotemi Jerry Oni, Joy Oziohu James, Zawu Yoowa Duyann, Samuel Adeniyi Oyegbade, and Oluwafemi Adebayo Oyewole

9.1 INTRODUCTION

A core basis of microbiology in the modern era is metagenomics. This is linked to the fact that metagenomics fosters the culture-autonomous examination of multifaceted microbial communities across host and environment-linked ecosystems (Handelsman *et al.*, 1998). Advances in metagenomics have paved the way for scientists to surmount limitations and to locate vastly diverse microbes which are not culturable using conventional microbiological approaches.

Metagenomics is generally classified into shotgun and amplicon-grounded. The latter depends on conserved marker genes' sequencing and amplification (particularly bacterial 16S rRNA gene and fungal internal transcribed spacer (ITS) regions). Contrarily, the former, shotgun metagenomics, entails sequencing of all extracted nucleic acids, offering details into taxonomic constitution, determinants of AMR, and virulence-linked genes concurrently (Quince et al., 2017). This technique fosters comparative and functional genomic analyses on a high-resolution level. Metagenomic analyses are routinely adopted on a vast array of marine, freshwater, and soil environments and host-linked microbiota (Mackelprang *et al.,* 2011). Furthermore, the series of evolutions surrounding the technologies and protocols of DNA extraction have enhanced the genomic material fragment recovery from complex samples, fostering the reconstruction of genome and the detection of mobile genetic elements that foster horizontal gene transfer and microbial adaptation.

However, the vast diversity of metagenomic datasets introduces critical analytical constraints in the course of annotation, binning, and assembling of genomes (Wooley et al., 2010). The selection of sequencing technology substantially affects the accuracy and resolution of metagenomes. Even though short-read sequencing platforms demonstrate a high degree of accuracy, they lack the capacity to resolve broad insertions, repetitive regions, and complicated structural variants.

Long-read sequencing technologies, especially those created by Oxford Nanopore Technologies (ONT) and Pacific Biosciences, have substantially enabled the generation of elongated read lengths,

DOI: 10.1201/9781003727514-11

which brings about enhanced metagenomic assemblies and lowered segmentation (Bleidorn, 2016), and are crucial in the resolution of poor assemblies, repetitive genomic architectures and high-GC regions (Tørresen et al., 2019).

Metagenomic data analysis is still limited despite technological advancements, due to sparseness of data, inaccurate taxonomic resolution, complexity of the algorithm, and extended dimensionality. These, in turn, continue to bring about non-reliability of traditional bioinformatics pipelines (Wooley et al., 2010). These constraints further constrain the availability of curated functional databases for environmental microbes, which distorts comparative investigation between different samples as well as functional annotation (New and Brito, 2020).

Expansive incorporation of multi-omics and rising complications of metagenomic ecosystems have heightened demands for post-processing and brought to bare the drawbacks of classical computational methods (Teeling and Glöckner, 2012). AI and ML have emerged as potent metagenomic data analysis tools to respond to and surmount the limitations by unveiling improved taxonomic categorisation, automated feature extraction, and enhanced genome binning. In most recent times, AI-grounded frameworks showcase higher and excellent management of diversely dimensional metagenomic datasets, bringing about a more superior computational efficiency, effectiveness, scalability and preciseness than analytical models (Qayyum et al., 2025).

This chapter aims to deeply investigate the incorporation of machine learning (ML) and artificial intelligence (AI) in metagenomic analysis, establishing their function in advancing microbiological research within the clinical and environmental domains, improving the interpretation of data, and addressing analytical bottlenecks.

9.2 HUMAN MICROBIOME AND HEALTH

The gut part of the human body is the most robustly diverse microbial ecosystem, accompanied by the mouth and respiratory tract (Figure 9.1). The microbiome communities offer xenobiotic metabolism, immune education, and nutrition. Dysbiosis has been linked to neuropsychiatric and inflammatory conditions, cancer, and metabolic diseases (Aggarwal *et al.*, 2022; Ma *et al.*, 2024). The human microbiome encompasses a complicated ecosystem constituting several billions of microbes such as fungi, archaea, viruses, and bacteria, which are responsible for colonisation of several body parts, while this community offers key roles such as immune modulation, metabolism of nutrients, and safeguarding against pathogens and epithelial barrier integrity maintenance (Pasolli *et al.*, 2019; Lloyd-Price et al., 2019). Robust investigation of microbiome's characteristics is now attainable for increasing progression in metagenomic research endeavours far more than culture-linked techniques. Oral, dermal, and gut cavities house numerous microbial communities impacted by releases by the host, pH, and availability of oxygen (Quince et al., 2017). Dysbiosis refers to the loss of balance in the role of microbes and their constitution. It is particularly associated with pathological factors such as respiratory infections, cancer, obesity, diabetes, and neurological imbalances (Shreiner et al., 2015).

The World Health Organisation (WHO) set up the global genomic surveillance strategy for pathogens with pandemic and epidemic potential from 2022 to 2032 to enable countries to establish their own genomic surveillance strategy for important pathogens (World Health Organisation, 2022). WHO further stresses that it is in the best interests of all Member States to share such data since it is essential for preventing, identifying, and promptly responding to pandemics and epidemics at all levels (Aborode et al., 2021). Aside from the high prevalence of medication resistance, which is thought to have contributed to at least 1.05 million fatalities in 2019 (Aguilar et al., 2023), Africa continues to be more

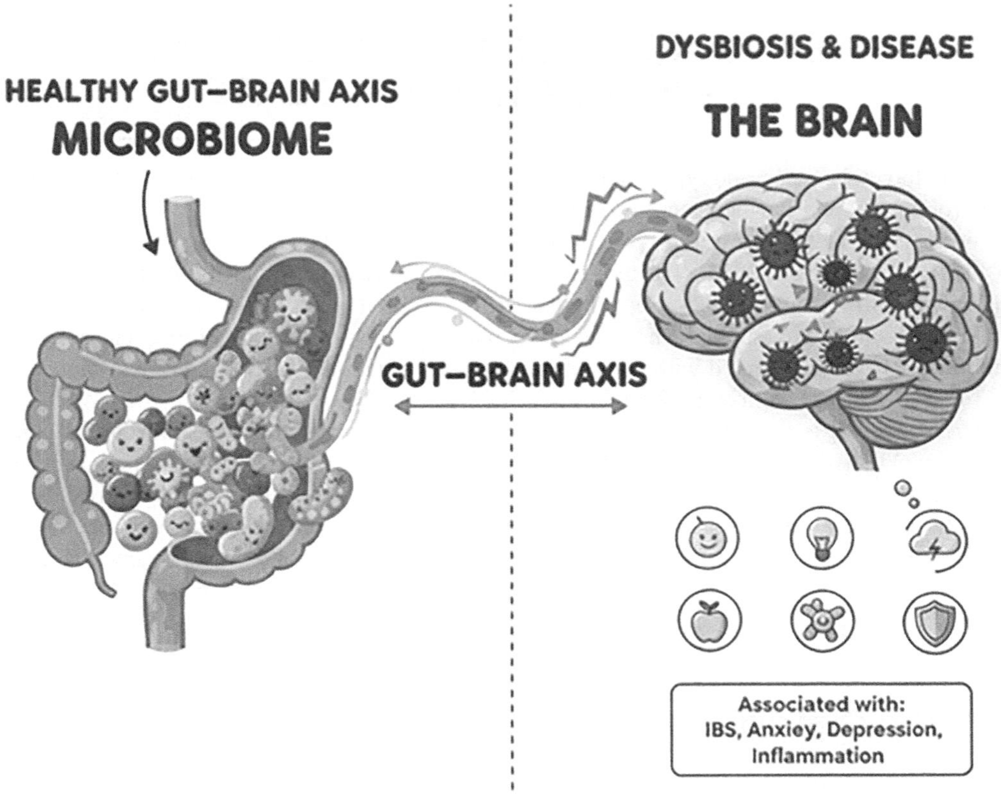

FIGURE 9.1 The gut–brain axis, exhibiting bidirectional communication between the brain and a varied microbiome in the gut. Health-advantageous bacteria produce neurotransmitters and short-chain fatty acids, fostering immune roles, metabolism, and mood support. Dysbiosis can yield systemic inflammation, depression, and anxiety.

vulnerable to infectious disease risks, with at least 140 disease outbreaks reported each year (Makoni, 2020). This necessitates the immediate development and maintenance of capacity for routine pathogen genomic surveillance, near real-time epidemic identification, and characterisation.

Microbiome interactions are important for human health because of this relationship, which can be neutral, harmful, or favourable. By having different kinds of enzymatic proteins, these additional genes help to regulate host physiology by changing the metabolites that are created and, consequently, host metabolism (Hooper and Gordon, 2001). Over time, an all-encompassing method based on the halobiont theory has been used rather than examining the interaction between a single bacterium and its host (Theis et al., 2016). Numerous microbes serve crucially in ecosystem functioning and have mutually beneficial microbe–microbe connections, which affect its functionalities as well as dynamics in population.

Differentiating between a temporal microbiome shaped by peculiar environmental influences is made much less difficult when the main microbiota is determined (Berg et al., 2020). While transient microbiota varies over time, core microbiota is the microbial population that is consistently linked to a particular host genotype or environment. By recognising these variations, the methodology used in microbiome research for therapeutic purposes can be improved using a suitable experimental, methodological, and statistical design.

9.3 LINKS BETWEEN MICROBIOME COMPOSITION AND DISEASE

Obesity, colorectal cancer, and type 2 diabetes have been substantially linked to the constitution of the gut microbiome. Cohesive metagenomic sequencing has enabled better recognition of lesions far above conventional risk factors. Respiratory microbiome dysbiosis is linked with lingering lung contagions (de Córdoba-Ansón *et al.,* 2025). The gut microbiome hosts a vast microbial community in the body of humans with trillions of microbes (Sender et al., 2016). Proof abounds of the substantial contributions of the gut microbiome to metabolic homeostasis. Changes in the population of Firmicutes in relation to Bacteroidetes have been linked with obesity as a result of variation in polysaccharides' energy extraction efficiencies (Turnbaugh et al., 2006).

Carcinogenesis is modulated by the gut microbiota via diverse mechanisms such as genotoxic metabolite synthesis and modulation of the immune system. *Fusobacterium nucleatum* has been associated with colorectal cancer (Kostic et al., 2012), while *Bifidobacterium longum* improves immune checkpoint inhibitors in cancer therapy (Routy et al., 2018). The gut–brain region connecting microbial metabolites to neurological consequences has gained substantial attention; meanwhile, dysbiosis (Figure 9.1) is implicated in depression, Parkinson's disease, and autism spectrum disorder (Cryan et al., 2019).

9.3.1 The Oral Microbiome

The oral cavity constitutes the second most abundantly diverse microbes comprising the teeth, gingiva, and tongue surfaces, which are colonised by hundreds of bacterial species (Dewhirst et al., 2010), significantly including the *Actinomyces*, *Veillonella*, and *Streptococcus* genera. Aside from contributing to systemic conditions, periodontal pathogens also foster oral diseases, such as those seen in *Porphyromonas gingivalis*, which is caught up in atherosclerosis through molecular mimicry and systemic inflammation (Hajishengallis, 2015). Oral dysbiosis is also associated with a higher risk and occurrence of Alzheimer's disease and pancreatic cancer (Michaud et al., 2013).

9.3.2 The Skin Microbiome

The skin houses microbial communities that are specific to sites, as influenced by exposure, moisture, and sebum. Major taxa are such as *Staphylococcus and Propionibacterium* species (Byrd et al., 2018). Skin conditions, including psoriasis, atopic dermatitis, and acne vulgaris, are consequences of dysbiosis. This has been reported by Nakatsuji et al. (2017) through their study investigating the overgrowth of *Staphylococcus aureus*, which aggravates atopic dermatitis via activation of toxin-mediated immunity.

9.3.3 The Respiratory Microbiome

A typical sterile respiratory tract houses a peculiar microbiome which encompasses *Prevotella*, *Streptococcus*, and *Veillonella* genera, which substantially impact pathogen resistance and mucosal immunity (Man et al., 2017). Respiratory conditions and infections, such as pulmonary infection and asthma, are triggered by dysbiosis (Figure 9.2). Altered microbial aggregation is likely to be

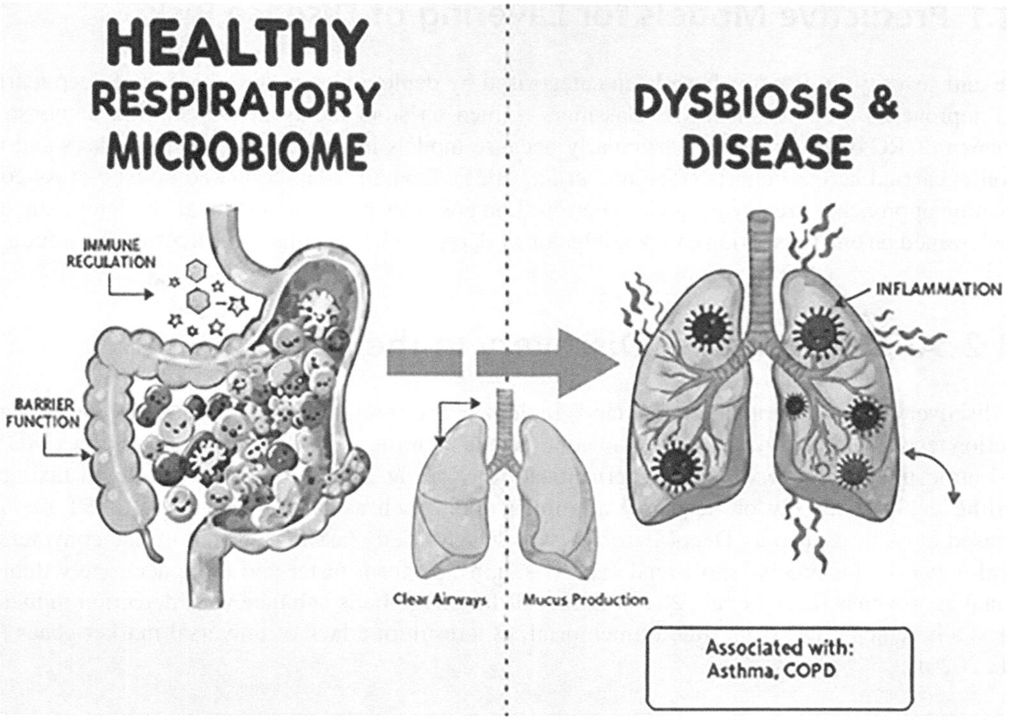

FIGURE 9.2 This figure establishes dissimilarities between barrier function with dysbiosis and function of healthy gut microbiome in controlling the immune system. Useful signals are sent from the healthy gut (as shown by the arrow) to the lungs. A microbiome that is imbalanced could facilitate lung inflammation, compounding diseased conditions such as asthma.

more vulnerable (Hilty et al., 2010). Based on several research evidences and investigations, the microbiome brings about diverse health and disease conditions. Metabolites from the gut blow-out and mouth-associated pathogens influence neurological dysfunction and systemic inflammation, respectively.

9.4 AI-BOOSTED METAGENOMIC INSIGHTS

Metagenomics has stretched microbiome studies forward and, in turn, enhanced functionality profiling and diversity in culture-independent microorganisms. PacBio and Illumina are prominent high-throughput sequencing platforms which create enormous raw sequencing data in a single experiment (Quince et al., 2017).

Although diverse technologies have fostered robust comprehension of microbial ecology, the complexity of the generated datasets brings about substantial analytical challenges. Deep and ML methods contribute significantly to translating metagenomic datasets and fostering disease prediction, functional annotation, automated taxonomic categorisation, and biomarker discovery. Notably, AI-driven methods surmount the drawbacks of conventional bioinformatics pipelines (Noé & Clementi, 2017).

9.4.1 Predictive Models for Layering of Disease Risk

State and severity of diseases have been categorised by deploying supervised ML and deep learning (DL) approaches. For instance, ML classifiers trained on stool metagenomes showed demonstrable changes in CRC discrimination, particularly because models integrate strain-level markers and validation is carried across cohorts (Piccinno *et al.,* 2025). Domain adaptation and holistic cross-cohort validation approaches are key because generalisation poses a significant technical challenge, since one model trained on one population can possibly not work on another population as a result of confounders.

9.4.2 AI for Biomarker Discovery in the Microbiome

The discovery of novel biomarkers demands biological interpretability and predictive power. Feature selection techniques are widely utilised in gene or taxa identification. Recent work integrates ML with multi-omic integration for biomarker prioritisation (Xiao & Zhao, 2024). Conventional taxonomic profiling depends largely on sequence alignment tools such as marker gene or BLAST methods. AI-based classifiers such as DeepMicrobes, typical AI-based classifiers which utilise convolutional neural networks for precise short-read categorisations, perform faster and more accurately than traditional approaches (Liang et al., 2020). Also, AI-driven methods enhance viral detection in metagenomes, a herculean task from time immemorial, as a result of a lack of universal marker genes (Ren et al., 2020).

9.4.3 Personalised Medicine: Microbiome-Informed Diagnostics and Therapeutics

ML models bring about patients' stratification for therapy response and microbiome management approaches. Translational uses are such as microbial panels for CRC screening, antibiotic resistance prediction, and identification of patients (Ma *et al.,* 2024). Microbiome conformation impacts the effectiveness of immune checkpoint inhibitors (Routy et al., 2018).

The high-dimensional microbiome datasets make them suitable for the discovery of AI-driven biomarkers. ML algorithms such as support vector machines and random forests have been deployed for microbial signatures identification of microbial signatures linked with disease conditions. For instance, Pasolli et al. (2016) advanced an ML framework that accurately differentiated healthy and diseased gut microbiomes within numerous datasets. DL models exceed linear associations, encompassing the complexity of interactions between microbial taxa.

DeepMicro framework has exhibited higher performance in inflammatory bowel disease classification, such as type 2 diabetes and colorectal cancer, from metagenomic data (Oh *et al.,* 2021). AI fosters metagenomics' incorporation with proteomics, transcriptomics, and metabolomics. Integrative DL frameworks facilitate host–microbiome connections' reconstruction (Zhernakova et al., 2016).

9.4.4 AI in Antimicrobial Resistance and Drug Discovery

Metagenomic sequencing has revealed varied reservoirs of ARG genes. Conventional homology-based methods often underestimate the diversity of antimicrobial-resistant genes. AI-based models utilise DL to ARG prediction (Arango-Argoty et al., 2018). Microbiomes harbour enormous biosynthetic gene clusters encoding non-artificial and therapeutic-based products. AI-based tools predict BGC types and linked metabolites from metagenomic data (Hannigan *et al.,* 2019). This fast-tracks the finding of novel antifungals, antibiotics, and anticancer agents, particularly when AMR serves as a threat to global health.

9.5 CASE STUDIES AND APPLICATIONS

Metagenomics incorporated with AI is a practical method that brings about clear-cut outcomes in biomedical and clinical settings. Case studies highpoint AI-powered evaluation of metagenomic data has improved predicted therapeutic consequences, public health interventions, and disease diagnostics.

AI applications in the identification of microbial signatures of colorectal cancer and AMR detection from metagenomic datasets. Multiple studies revealed microbial signatures for colorectal cancer, which is a key cancer-causative factor leading to mortality globally. Incorporated metagenomic evaluations across organisations depict strain-level markers; AI models bring together functional and taxonomic characteristics and can be incorporated into non-invasive screening workflows (Piccinno *et al.,* 2025; Sung *et al.,* 2021). Detecting CRC on time substantially enhances survival rates. AMR prediction from genomes and metagenomes influences sequence characteristics and ML models. Ren et al. (2020) conducted a review on ML methods for AMR prediction and asserted that while gene-based models are potent, incorporating mobile genetic elements on enhance precision (Valavarasu *et al.,* 2025). ML techniques have been applied to vast metagenomic datasets for CRC-linked microbial signatures identification.

AMR, if unmanaged, is likely to cause 10 million mortalities yearly by 2050 (O'Neill, 2016). Conventional culture-grounded susceptibility testing operates at a fast rate and cannot embrace unculturable microbes' resistome. AI incorporated with metagenomics provides a robust strategy for AMR profile prediction from environmental and clinical samples. DeepARG, a DL model, has been skilled on vast databases of resistance genes for the detection of AMR determinants from metagenome-based data (Arango-Argoty et al., 2018). These AI approaches do better than traditional alignment-based methods by taking non-linear patterns in sequence data. Correspondingly, Rahaman *et al.,* (2018) utilised DL for metagenomic contigs categorisation into resistance groupings, fostering antimicrobial resistance surveillance in community environments.

9.5.1 AI-Driven Analysis of Microbiome Changes in COVID-19 Patients

The COVID-19 instigated by SARS-CoV-2 was discovered that the virus affects the oral and gut microbiomes. Dysbiosis in these microbiomes has been associated with immune dysregulation and disease severity (Zuo et al., 2020). AI is crucial for the analysis of abundant metagenomic datasets obtained from patients with COVID-19.

9.5.2 Data Heterogeneity and Standardisation Issues

In many different fields, problems with data uniformity and variability present formidable obstacles. Heterogeneous product descriptions hinder effective buyer-vendor interactions in electronic commerce, necessitating both standardisation and integration strategies because of the possible variety of standards (Ng, 2008). Correspondingly, in nanomedicine, disparities in model preparation procedures, LC-MS workflows, and data processing lead to significant heterogeneity in protein corona characterisation among proteomics facilities; typical approaches increase protein ID overlaps from 11% to 40% over time, which needs to be looked into (Ashkarran et al., 2024).

9.5.3 Artificial Intelligence (AI) in Big Data Interpretation

The amount of digital data is increasing quickly due to the quick advancements in digital technologies (Klein, 2017). As a result, a lot of data is generated from various sources, including social networks, smartphones, sensors, etc. Big data is the term used to describe the enormous volumes of data that traditional relational databases and analytical methods cannot handle or store. To find patterns in big datasets, new tools and analytical methods must be developed. Big data is rapidly generated in a variety of formats from a wide range of sources. In order to properly take advantage of them, the new analytical tools should now be able to identify correlations between quickly changing data. Utilising cutting-edge AI methods to glean knowledge and insights from massive, intricate datasets is known as data analytics with AI (Gandomi et al., 2022). This entails using natural language processing methods, DL models, and ML algorithms to find patterns and relationships in large datasets that can guide choices and spur creativity. AI-powered big data analytics aims to automate data analysis and improve its speed, accuracy, and scalability so that businesses can fully utilise their data and obtain a competitive edge. Big data analytics in healthcare enables the analysis of intricate and varied datasets, offering insights into patient behaviours, treatment outcomes, and healthcare trends at the individual and population levels, as stated by Bag *et al.* (2023). Predictive healthcare, better patient outcomes, and resource optimisation all depend on this analysis.

9.6 CHALLENGES IN INTEGRATING AI IN METAGENOMIC DATA ANALYSIS

A metagenome is a grouping of many complete genomes. In theory, comprehensive DNA sequencing should produce draft-quality, entire genome sequences for each member in order to completely comprehend an ecosystem. Nevertheless, even with the advent of high-throughput second-generation sequencers, this remains an extremely costly and difficult undertaking. Because metagenomic datasets are a complex collection of sequences from multiple species, their analysis and annotation differ from whole-genome investigations.

Determining what is actually discovered is another drawback of the system (López-Labrador et al., 2021). The actual identification of the pathogen following sequencing, for instance, is impacted by incomplete databases, incorrectly annotated sequences, databases that contain contaminating organisms, bias in databases, and incorrect classification of organisms (Diao *et al.,* 2022; López-Labrador et al., 2021). Furthermore, hospitals and commercially available mNGS systems may identify pathogens differently and inaccurately due to variances in pipelines, repeatability, quality control (QC), and workflow (López-Labrador et al., 2021). Lack of consistency, bioinformatic data storage, and patient privacy are further logistical issues (Diao *et al.,* 2022). Additionally, the detection of exogenous DNA contamination from reagents and the laboratory environment can be decreased with the use of negative controls, strict decontamination techniques during nucleic acid processing, and unidirectional workflow, among other strict QCs in the laboratory (Chen *et aal.,* 20222).

9.6.1 Computational and Algorithmic Limitations

Fundamental restrictions in a variety of disciplines are revealed by research on computational and algorithmic limitations (Love et al., 2015). In addition, it contends that working at different levels of Marr's computational framework has theoretical constraints and supports an integrated "inside-out" strategy that uses algorithmic modelling to bridge the computational, algorithmic, and implementation levels. Moreover, in computational biology, the outlines of objectives that bound the algorithmic compression in computer science, pointing out that areas that are too complex for computation, the limitations of the human brain, and intrinsic computational constraints make full mastery of complexity unachievable (Cirillo, 2018; Love et al., 2015).

9.6.2 Interpretability and Reproducibility of AI Models

Nowadays, the interpretability and reproducibility of AI models in high-stakes applications like healthcare have become crucial issues. At the same time, in order to assure practical viability, interpretability techniques in production ML systems must address additional difficulties that call for thorough repeatability solutions, such as time-ordered event sequences (Dokshin et al., 2018). While the growing number of publications is commendable, the lack of methodological thoroughness has resulted in misinterpretability and disasters linked to reproducibility within the medical space (Ciobanu-Caraus et al., 2024). There is a looming crisis surrounding AI models, particularly their severe impacts on patients globally, due to high accuracy and interpretability requirements (Leventi-Peetz and Östreich, 2022).

9.6.3 Ethical, Legal, and Social Implications (ELSI)

The ELSI area was initiated to investigate the social influence of biomedical development endeavours in the human genome project around the 1900s (Guerrini et al., 2022; Ogbogu and Ahmed, 2022). A unique research programme in NIH-funded biomedical research is the ELSI programme at the National Human Genome Research Institute, which supports investigator-initiated studies aimed at anticipating and addressing the implications of human genetics and genomics research for individuals, families, communities, and society as a whole (McEwen, 2012; McEwen et al., 2014). The initiative has had a major influence on the application of genetic medicine, the conduct of genomics research, and general public policies (McEwen et al., 2014).

9.6.4 Data Privacy and Ownership in Metagenomic Studies

Because biomedical and genomic data are delicate, data privacy and ownership are important issues in metagenomic research (Mittelstadt and Floridi, 2016). The process of name ownership, privacy, and informed permission is the primary ethical issue with biological big data, highlighting the challenges of data protection, anonymisation, and intellectual property rights in metagenomic exploration (Singh et al., 2012). Meta-analyses of sequencing data carry privacy risks, such as the potential for personal information to be exposed through information about uncommon variations. A cryptographic method called MetaSeq makes it possible to conduct safe meta-analysis without disclosing the identities of carriers or sensitive variations (Tang et al., 2012). Pasolli et al. (2016) do not specifically address privacy or ownership issues; instead, they concentrate on ML meta-analyses of sizable metagenomic datasets, stressing the necessity of reliable computational methods and data sharing. Data sharing has been one of the challenging means of information dissemination.

9.6.5 Ethical Challenges of Microbiome-Based Predictions

Predictions based on microbiomes pose serious ethical problems in a variety of fields. It will interest you to know that there are serious privacy concerns because the human microbiome contains private information about a person's lifestyle, medical history, prescription drugs, and mental state that helps to prevent future problems from taking place (Ahmed-Farid et al., 2021; Ma et al., 2021). Risks to personal identification are among the main ethical concerns because microbiome research may cast doubt on conventional notions of selfhood and individual identity (Ahmed-Farid et al., 2021; Rhodes, 2016). Commercialisation and overhype are other issues that might hurt people and jeopardise effective medicines (Ma et al., 2021). In my mind, careful regulatory actions and constant communication between scientists, medical experts, and the general public are required to address these issues (Ma et al., 2021).

9.6.6 Risk of Bias in AI-Driven Health Predictions

There is a considerable chance that AI-driven health projections will be biased, which could have detrimental clinical effects. Throughout the whole algorithm development process, from gathering training data to implementation and post-deployment monitoring, bias might appear (Mihan et al., 2024). AI algorithms tend to reinforce pre-existing prejudices when training datasets are homogeneous or inaccurately reflect population variability. This is especially true for under-represented demographic groups based on age, sex, colour, ethnicity, and socioeconomic status (Mihan et al., 2024). AI is reshaping the healthcare sector and clinical applications beyond imagination. However, algorithmic bias still poses a core constraint in the full deployment of AI for clinical operations. For the human population to be fully captured, AI systems need to be exposed to enormous biological datasets with minimal or no exclusion of the human population (Norori et al., 2021). Resolving these issues calls for all-encompassing mitigation techniques across the AI lifecycle, such as enhanced data representation, legislative frameworks, and cooperative initiatives between healthcare professionals and AI developers (Abdelwanis and Elmezain, 2024).

9.7 MACHINE LEARNING AND DEEP LEARNING IN GENOMICS

ML is the development of computer programs that can receive data and use it to learn on their own. The standard ML approach includes the following steps: data collection, feature selection, training, model evaluation, and testing. In general, supervised, unsupervised, and reinforcement learning are the three main categories of ML. The term "deep learning" describes a group of novel approaches that, when combined, have demonstrated groundbreaking improvements over current ML methods in a number of domains. A subset of ML algorithms known as DL algorithms is capable of forecasting and effectively dimensionally reducing. The primary distinction between DL and conventional ML techniques used in genomics is the latter's superior capacity to handle massive amounts of genetic data and its quick adaptability. QC is the initial stage of processing next-generation sequencing (NGS) data; however, even this step has speed constraints and challenges with contaminant screening. If precise and efficient data QC is not possible, metagenomic data, which is made up of NGS data from several genomes that are typically unknown beforehand, faces a more significant issue.

Various sequencing artefacts, including contaminated and low-quality reads, may be present in raw metagenomic NGS reads: (1) Reads with poor sequencing quality can seriously impair subsequent analyses. They are the outcome of sample preparation tests like emPCR3 and flawed sequencing instrument properties. The accuracy of metagenomic sequencing data on microbial diversity can be significantly increased through quality screening. (2) The metagenomic samples may contain genomes other than bacteria due to contaminations brought on by contaminated samples or improper sample preparation. High eukaryotic species are typically regarded as contaminants for metagenomic data, and they must be detected and filtered before additional analysis in order to avoid inaccurate findings. Furthermore, contaminating genomes from other microbial communities may potentially be present in metagenomic data.

9.8 AI-DRIVEN TAXONOMIC CLASSIFICATION AND FUNCTIONAL PREDICTION

Workflows for metagenomic analysis fall into one of two categories: (1) de novo assembly and binning tools, which assist in identifying genomes for additional research when no prior knowledge of a microbial community is available, or (2) taxonomic classification tools, which compare reads to a reference database and are employed in situations where appropriate references are available (like pathogen detection) or where sequencing and analysis expenses must be minimised (like complex communities or extensive clinical studies) (Breitwieser & Salzberg, 2020). The sequencing technologies used also influence the usage of these paradigms to some degree; taxonomic classification is more common for short reads (<1 kbp) using methods created by ONT and Pacific Biosciences, and it is preferred for de novo assembly (Amarasinghe *et al.*, 2020). Recent research has also examined the potential of long-read technologies for extremely sensitive and accurate taxonomic classification as they become more generally available and affordable. To lessen the impact of greater sequencing error rates (usually >1%), they have adopted an alignment-based strategy (Wang *et al.*, 2021). The prediction of protein function has changed dramatically in recent years, moving away from simple sequence alignment and towards methods based on ML, natural language processing, or biological network analysis. Despite these developments, not many techniques have been created for and tested on metagenome or amplicon sequencing data, primarily due to the lack of a "ground truth" unless the techniques are used on "mock communities," which are extremely simplified representations of real microbial communities and do not accurately reflect the complexity found in real-world situations.

9.9 FUTURE PERSPECTIVES AND CLINICAL TRANSLATION

The research articles explore future possibilities in a variety of technical and medical domains, emphasising recurring themes of changing difficulties and advancements in methodology. Tackling global challenges such as population rise, increasing sea levels, and resource depletion stirs the need towards adopting robust, future-ready, and collaborative research initiatives and approaches. (Chowdhury and Flentje, 2007). The past decade has been marked with vivid shifts in advances feared towards managing chronic obstructive pulmonary disease amidst undefined and hidden issues (Agostí and Vestbo, 2011).

AI-driven multi-omics technologies incorporation is gradually brewing up enhanced drug development and precision medicine. DL algorithms, such as transformers, convolutional neural networks, and deep neural networks, make it possible to identify intricate biological relationships and patterns across several omics layers that are impossible to find with conventional single-omics methods (Periyasamy, 2025).

Applications for digital health technologies and mobile health (mHealth) that include features for symptom management, provider communication, improved healthcare access, and health information searches can help those living in remote and underserved areas (Deitenbeck et al., 2018). Incorporating telemedicine, remote monitoring, and health education elements that lower technology barriers in underserved areas, cross-platform health applications further improve accessibility by functioning across many devices and systems (Owoade, 2025). The ability of digital health solutions to improve equity may be limited, though, as research shows that they have disproportionately concentrated on downstream access facilitators rather than upstream determinants like health literacy and care-seeking habits (Shaw et al., 2024).

9.10 CONCLUSION

Metagenomic studies have been transformed and advanced substantially into clinically useful tools with the advent of AI by fostering disease prediction, precise taxonomic sorting, and functional annotation from enormous datasets. ML and DL models enhance the development of drugs and the discovery of biomarkers and also unravel unseen microbial links. Incorporation of AI into microbiome science is revolutionising precision medicine in spite of ethics, interpretation, reproducibility, and data heterogeneity issues.

As depicted in various case studies, such as AMR prediction, diagnosis of colorectal cancer, and COVID-19 microbiome investigation, AI fosters the capacity for microbial signatures identification and emerging disease monitoring. The evolving growth of datasets and the maturity of AI algorithms are also pointers to the fact that their incorporation into healthcare will bring about precision in personalised human health interventions. Immunity, health, and vulnerability to disease are substantially influenced by the oral, respiratory, gut, and dermal microbiomes. As metagenomics research progresses, there is improved mechanistic understanding of disease conditions and lessens the complexities that exist between the microbiota and the host.

REFERENCES

Abdelwanis, M. I., & Elmezain, M. I. (2024). A comprehensive review of hybrid AC/DC networks: Insights into system planning, energy management, control, and protection. *Neural Computing and Applications*, *36*(29), 17961–17977.

Aborode, A. T., Hasan, M. M., Jain, S., Okereke, M., Adedeji, O. J., Karra-Aly, A., & Fasawe, A. S. (2021). Impact of poor disease surveillance system on COVID-19 response in Africa: Time to rethink and rebuild. *Clinical Epidemiology and Global Health*, *12*, 100841. https://doi.org/10.1016/j.cegh.2021.100841

Aggarwal, V., Sunder, S., & Verma, S. R. (2022). Disease-associated dysbiosis and potential therapeutic role of *Akkermansia muciniphila*, a mucus-degrading bacteria of gut microbiome. *Folia Microbiologica*, *67*(6), 811–824. https://doi.org/10.1007/s12223-022-00973-6

Agostí, A., & Vestbo, J. (2011). Current controversies and future perspectives in chronic obstructive pulmonary disease. *American Journal of Respiratory and Critical Care Medicine*, *184*(5), 507–513.

Aguilar, G. R., Swetschinski, L. R., Weaver, N. D., Ikuta, K. S., Mestrovic, T., Gray, A. P., & Naghavi, M. (2023). The burden of antimicrobial resistance in the Americas in 2019: A cross-country systematic analysis. *The Lancet Regional Health–Americas*, *25*, 100561.

Ahmed-Farid, O. A., Salah, A. S., Nassan, M. A., & El-Tarabany, M. S. (2021). Performance, carcass yield, muscle amino acid profile, and levels of brain neurotransmitters in aged laying hens fed diets supplemented with guanidinoacetic acid. *Animals*, *11*(11), 3091.

Amarasinghe, S. L., Su, S., Dong, X., Zappia, L., Ritchie, M. E., & Gouil, Q. (2020). Opportunities and challenges in long-read sequencing data analysis. *Genome Biology*, *21*, 30. https://doi.org/10.1186/s13059-020- 1935-5

Arango-Argoty, G., Garner, E., Pruden, A., Heath, L. S., Vikesland, P., & Zhang, L. (2018). DeepARG: A deep learning approach for predicting antibiotic resistance genes from metagenomic data. *Microbiome*, *6*, 23. https://doi.org/10.1186/s40168-018-0401-z.

Ashkarran, A. A., Sharifi, S., LaRock, C. N., & Mahmoudi, M. (2024). Improved methodology for evaluating nanomedicine antibacterial properties in biological fluids. *ACS Pharmacology & Translational Science*, *7*(3), 855–862.

Bag, S., Dhamija, P., Singh, R., Rahman, M., & Sreedharan, V. R. (2023). Big data analytics and artificial intelligence technologies based collaborative platform empowering absorptive capacity in health care supply chain: An empirical study. *Journal of Business Research*, *154*, 113315. https://doi.org/10.1016/j.jbusres.2022.113315

Berg, G., Rybakova, D., Fischer, D., Cernava, T., Verges, M. C., Charles, T., Chen, X., Cocolin, L., Eversole, K., & Corral, G. H. et al (2020). Microbiome definition re-visited: Old concepts and new challenges. *Microbiome*, *8*, 103.

Bleidorn, C. (2016). Third-generation sequencing: Technology and its potential impact on evolutionary biodiversity research. *Systematics and Biodiversity*, *14*(1), 1–8.

Breitwieser, F. P., & Salzberg, S. L. (2020). Pavian: Interactive analysis of metagenomics data for microbiome studies and pathogen identification. *Bioinformatics*, *36*(4), 1303–1304. https://doi.org/10.1093/bioinformatics/btz715

Byrd, A. L., Belkaid, Y., & Segre, J. A. (2018). The human skin microbiome. *Nature Reviews Microbiology*, *16*(3), 143–155. https://doi.org/10.1038/nrmicro.2017.157

Chen, F., Xue, J., Bai, M., Fan, C., & Zhao, Y. (2022). Lighting up nucleic acid modifications in single cells with DNA-encoded amplification. *Accounts of Chemical Research*, *55*(16), 2248–2259. https://doi.org/10.1021/acs.accounts.2c00269

Chowdhury, R., & Flentje, P. (2007). Perspectives for the future of geotechnical engineering.

Ciobanu-Caraus, O., Aicher, A., Kernbach, J. M., Regli, L., Serra, C., & Staartjes, V. E. (2024). A critical moment in machine learning in medicine: On reproducible and interpretable learning. *Acta Neurochirurgica*, *166*(1), 14.

Cirillo, V. (2018). Job polarization in European industries. *International Labour Review*, *157*(1), 39–63.

Cryan, J. F., O'Riordan, K. J., Cowan, C. S. M., Sandhu, K. V., Bastiaanssen, T. F. S., Boehme, M., & Dinan, T. G. (2019). The microbiota-gut-brain axis. *Physiological Reviews*, *99*(4), 1877–2013. DOI: 10.1152/physrev.00018.2018.

de Córdoba-Ansón, P. F., Linares-Ambohades, I., Baquero, F., Coque, T. M., & Pérez-Cobas, A. E. (2025). The respiratory tract microbiome and human health. *Microbial Biotechnology*, *18*(5), e70147. https://doi.org/10.1111/1751-7915.70147

Deitenbeck, B., Qureshi, S., & Xiong, J. (2018). The role of mHealth for equitable access to healthcare for rural residents.

Dewhirst, F. E., Chen, T., Izard, J., Paster, B. J., Tanner, A. C. R., Yu, W.-H., & Wade, W. G. (2010). The human oral microbiome. *Journal of Bacteriology*, *192*(19), 5002–5017. https://doi.org/10.1128/JB.00542-10.

Diao, J., Zhou, Z., Xue, X., Zhao, D., & Chen, S. (2022). Bioinformatic workflow fragment discovery leveraging the social-aware knowledge graph. *Frontiers in Genetics*, *13*, 941996. https://doi.org/10.3389/fgene.2022.941996

Dokshin, G. A., Ghanta, K. S., Piscopo, K. M., & Mello, C. C. (2018). Robust genome editing with short single-stranded and long, partially single-stranded DNA donors in *Caenorhabditis elegans*. *Genetics*, *210*(3), 781–787.

Gandomi, A. H., Chen, F., & Abualigah, L. (2022). Machine learning technologies for big data analytics. *Electronics*, *11*(3), 421. https://doi.org/10.3390/electronics11030421

Guerrini, G., Magrì, D., Gioria, S., Medaglini, D., & Calzolai, L. (2022). Characterisation of nanoparticle-based vaccines for COVID-19. *Nature Nanotechnology*, *17*(6), 570–576.

Hajishengallis, G. (2015). Periodontitis: From microbial immune subversion to systemic inflammation. *Nature Reviews Immunology*, *15*(1), 30–44. https://doi.org/10.1038/nri3785.

Handelsman, J., Rondon, M. R., Brady, S. F., Clardy, J., & Goodman, R. M. (1998). Molecular biological access to the chemistry of unknown soil microbes: A new frontier for natural products. *Chemistry & Biology*, *5*, R245–R249.

Hannigan, G. D., Prihoda, D., Palicka, A., Soukup, J., Klempir, O., Rampula, L., Durcak, J., Wurst, M., Kotowski, J., Chang, D., Wang, R., Piizzi, G., Temesi, G., Hazuda, D. J., Woelk, C. H., & Bitton, D. A. (2019). A deep learning genome-mining strategy for biosynthetic gene cluster prediction. *Nucleic Acids Research*, *47*(18), e110. https://doi.org/10.1093/nar/gkz654

Hilty, M., Burke, C., Pedro, H., Cardenas, P., Bush, A., Bossley, C., & Moffatt, M. F. (2010). Disordered microbial communities in asthmatic airways. *PLoS ONE*, *5*(1), e8578. https://doi.org/10.1371/journal.pone.0008578.

Hooper, L. V., & Gordon, J. I. (2001). Commensal host-bacterial relationships in the gut. *Science*, *292*, 1115–1118.

Klein, S. (2017). *The world of big data and IoT.* In *IoT Solutions in Microsoft's Azure IoT Suite* (pp. 3–13). Berkeley, CA: Apress.

Kostic, A. D., Gevers, D., Pedamallu, C. S., Michaud, M., Duke, F., Earl, A. M., & Meyerson, M. (2012). Genomic analysis identifies association of *fusobacterium* with colorectal carcinoma. *Genome Research*, *22*(2), 292–298. https://doi.org/10.1101/gr.126573.111.

Leventi-Peetz, A. M., & Östreich, T. (2022). Deep learning reproducibility and explainable AI (XAI). *arXiv preprint arXiv:2202.11452.*

Liang, Q., Bible, P. W., Liu, Y., Zou, B., & Wei, L. (2020). DeepMicrobes: Taxonomic classification for metagenomics with deep learning. *NAR Genomics and Bioinformatics*, *2*(1), lqaa009. https://doi.org/10.1093/nargab/lqaa009.

Lloyd-Price, J., Arze, C., Ananthakrishnan, A. N., Schirmer, M., Avila-Pacheco, J., & Poon, T. W. et al (2019). Multi-omics of the gut microbial ecosystem in inflammatory bowel diseases. *Nature*, *569*, 655–662. https://doi.org/10.1038/s41586-019-1237-9

López-Labrador, F. X., Brown, J. R., Fischer, N., Harvala, H., Van Boheemen, S., Cinek, O., & de Vries, J. J. (2021). Recommendations for the introduction of metagenomic high-throughput sequencing in clinical virology, part I: Wet lab procedure. *Journal of Clinical Virology*, *134*, 104691.

Love, T., Laier, C., Brand, M., Hatch, L., & Hajela, R. (2015). Neuroscience of internet pornography addiction: A review and update. *Behavioral Sciences*, *5*(3), 388–433.

Ma, J., Li, Y., Grundish, N. S., Goodenough, J. B., Chen, Y., Guo, L., & Wan, L. J. (2021). The 2021 battery technology roadmap. *Journal of Physics D: Applied Physics*, *54*(18), 183001.

Ma, J., Al Moussawi, K., Lou, H., Wang, Y., Chadwick, J., Phetsouphanh, C., Slee, E., Zhong, S., Leissing, T., Roth, A., Qin, X., Chen, S., Yin, J., Ratnayaka, I., Hu, Y., Louphrasitthiphol, P., Taylor, L., Bettencourt, P., Muers, M., Greaves, D., McShane, H., Goldin, R., Soilleux, E., Coleman, M., Ratcliffe, P., & Lu, X. (2024). *Ma et al., PNAS 2024 "Deficiency of factor-inhibiting HIF creates a tumor-promoting immune microenvironment" main figures* (Version 1) [Data set]. Mendeley Data. https://doi.org/10.17632/y7s8hf6msw.1

Mackelprang, R., Waldrop, M. P., DeAngelis, K. M., David, M. M., Chavarria, K. L., Blazewicz, S. J., Rubin, E. M., & Jansson, J. K. (2011). Metagenomic analysis of a permafrost microbial community reveals a rapid response to thaw. *Nature*, *480*(7377), 368–371. https://doi.org/10.1038/nature10576

Makoni, M. (2020). Africa's $100-million pathogen genomics initiative. *The Lancet Microbe*, *1*(8), e318. https://doi.org/10.1016/S2666-5247(20)30206-8

Man, W. H., de Steenhuijsen Piters, W. A. A., & Bogaert, D. (2017). The microbiota of the respiratory tract: Gatekeeper to respiratory health. *Nature Reviews Microbiology*, *15*(5), 259–270. https://doi.org/10.1038/nrmicro.2017.14.

McEwen, A. S., Dundas, C. M., Mattson, S. S., Toigo, A. D., Ojha, L., Wray, J. J., & Thomas, N. (2014). Recurring slope lineae in equatorial regions of mars. *Nature Geoscience*, *7*(1), 53–58.

McEwen, B. S. (2012). Brain on stress: How the social environment gets under the skin. *Proceedings of the National Academy of Sciences*, *109*(Supplement_2), 17180–17185.

Michaud, D. S., Izard, J., Wilhelm-Benartzi, C. S., You, D., Grote, V., Tjønneland, A., & Riboli, E. (2013). Plasma antibodies to oral bacteria and risk of pancreatic cancer in a large European prospective cohort study. *Gut*, *62*(12), 1764–1770. https://doi.org/10.1136/gutjnl-2012-303006.

Mihan, A., Pandey, A., & Van Spall, H. G. (2024). Mitigating the risk of artificial intelligence bias in cardiovascular care. *The Lancet Digital Health, 6*(10), e749–e754.

Mittelstadt, B. D., & Floridi, L. (2016). The ethics of big data: current and foreseeable issues in biomedical contexts. *The Ethics of Biomedical Big Data,* 445–480.

Nakatsuji, T., Chen, T. H., Narala, S., Chun, K. A., Two, A. M., Yun, T., & Gallo, R. L. (2017). Antimicrobials from human skin commensal bacteria protect against *Staphylococcus aureus* and are deficient in atopic dermatitis. *Science Translational Medicine, 9*(378), eaah4680. https://doi.org/10.1126/scitranslmed.aah4680.

New, F. N., & Brito, I. L. (2020). What is metagenomics teaching us, and what is missed? *Annual Review of Microbiology, 74*(1), 117–135.

Ng, W. L. (2008). An efficient and simple model for multiple criteria supplier selection problem. *European Journal of Operational Research, 186*(3), 1059–1067.

Noé, F., & Clementi, C. (2017). Collective variables for the study of long-time kinetics from molecular trajectories: Theory and methods. *Current Opinion in Structural Biology, 43,* 141–147. https://doi.org/10.1016/j.sbi.2017.02.006

Norori, N., Hu, Q., Aellen, F. M., Faraci, F. D., & Tzovara, A. (2021). Addressing bias in big data and AI for health care: A call for open science. *Patterns, 2*(10), 100347.

O'Neill, J. (2016). Tackling drug-resistant infections globally: final report and recommendations (AMR Review). *Review commissioned by the UK Government.*

Oh, Y.-J., Park, Y.-R., Hong, J., & Lee, D.-Y. (2021). Metagenomic, metabolomic, and functional evaluation of kimchi broth treated with light-emitting diodes (LEDs). *Metabolites, 11*(8), 472. https://doi.org/10.3390/metabo11080472

Ogbogu, U., & Ahmed, N. (2022). Ethical, legal, and social implications (ELSI) research: Methods and approaches. *Current Protocols, 2*(1), e354.

Owoade, A. A. (2025). A classification model for sentiment analysis of depression using Nigerian tweets. *Science World Journal, 20*(2), 647–658.

Pasolli, E., Asnicar, F., Manara, S., & Zolfo, M., et al. (2019). Extensive unexplored human microbiome diversity revealed by over 150,000 genomes from metagenomes. *Cell, 176*(3), 649–662. https://doi.org/10.1016/j.cell.2018.12.028.

Pasolli, E., Truong, D. T., Malik, F., Waldron, L., & Segata, N. (2016). Machine learning meta-analysis of large metagenomic datasets: Tools and biological insights. *PLoS Computational Biology, 12*(7), e1004977. https://doi.org/10.1371/journal.pcbi.1004977.

Periyasamy, M. (2025, January). AI-driven multi-omics integration for enhanced drug discovery pipelines. In *2025 International Conference on Multi-Agent Systems for Collaborative Intelligence (ICMSCI)* (pp. 1553–1558). IEEE.

Piccinno, G., Thompson, K. N., Manghi, P., Ghazi, A. R., Thomas, A. M., Blanco-Míguez, A., Asnicar, F., Mladenovic, K., Pinto, F., Armanini, F., Punčochář, M., Piperni, E., Heidrich, V., Fackelmann, G., Ferrero, G., Tarallo, S., Nguyen, L. H., Yan, Y., Keles, N. A., Tuna, B. G., Vymetalkova, V., Trompetto, M., Liska, V., Hucl, T., Vodicka, P., Bencsiková, B., Čarnogurská, M., Popovici, V., Marmorino, F., Cremolini, C., Pardini, B., Cordero, F., Song, M., Chan, A. T., Derosa, L., Zitvogel, L., Huttenhower, C., Naccarati, A., Budinska, E., & Segata, N. (2025). Pooled analysis of 3,741 stool metagenomes from 18 cohorts for cross-stage and strain-level reproducible microbial biomarkers of colorectal cancer. *Nature Medicine, 31*(7), 2416–2429. https://doi.org/10.1038/s41591-025-03693-9

Qayyum, A., Muhammad, F., Bawazir, A. A., & Qatawneh, M. A. S. (2025). Exploring the environmental trade-offs of renewable energy and green technology innovation on Pakistan's ecological footprint. *International Journal of Energy Economics and Policy, 15*(3), 595.

Quince, C., Walker, A. W., Simpson, J. T., Loman, N. J., & Segata, N. (2017). Shotgun metagenomics, from sampling to analysis. *Nature Biotechnology, 35*(9), 833–844. https://doi.org/10.1038/nbt.3935.

Rahman, S. F., Olm, M. R., Morowitz, M. J., & Banfield, J. F. (2018). Machine learning leveraging genomes from metagenomes identifies influential antibiotic resistance genes in the infant gut microbiome. *mSystems, 3*(1), e00123-17. https://doi.org/10.1128/mSystems.00123-17

Ren, J., Song, K., Deng, C., Ahlgren, N. A., Fuhrman, J. A., Li, Y., & Sun, F. (2020). Identifying viruses from metagenomic data using deep learning. *Quantitative Biology, 8*(1), 64–77. https://doi.org/10.1007/s40484-019-0187-4.

Rhodes, M. G. (2016). Judgments of learning: Methods, data, and theory. *The Oxford Handbook of Metamemory, 1,* 65–80.

Routy, B., Le Chatelier, E., Derosa, L., Duong, C. P., Alou, M. T., Daillère, R., & Zitvogel, L. (2018). Gut microbiome influences efficacy of PD-1-based immunotherapy against epithelial tumors. *Science, 359*(6371), 91–97. https://doi.org/10.1126/science.aan3706.

Sender, R., Fuchs, S., & Milo, R. (2016). Revised estimates for the number of human and bacteria cells in the body. *PLoS Biology, 14*(8), e1002533. https://doi.org/10.1371/journal.pbio.1002533.

Shaw, T. A., Arias, P. A., Collins, M., Coumou, D., Diedhiou, A., Garfinkel, C. I., & Wang, L. (2024). Regional climate change: Consensus, discrepancies, and ways forward. *Frontiers in Climate, 6*, 1391634.

Shreiner, A. B., Kao, J. Y., & Young, V. B. (2015). The gut microbiome in health and in disease. *Current Opinion in Gastroenterology, 31*(1), 69–75. https://doi.org/10.1097/MOG.0000000000000139.

Singh, N. K., Gupta, D. K., Jayaswal, P. K., Mahato, A. K., Dutta, S., Singh, S., & Sharma, T. R. (2012). The first draft of the pigeonpea genome sequence. *Journal of Plant Biochemistry and Biotechnology, 21*(1), 98–112.

Sung, H., Ferlay, J., Siegel, R. L., Laversanne, M., Soerjomataram, I., Jemal, A., & Bray, F. (2021). Global cancer statistics 2020: GLOBOCAN estimates of incidence and mortality worldwide for 36 cancers in 185 countries. *CA: A Cancer Journal for Clinicians, 71*(3), 209–249. https://doi.org/10.3322/caac.21660

Tang, W. W., Zeng, G. M., Gong, J. L., Liu, Y., Wang, X. Y., Liu, Y. Y., & Tu, D. Z. (2012). Simultaneous adsorption of atrazine and cu (II) from wastewater by magnetic multi-walled carbon nanotube. *Chemical Engineering Journal, 211*, 470–478.

Teeling, H., & Glöckner, F. O. (2012). Current opportunities and challenges in microbial metagenome analysis—A bioinformatic perspective. *Briefings in Bioinformatics, 13*(6), 728–742.

The Royal Society. (2018). Royal Society ACME Conference 2018: Working together—Mathematics education in a changing landscape. https://royalsociety.org

Theis, K. R., Dheilly, N. M., Klassen, J. L., Brucker, R. M., Baines, J. F., Bosch, T. C., & Lloyd, E. A. (2016). Getting the hologenome concept right: An eco-evolutionary framework for hosts and their microbiomes. *mSystems, 1*, e00028–16.

Tørresen, O. K., Star, B., Mier, P., Andrade-Navarro, M. A., Bateman, A., Jarnot, P., Gruca, A., Grynberg, M., Kajava, A. V., Promponas, V. J., Anisimova, M., & Jakobsen, K. S. (2019). Tandem repeats lead to sequence assembly errors and impose multi-level challenges for genome and protein databases. *Nucleic Acids Research, 47*(21), 10994–11006. https://doi.org/10.1093/nar/gkz841

Turnbaugh, P. J., Ley, R. E., Mahowald, M. A., Magrini, V., Mardis, E. R., & Gordon, J. I. (2006). An obesity-associated gut microbiome with increased capacity for energy harvest. *Nature, 444*(7122), 1027–1031. https://doi.org/10.1038/nature05414

Valavarasu, S., Sangu, Y., & Mahapatra, T. (2025). Prediction of antibiotic resistance from antibiotic susceptibility testing results from surveillance data using machine learning. *Scientific Reports, 15*, 30509. https://doi.org/10.1038/s41598-025-14078-w

Wang, H., Wang, M., Fan, S., Lu, J., Lan, Y., Li, M., Li, J., Liu, R., Sun, J., Fang, J., Qian, P.-Y., Zhang, Y.-Z., & Zhang, W. (2021). Culture enrichment combined with long-read sequencing facilitates genomic understanding of hadal sediment microbes. *Frontiers in Marine Science, 8*, 754332. https://doi.org/10.3389/fmars.2021.754332

Wooley, J. C., Godzik, A., & Friedberg, I. (2010). A primer on metagenomics. *PLoS Computational Biology, 6*(2), e1000667. https://doi.org/10.1371/journal.pcbi.1000667

World Health Organisation. (2022). *WHO Guiding Principles for Pathogen Genome Data Sharing*. World Health Organization.

Xiao, L., & Zhao, F. (2024). *Exploring the frontier of microbiome biomarker discovery with artificial intelligence.* National Science Review, *11*(11), nwae325. https://doi.org/10.1093/nsr/nwae325

Zhernakova, A., Kurilshikov, A., Bonder, M. J., Tigchelaar, E. F., Schirmer, M., Vatanen, T., & Fu, J. (2016). Population-based metagenomics analysis reveals markers for gut microbiome composition and diversity. *Science, 352*(6285), 565–569. https://doi.org/10.1126/science.aad3369.

Zuo, T., Zhang, F., Lui, G. C., Yeoh, Y. K., Li, A. Y. L., Zhan, H., & Ng, S. C. (2020). Alterations in gut microbiota of patients with COVID-19 during time of hospitalization. *Gastroenterology, 159*(3), 944–955.e8. https://doi.org/10.1053/j.gastro.2020.05.048.

AI in Drug Discovery, Development, and Diagnostics

10

Mayowa Agbi, Samuel Adeniyi Oyegbade, Daniel Oluwatobiloba Balogun, Samson Olusola Adeboye, and Oluwafemi Adebayo Oyewole

10.1 INTRODUCTION

The pathway from a new drug's laboratory concept to the market is a time-consuming, costly, and extraordinarily inefficient process. The conventional drug development routes are linked with long durations that often exceed ten years and expenses that may surpass $2.6 billion for each approved drug. Besides, there is a very small probability of success; nearly 90% of drug candidates that reach clinical trials will be denied by the government because of the occurrence of unexpected safety or efficacy problems (Brouwer et al., 2021). The pharmaceutical industry, thus, is slowly but surely finding its way out of these chronic issues, and this breakthrough is mainly attributed to AI and ML adoption.

AI signifies a revolution in the way of thinking that no longer relies upon the traditional, often serendipitous, trial-and-error method but rather through a very rational, predictive, and data-driven one. AI is prepared to enhance efficiency, lower costs, and increase the success rate at all steps of the drug development cycle by means of the capability to analyze vast and varied datasets, which cover genomics, proteomics, chemical structures, and clinical records (Figure 10.1). The AI application in drug development and optimization is the subject of the report, and the report is providing a comprehensive, expert-level analysis. The report goes into the role AI plays in lead optimization, discusses the impact it is having on medical development and trials, gives examples from the real world, and finally, outlines and estimates the challenges, ethical considerations, and future trajectories that will characterize this new frontier in medicine (Gao et al., 2023).

The use of AI in various steps of drug development and optimization has impacted the pharmaceutical industry by opening up new and previously non-existent ways of speeding up the discovery process, increasing the effectiveness, and reducing the risks associated with these products. AI has played a role in different areas, from the refinement of the promising compounds to the conduction of the clinical trials, and thus its potential is changing the manner in which new treatments are made available to the patients.

DOI: 10.1201/9781003727514-12

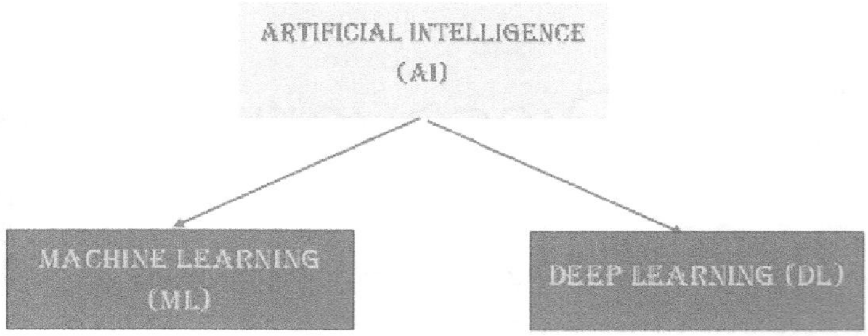

FIGURE 10.1 AI-powered pipeline illustrating artificial intelligence tools (machine learning, deep learning, NLP, and graph neural networks) integration across major stages of drug discovery and development, spanning from data input to post-market surveillance.

10.2 AI IN LEAD OPTIMIZATION

Lead optimization is a key step in the drug discovery process where the compound that shows the greatest potential is gradually changed into the drug candidate that has the best potency, selectivity, and pharmacokinetic features. AI is giving a radical makeover to the sophisticated chemical process and biological testing, which has been the long-standing method of the industry and is no longer so resource-consuming. Not only is AI accelerating this process, but it is also changing the whole idea of the best lead by merging the criteria of downstream developability, such as manufacturability and toxicity, at the very beginning of the design. This proposed all-in-all, multiparameter optimization method is making the process of lead optimization to be viewed as a predictive and de-risked engineering discipline rather than a sequential, problem-solving one. Lead optimization is the most important stage, at which starting hit compounds are improved to drug candidates with better potency, selectivity, pharmacokinetics, and less toxicity. AI techniques are becoming indispensable in this difficult task.

10.2.1 Structure-Activity Relationship (SAR) Prediction

The core principle that drives medicinal chemistry is the SAR, which bridges the chemical structure of a compound and its biological efficacy. For this purpose, quantitative structure-activity relationship (QSAR) models have been the leading computational tool for many years. However, conventional QSAR techniques largely relied upon linear models and human-designed molecular descriptors that limited their ability to characterize the intricate and nonlinear interactions that predominate in biological systems.

Artificial intelligence (AI) has led to a revolution in SAR modeling, with incredible increase in predictive power. Cutting-edge ML techniques such as random forests (RF) and deep learning (DL) frameworks with deep neural networks (DNNs) and graph neural networks (GNNs) can hierarchically learn very abstract feature representations directly from the input molecular structures. The ability to automatically extract attributes from raw input data eliminates the need for human-defined descriptors and thus makes it possible for models to find some subtle, nonlinear input-output relationships that were earlier inaccessible.

Despite the fact that some comparative studies have concluded that the overall performance of DNNs can be at par with that of traditional shallow learning methods such as RF, the real benefit of DL is in the creation of special architectures that are specific to the chemical task. For example, the models that are based on the molecular graph encoding convolutional neural network (MGE-CNN) architecture have been proven to give better results than shallow learning models in both quantitative toxicity prediction and toxicant classification. There is a growing trend toward specialized and high-performing architectures, which is increasing the accuracy of SAR modeling, thus allowing researchers to predict a compound's effectiveness with higher accuracy before it is synthesized.

The identification of the link between a compound's chemical structure and its biological activity is the backbone of lead optimization. The classic SAR investigations are usually very laborious and time-consuming. With AI, and especially ML algorithms, the learning of intricate SAR patterns from extensive datasets can be done in no time. QSAR models using algorithms such as random forests, support vector machines (SVM), and neural networks advise the new compound's activity based on their molecular descriptors (molecular weight, logP, etc.) (Yan et al., 2021). The deep learning models that can automatically capture features from molecular structures confer to the SAR predictions even more sophisticated level, thus allowing medicinal chemists to prioritize the synthesis and testing of compounds possessing the highest therapeutic potential (Kim et al., 2023).

10.2.2 Drug-Drug Interaction (DDI) and Toxicity Prediction

It is very important to predict potential DDIs and toxicity early during drug development to avoid undesirable events and very expensive failures in the late stages of development. AI models are being fed with vast amounts of data from already known DDIs, adverse drug reactions (ADRs), and in vitro and in vivo toxicity studies. For example, graph neural networks (GNNs) are capable of performing very accurate analyses of drug-target networks, thus determining the probability of DDIs through the interaction of the molecules (Rao et al., 2020). In the same way, deep learning architectures can not only detect structural alerts that are associated with toxicity but also forecast different toxicological endpoints (hepatotoxicity, cardiotoxicity, etc.) by recognizing chemical structures and correlating them with omics data (Xu et al., 2022). Reacting to the possible problems in advance, the drug companies will be able to either completely get rid of the lead compounds or modify them in the early stages.

10.2.3 AI for Optimizing Pharmacokinetics (PK) and Pharmacodynamics (PD)

The absorption, distribution, metabolism, and excretion (ADME) properties (PK) and the mechanism and magnitude of drug action (PD) must be fully optimized for a drug to be considered successful. To this end, AI models come in as valuable tools that allow the prediction of these complicated behaviors with more accuracy than before. The use of AI in PK can result in the prediction of factors like bioavailability, half-life, and clearance by integrating chemical properties with biological data (Brouwer et al., 2021). In terms of PD, AI can assist in predicting the drug's potency as well as its receptor-binding affinity, thus allowing the selection of the most engaging compounds with the target. Furthermore, it is possible that machine learning techniques will be employed to create more metabolically stable compounds or those with less efflux tendency, which, in turn, might lead to increased oral bioavailability and prolonged effect (Gao et al., 2023).

10.3 AI IN PRECLINICAL AND CLINICAL DEVELOPMENT

The process of transforming lead compounds into commercial drugs is a long, expensive, and highly uncertain venture. Nevertheless, AI is capable of doing a complete turnaround in the whole process through its ability to make the preclinical and clinical phases less risky and faster.

10.3.1 The Use of Predictive Models for the Translation of Animal-to-Human

It is still the case that translating results from preclinical animal models to human clinical outcomes is a major hurdle in drug development. One solution to this problem is coming from AI in the form of predictive models that will help in the translation of drug development. AI is capable of pinpointing biomarkers and creating models that will forecast human response, relying on the preclinical data, through the analysis of large datasets that comprise meta-analyses of animal study data, human clinical trial data, and various omics data (e.g., genomics and proteomics) (Daneshjou et al., 2021). This not only helps to make more informed decisions about which compounds to advance to human trials, but it also minimizes the risk of failures caused by differences among species.

10.3.2 Patient Stratification for Clinical Trials

One of the major reasons that clinical trials do not succeed is patient heterogeneity, whereby a medication might be beneficial in one particular subgroup of patients only. The strength of AI in this regard is that, through the so-called clinical trial patient stratification, it is able to pinpoint these subgroups. AI algorithms are capable of processing and analyzing different patient data sources like genomic profiles, electronic health records (EHRs), imaging data, and biomarkers to discover the exact patient populations most likely to react positively to a specific treatment (Rajpurkar et al., 2022). Thus, it becomes feasible to conduct clinical trials that are very precise and effective, thus increasing the success rate and reducing the patient count.

10.3.3 Artificial Intelligence in Trial Design, Monitoring, and Adaptive Trials

AI is transforming the management of clinical trials in multiple ways. At the very beginning of the clinical trial process, the AI can predict the trial outcomes based on historical data and, thus, fine-tune the parameters like sample size, endpoints, and the criteria for the eligibility of the patients (Topol, 2019). Moreover, AI will be able to analyze the real-time data that has been collected during the trial to give insights regarding safety, identify any deviation from the protocol, and predict the rate of patient dropouts, leading to interventions being made at the right time. Moreover, AI is instrumental in adaptive design of trials, which means trial parameters can be changed in accordance with the results of interim data analysis. AI, for instance, can play the role of expediting the drug development process by helping in the modification of the dose levels, switching of treatment arms, and alteration of patient recruitment strategies for better efficiency (Schork, 2019).

10.4 CASE STUDIES IN DRUG DEVELOPMENT

The role of AI in drug development is no longer a matter of speculation but rather a fact confirmed by several successful applications.

10.4.1 AI in Oncology Drug Development

Oncology is a predominant field for the implementation of AI technologies. Insilico Medicine is one of those companies that has successfully applied AI techniques to determine new drug targets and engineer new molecules. For example, Insilico Medicine has remarkably spotted a new target for idiopathic pulmonary fibrosis and developed a very effective inhibitor in just 18 months, which is now undergoing clinical trials (Popova et al., 2021). The speed attained by AI is an evident demonstration of its potential to cut down the discovery process by a substantial margin.

10.4.2 Employment of AI in Drug Repositioning

Drug repurposing, meaning finding new applications for already approved pharmaceuticals, is a method that is faster and less expensive when it comes to the market. AI algorithms can discover potential new indications by searching vast biological information networks such as gene expression profiles, disease pathways, and drug-target interactions. During the COVID-19 pandemic, AI was instrumental in the identification of potential antiviral agents from existing drug libraries (Guan et al., 2020). This approach is mainly large-scale discovery and preclinical testing elimination, since the safety profiles of the existing drugs are known.

10.4.3 Success Stories from Pharma/Biotech Industries

The integration of AI firms with pharmaceutical giants is proving to be an excellent technology. For example, BenevolentAI has collaborated with AstraZeneca to find new targets and treatment options for chronic kidney disease as well as idiopathic pulmonary fibrosis (Marks et al., 2021). The Recursion Pharmaceuticals story is another successful example that combines AI and automated wet-lab biology to merge disease biology and speed up the introduction of new medicines to the market across multiple indications. Such examples show how much AI is becoming more and more important in the drug development process.

10.5 AI IN DIAGNOSTIC TOOL DEVELOPMENT

AI technology paradigm shift in diagnostic tool development is nothing short of miraculous, as it is giving disease detection a really precise and quick way. In contrast to manual diagnostics that need human interpretation, AI models are taught through huge amounts of data to extract the same type of patterns, which can then be used in different areas such as radiology, genomics, and pathology (Kumar et al., 2023). AI diagnostics not only offer high precision and quickness as a plus but also eliminate the necessity of expensive reagents and, at the same time, keep on learning from new data. On the other hand, such progress brings along some issues like the quality of data and privacy of patients. Thus, a concerted effort by the industry stakeholders is required in order to convert their research into trustworthy products that consumers will accept. Once all this is done, the potential of AI would be more than just early disease progression detection (Alowais et al., 2023) (Figure 10.2).

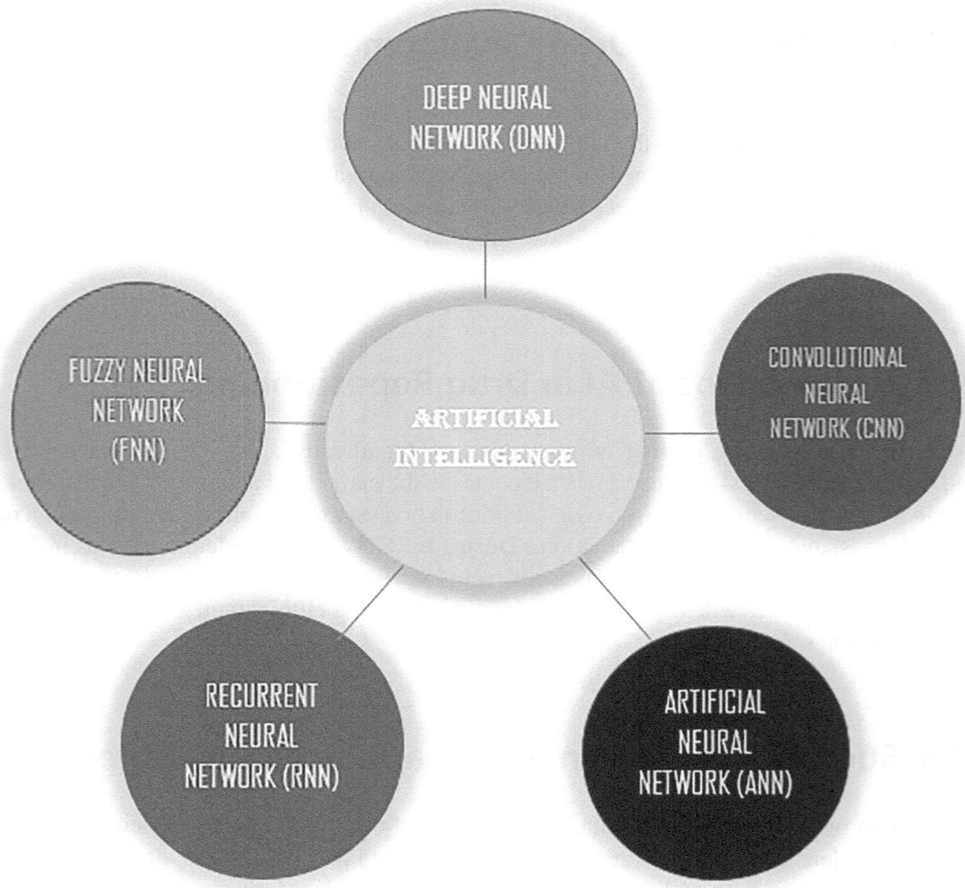

FIGURE 10.2 AI integration across diagnostic modalities and precision medicine. Data sources (imaging, genomics, EHRs, and sensors) are processed through AI layers for deep learning, data fusion, and pattern recognition to drive imaging, genomic, and point-of-care diagnostics, resulting in early detection and personalized patient care.

10.5.1 AI for Imaging-Based Diagnostics

In the last few years, the calling of medical imaging has reached the highest point of modern diagnosis, as it makes it possible for doctors to see the changes in function and structure caused by diseases, and thus, the doctors can make better decisions. Nevertheless, reading images from radiology and pathology is subject to human interpretation and, therefore, the inherent limitations of readers' fatigue plus the shortage of qualified people (Rong & Liu, 2024). AI, in particular, deep-learning methods in the analysis of images, is removing these barriers by offering quick and consistent evaluation of these intricate visual data. For instance, in radiology, the AI systems have proved their worth by detecting very small abnormalities with the same accuracy as human experts in the areas of X-ray, magnetic resonance imaging, and computed tomography (Bhandari, 2024). Large datasets have been used to train neural networks that are capable of identifying lung nodules, intracranial hemorrhage, and stroke (Wang et al., 2023). Moreover, in places where resources are scarce, commercial tools for chest imaging may enhance the diagnosis of tuberculosis, thus proving AI's utmost importance in the areas with the least trained personnel (John et al., 2023). On the other hand, pathology has experienced the benefits of AI through the use of whole-slide imaging that allows for the digitization of histological specimens and thus creates data volumes that can be analyzed by AI (Verma et al., 2025). The situation has now reached a point where the use of AI in the scoring of prognostic biomarkers like Ki-67 and PD-L1 has led to consistent results across numerous laboratories (Rigby et al., 2025). AI has been mentioned in the context of precision pathology not only in the case of cancer but also in other areas beyond that. However, these developments still face a number of very serious hurdles, including the problem of models being trained on very narrow or homogenous datasets. Such models often fail dramatically when they meet variations in patient populations or staining protocols, leading to the question of whether AI is indeed biased or not (Ohia & Daniel, 2024). Apart from the language models, regulatory approval and data privacy in clinical systems are some of the issues that must be resolved before these models can be deployed on a large scale.

Presently, the focus of research is on developing model frameworks that integrate clinical data with imaging from proteomic and genomic perspectives to solve this issue (Li et al., 2024). In such a scenario, successful implementation would still demand the interpretive skills of researchers and the clinical judgment of professionals such as pathologists and radiologists.

10.5.2 Genomic and Biomarker-Based Diagnostics

High-throughput sequencing and molecular methods have progressed so much that they have brought clinical diagnostics to such a level where disease detection can be done not only based on morphology but also genetically underneath of pathology (Dongare et al., 2025). Genomic and biomarker-based diagnostics utilize the trends of DNA, RNA, proteins, and metabolites to spot diseases at an early stage for treatment. AI takes this technology to another level by recognizing patterns, which is also made possible by enormous datasets. In the case of cancer treatment, AI algorithms are capable of studying next-generation sequencing data to recognize copy number variations and somatic mutations responsible for tumor growth. This is done through machine learning models that forecast tumor mutational burden (Srivastava, 2025). Not only in cancer but also in other infectious diseases, AI is revolutionizing pathogen detection and rare disease diagnostics. Deep neural network models programmed with metagenomic sequencing data are able to classify unknown microbial reads almost instantly, thus allowing the identification of infectious microbes in the very beginning of the outbreak (Roy et al., 2024). In the same manner, AI has proved to be the fastest way to determine the causes of Mendelian disorders by merging whole-genome data.

AI has the capability to lead the discovery of proteomic and metabolomic biomarkers. For instance, using a multi-omics approach that integrates genomic, metabolomic, and proteomic sets of data, researchers will be able to create predictive models by AI and even simulate responses to disease risk profiles (Sanches et al., 2024). The application of AI in this area still faces the challenge of different data formats, which would otherwise allow for the synchronization of results among labs. The minute predictions of the future point to real-time genomic monitoring and diagnosis, where small sequencing gadgets will be attached to AI machines to conduct rapid genomic analysis even in rural or disaster areas (Struelens et al., 2024). This will also pave the way for not only the inclusion of patients' EHRs and lifestyle data for disease management but also, soon enough, AI-based genomic diagnostics being a cornerstone for precision medicine, or rather, a personalization of every patient's treatment.

10.5.3 Diagnosis at the Point of Care with AI

Lastly, AI is changing the course of action related to the patients' health through its application in diagnostics at the point of care (Pillay et al., 2025). The usage of traditional point-of-care tests like rapid antigen assays in the clinical settings has been due to their easy and fast results. However, drawback is that human bias plays a role in the interpretation of results. In this scenario, incorporation of AI models in the shape of algorithms might give rise to precision. A case in point is the smartphone-enabled reader with integrated models for reading and colorimetric test strips, thus minimizing the error of false negatives, no matter the level of skill of the operator (Geng et al., 2023). Besides, machine learning models have been utilized along with wearable sensor signals and patient-reported symptoms to provide real-time diagnostics of clinical settings (Mohammadian Rad & Marchiori, 2021). These smart chips are becoming quite a lot more useful in-screening infectious diseases like malaria and tuberculosis as well as chronic disease management, for instance, glucose monitoring (Shang et al., 2023). AI has also been very helpful in predicting models that have resulted in the decrease of time from sample collection to clinical actions in healthcare delivery. Henceforth, the partnership between AI and point-of-care diagnostics is going to be a powerful weapon in the fight for global health as the reaping of benefits from multi-test, rather than single-analyte tests is made.

10.6 AI IN PERSONALIZED AND PRECISION MEDICINE

In a very short period of time, AI has become the key or even the main supporting factor of personalized and precision medicine that allows the doctors and researchers to customize the prevention and treatment of diseases according to the unique clinical profile of each patient. You could say that the traditional methods often neglected the genetic diversity as well as the lifestyle factors that are involved in the disease's progression. AI is picking up on this limitation, though it is a difficult challenge, by teasing out the crucial patterns from the entire set of data to support laboratory results (Chen et al., 2024). The healthcare industry is on the verge of being tipped by these AI-driven insights from the generic treatment protocols to individual care that not only improves but also minimizes the side effects of treatment for patients around the globe.

10.7 INTEGRATING DRUG RESPONSE PREDICTIONS WITH PATIENT-SPECIFIC DATA

The term "precision medicine" refers to the process of determining the exact right therapy for the exact right patient at the exact right time (Akhoon, 2021). Not to be overlooked, AI plays an important role by making personalized therapy possible, wherein the whole range of patient-specific data, including genomic sequences, lifestyle information, and clinical history, is put together into a single predictive model for drug response (Taherdoost & Ghofrani, 2024). This is feasible owing to the fact that the AI systems are able to comprehend the intricate, nonlinear interactions between the different layers of data, which subsequently makes it possible to predict the individual patient's response. Machine learning algorithms have the capability, through the use of EHRs and longitudinal data, to consider both previous medication exposure and the dynamic changes of the disease state that predict how likely a certain patient is to respond to a specific drug or dosage (Swinckels et al., 2024).

This method enables the repurposing of drugs and the optimization of clinical trial design by the classification of patients based on their predicted reactions. An instance of such a scenario is the area of oncology, where the AI systems are being trained on tumor genomics and treatment histories, which then can make it possible for the oncologists to know the patients who are probably going to get the most benefit from the targeted kinase inhibitors (Lotter et al., 2024). The area of therapy around this whole patient-centered approach is not just confined to cancer, but it also extends to heart diseases and autoimmune conditions. To put it differently, the predictions for drug responsiveness can really influence the way clinicians prescribe medications based on the patients' data as the health systems gradually update their clinical databases with AI assimilation, hence going from trial-and-error approaches to personalized therapy that maximizes the effect of treatment.

10.7.1 AI in Pharmacogenomics and Individualized Therapy

Pharmacogenomics studies the role of genetic alteration in drug uptake, distribution, metabolism, and drug response (Qahwaji et al., 2024). AI is steadily becoming the go-to technology for transforming such genomic data into personalized treatments. The analysis of huge amounts of data that are comprised of patient genotypes and transcriptomic profiles by AI has the capability of revealing the connection between the genotype and the phenotype of a person (Wu & Xie, 2025).

For instance, machine learning algorithms can indicate the level of influence that single-nucleotide polymorphisms or uncommon mutations exert on the functioning of drug-metabolizing enzymes like CYP450. Based on such information, drug dosage adjustments and selection can be done accordingly, thereby preventing negative drug reactions (Muhammad et al., 2024). Moreover, deep learning by AI can combine various types of data to create risk profiles for each patient. AI-powered pharmacogenetics may boost the drug development process, as scientists will be able to plan and carry out clinical trials by selecting the populations with the right genetic makeup that are more likely to be responsive to certain compounds (Taherdoost & Ghofrani, 2024). The convergence of AI and pharmacogenomics is steadily transforming individualized therapy that maximizes genome-informed prescribing and proactive adjustments.

10.7.2 Companion Diagnostics and Co-Development with Drugs

Companion diagnostics are lab tests that help to find out if a person will profit from a specific drug. These diagnostics are a major component of precision medicine since they not only aid and/or assure the molecular matching of the drug to the individual's disease but also the opposite; they ensure that the right drug is prescribed based on the molecular characteristics of each individual's disease (Agarwal et al., 2015). AI is already in play in every aspect of companion diagnostics mainly for understanding the interaction of diagnostics and therapeutics. One of the important applications of AI is in developing machine learning models for analyzing multi-omics datasets in order to find potential biomarkers predicting adverse effects in certain patients. Biomarkers are then used to create diagnostic tests that are associated with new drugs. Besides, AI helps in making clinical and regulatory trial processes more efficient. Consequently, trials become not only faster but also cheaper while producing high-quality evidence, which is needed for giving regulatory approval to a companion diagnostic (Askin et al., 2023).

When it comes to clinical applications, AI-based companion diagnostics turn the interpretation of diagnostic outputs in real-time, like whole genome sequencing, into a recommendation at the point of care (Fahim et al., 2025). This effectively decreases the time period from the discovery of a biomarker to a treatment decision; thus, patients with rapidly progressive diseases reap the benefits of it.

10.8 INTEGRATION OF AI IN CLINICAL DECISION SUPPORT SYSTEMS

The discovery of the application of AI in clinical decision-making has altered the whole scenario when it comes to data access by doctors and their treatment suggestions. The combination of ML algorithms and NLP can perform the task described above continuously, and that is why they are often referred to as "predictive" tools. The collaboration of the two methods not only increases the accuracy of the diagnosis but also supplies a more reliable foundation for the practice of personalized medicine, since it provides recommendations that are based on the individual characteristics of the patients. This paper provides an in-depth discussion on the adoption of AI in clinical decision support systems and its effect on practice.

10.8.1 AI-Enabled Diagnostic Platforms for Clinicians

The use of AI in the medical field is increasing and resulting in better diagnoses and the treatments that assist doctors in different specializations. For example, in radiology, the application of neural networks leads to X-ray and CT-scan interpretations with the same accuracy as those of specialists. In certain cases, the networks even detect very small lesions or diseases in the initial stages that are missed by human doctors in their hectic routine in clinics (Giordano et al., 2021). In family medicine, the implementation of natural language processing enables the extraction of key details from the massive data stored in the unstructured notes of clinicians. This not only highlights the issues related to diagnosis but also facilitates the continuous care of the patient. The AI integration into the clinical setup is slowly being recognized as not only a power for doctors and scientists but also a guiding factor that offers better specific scores without interfering with the physician's routine (Sezgin et al., 2023).

Data integration from genomics and laboratory tests is the main factor that makes such things possible. As a result, AI-diagnostic platforms become the first-class assistants to doctors as they somehow navigate through the very difficult and significant process of data interpretation, thus promoting a uniformity in the dialogue with the patients, which eventually results in trust.

10.8.2 Predictive Analytics for Early Disease Detection

The combination of predictive analytics and AI can pinpoint diseases' likelihood even before any clinical symptoms appear in hospitals. This cutting-edge idea could facilitate healthcare not only as a reactive treatment but also as a proactive prevention method (Dixon et al., 2024). Connecting EHRs and patients' lifestyle information with wearable sensors can assist in detecting the onset of a disease by the extremely subtle signals leading up to its progression (Dinh-Le et al., 2019). In the case of cardiology, for instance, the application of predictive models on multimodal datasets can notify physicians about the patients with the highest probability of heart failure or myocardial infarction, thus allowing for intervention with timely pharmacological therapy (DeGroat et al., 2024). Regarding metabolic disorders, not only the application of microbiome data and dietary pattern predictions might indicate the patients' progression of type 2 diabetes even before the biomarkers reach a diagnostic threshold (Rojas López & Barberis, 2024). The health consequences of this are at large, as predictive models are also able to follow changes in infectious diseases in the community and through wastewater surveillance to predict disease outbreaks. The establishment of such an early warning system puts health authorities in a position to prepare by deploying doctors and controlling the spread of the disease before it passes on to a large population (Desai et al., 2019). Nevertheless, relying too much on automated predictions without taking other factors into consideration could result in the wrong classification and ultimately discrimination. However, the case of predictive analytics that can enable healthcare providers to spot early risks in patients and then customize preventive measures accordingly is there, which will eventually lead to a decrease in the incidence of both chronic and acute diseases among all types of populations.

10.8.3 Case Studies: Cancer Diagnostics, Infectious Diseases, and Rare Diseases

The use of AI in clinical settings has the power to change the way treatment is done in various disease areas. In the case of cancer diagnostics, models based on deep learning that have been trained on huge datasets containing histopathology and different imaging are already outperforming radiologists in diagnosing breast, lung, and skin cancers (Unger & Kather, 2024). For example, when neural networks are analyzing whole slide pathology images, they can point out microscopic tumors that human observers might miss due to tiredness or being busy with other things (Hu et al., 2023). This makes it possible to carry out treatment that is exactly tailored to the patient and the disease at an early stage so that the patient's health can be maintained longer. Additionally, the oncologists have used the integration of the genomic profiles with the application of radiomic features to customize the chemotherapy and immunotherapy for specific patients (Li & Zhou, 2022).

AI has played a role in infectious diseases by speeding up the detection of outbreaks and the identification of pathogens. Models based on machine learning used in surveillance have signaled the emergence of hotspots for influenza and COVID-19 (Odone et al., 2025). Moreover, natural language processing has made it possible to develop rapid diagnostic tools that will help in the reading of chest X-rays for pneumonia and tuberculosis, thus helping treatment (Emma, 2025).

Lastly, in the field of rare diseases, the limited number of cases and their various presentations are a source of difficulty (Fehr & Prütz, 2023). But, still, AI can detect patterns in the data from different sources. The models have been first trained on sequencing data and global registries to make the discovery of new disease-gene associations and their early diagnosis possible (Kim et al., 2023). All these case studies taken together provide evidence that AI is an aid to clinical decision-making and has the ability to overcome the issues of speed, accuracy, and treatment personalization; hence, its importance in everyday clinical practice.

10.9 CHALLENGES, ETHICS, AND FUTURE PERSPECTIVES

Even with its huge potential, the incorporation of AI into the process of drug development still encounters a number of difficulties. First, data quality and availability are still critical problems; AI models may become faulty if their training data is biased or insufficient. On the other hand, if advanced AI techniques like deep learning are used, they might be regarded as "black boxes" that provide output but not the logic that leads to it, which eventually results in trust issues and non-acceptance by the authorities. Besides, ethical issues involving the privacy of data, the fairness of algorithms, and the accessibility of AI-researched drugs to all need to be taken into account and dealt with carefully (Vayena et al., 2018). But the future of AI in the field of drug development is very optimistic. To begin with, explainable AI (XAI) will provide a litter of transparency and interpretability around the AI models being implemented. Moreover, the pairing of AI with other next-gen technologies, such as quantum computing and advanced robotics, will not only boost but also expand AI's capabilities. Besides, personalized medicine, which is the outcome of AI's analyzing the patient's data, will become a more common practice, resulting in treatment that is highly customized and very effective. To sum up, AI is about to change the whole scenario of drug discovery and development from being slow, imprecise, and unfriendly to the patient to being fast, accurate, and patient-centric, which will help in making the life-saving drugs available to those in need quicker than before.

10.10 CONCLUSION

With the help of the newly constructed AI systems that facilitate the rapid drug-discovery process, a variety of flu medications are now available in the market. The utilization of AI systems in predictive modeling paves the way for the creation of precise diagnostics. Personalized medicine is a new trend in drug discovery due to AI involvement. The rise of predictive, data-oriented tools that are AI-driven will affect the workflows in hospitals, pharmacology, and clinical practices, as they will be regarded as a routine part of them. The healthcare industry is likely to change considerably as more AI systems will be redirected to the different predictive modeling paradigms within the healthcare framework.

REFERENCES

Agarwal, A., Ressler, D., & Snyder, G. (2015). The current and future state of companion diagnostics. *Pharmacogenomics and Personalized Medicine*, *8*, 99–110. https://doi.org/10.2147/PGPM.S49493

Akhoon, N. (2021). Precision medicine: A new paradigm in therapeutics. *International Journal of Preventive Medicine*, *12*, 12. https://doi.org/10.4103/ijpvm.IJPVM_375_19

Alowais, S. A., Alghamdi, S. S., Alsuhebany, N., Alqahtani, T., Alshaya, A. I., Almohareb, S. N., Aldairem, A., Alrashed, M., Bin Saleh, K., Badreldin, H. A., Al Yami, M. S., Al Harbi, S., & Albekairy, A. M. (2023). Revolutionizing healthcare: The role of artificial intelligence in clinical practice. *BMC Medical Education*, *23*(1), 689. https://doi.org/10.1186/s12909-023-04698-z

Askin, S., Burkhalter, D., Calado, G., & El Dakrouni, S. (2023). Artificial intelligence applied to clinical trials: Opportunities and challenges. *Health and Technology*, *13*(2), 203–213. https://doi.org/10.1007/s12553-023-00738-2

Bhandari, A. (2024). Revolutionizing radiology with artificial intelligence. *Cureus*, *16*(10), e72646. https://doi.org/10.7759/cureus.72646

Brouwer, K. L., Ramanathan, R., & Varma, M. V. (2021). Artificial intelligence and machine learning in drug discovery and development: Focus on ADME and pharmacokinetics. *Drug Metabolism and Disposition*, *49*(3), 227–238.

Chen, S., Yu, J., Chamouni, S., Wang, Y., & Li, Y. (2024). Integrating machine learning and artificial intelligence in life-course epidemiology: Pathways to innovative public health solutions. *BMC Medicine*, *22*(1), 354. https://doi.org/10.1186/s12916-024-03566-x

Daneshjou, R., Kim, M., & Topol, E. J. (2021). Artificial intelligence and machine learning for precision medicine in cardiology. *Nature Reviews Cardiology*, *18*(9), 675–691.

DeGroat, W., Abdelhalim, H., Peker, E., Sheth, N., Narayanan, R., Zeeshan, S., Liang, B. T., & Ahmed, Z. (2024). Multimodal AI/ML for discovering novel biomarkers and predicting disease using multi-omics profiles of patients with cardiovascular diseases. *Scientific Reports*, *14*(1), 26503. https://doi.org/10.1038/s41598-024-78553-6

Desai, A. N., Kraemer, M. U. G., Bhatia, S., Cori, A., Nouvellet, P., Herringer, M., Cohn, E. L., Carrion, M., Brownstein, J. S., Madoff, L. C., & Lassmann, B. (2019). Real-time epidemic forecasting: Challenges and opportunities. *Health Security*, *17*(4), 268–275. https://doi.org/10.1089/hs.2019.0022

Dinh-Le, C., Chuang, R., Chokshi, S., & Mann, D. (2019). Wearable health technology and electronic health record integration: Scoping review and future directions. *JMIR Mhealth and Uhealth*, *7*(9), e12861. https://doi.org/10.2196/12861

Dixon, D., Sattar, H., Moros, N., Kesireddy, S. R., Ahsan, H., Lakkimsetti, M., Fatima, M., Doshi, D., Sadhu, K., & Junaid Hassan, M. (2024). Unveiling the influence of AI predictive analytics on patient outcomes: A comprehensive narrative review. *Cureus*, *16*(5), e59954. https://doi.org/10.7759/cureus.59954

Dongare, D. B., Nishad, S. S., Mastoli, S. Y., Saraf, S. A., Srivastava, N., & Dey, A. (2025). High-throughput sequencing: A breakthrough in molecular diagnosis for precision medicine. *Functional & Integrative Genomics*, *25*(1), 22. https://doi.org/10.1007/s10142-025-01529-w

Emma, L. (2025). *AI-Powered Diagnostic Tools for Early Disease Detection.*

Fahim, Y. A., Hasani, I. W., Kabba, S., & Ragab, W. M. (2025). Artificial intelligence in healthcare and medicine: Clinical applications, therapeutic advances, and future perspectives. *European Journal of Medical Research*, *30*(1), 848. https://doi.org/10.1186/s40001-025-03196-w

Fehr, A., & Prütz, F. (2023). Rare diseases: A challenge for medicine and public health. *Journal of Health Monitoring*, *8*(4), 3–6. https://doi.org/10.25646/11826

Gao, D., Liu, W., & Zhang, R. (2023). Deep learning for predicting drug metabolism and pharmacokinetics. *Artificial Intelligence in Medicine*, *133*, 102404.

Geng, Z., Miao, Y., Zhang, G., & Liang, X. (2023). Colorimetric biosensor based on smartphone: State-of-art. *Sensors and Actuators A: Physical*, *349*, 114056. https://doi.org/10.1016/j.sna.2022.114056

Giordano, C., Brennan, M., Mohamed, B., Rashidi, P., Modave, F., & Tighe, P. (2021). Accessing artificial intelligence for clinical decision-making. *Frontiers in Digital Health*, *3*, 645232. https://doi.org/10.3389/fdgth.2021.645232

Guan, W. et al. (2020). AI-driven drug repurposing for COVID-19: A comprehensive review. *Briefings in Bioinformatics*, *22*(6), bbaa221.

Hu, W., Li, X., Li, C., Li, R., Jiang, T., Sun, H., Huang, X., Grzegorzek, M., & Li, X. (2023). A state-of-the-art survey of artificial neural networks for whole-slide image analysis: From popular convolutional neural networks to potential visual transformers. *Computers in Biology and Medicine*, *161*, 107034. https://doi.org/10.1016/j.compbiomed.2023.107034

John, S., Abdulkarim, S., Usman, S., Rahman, M. T., & Creswell, J. (2023). Comparing tuberculosis symptom screening to chest x-ray with artificial intelligence in an active case finding campaign in northeast Nigeria. *BMC Global and Public Health*, *1*(1), 17. https://doi.org/10.1186/s44263-023-00017-2

Kim, J., McFee, M., Fang, Q., Abdin, O., & Kim, P. M. (2023). Computational and artificial intelligence-based methods for antibody development. *Trends in Pharmacological Sciences*, *44*(3), 175–189. https://doi.org/10.1016/j.tips.2022.12.005

Kim, S. et al. (2023). Deep learning in drug discovery: Current applications and future directions. *Journal of Medicinal Chemistry*, *66*(2), 1145–1166.

Kumar, Y., Koul, A., Singla, R., & Ijaz, M. F. (2023). Artificial intelligence in disease diagnosis: A systematic literature review, synthesizing framework and future research agenda. *Journal of Ambient Intelligence and Humanized Computing*, *14*(7), 8459–8486. https://doi.org/10.1007/s12652-021-03612-z

Li, R., Romano, J. D., Chen, Y., & Moore, J. H. (2024). Centralized and federated models for the analysis of clinical data. *Annual Review of Biomedical Data Science*, *7*(1), 179–199. https://doi.org/10.1146/annurev-biodatasci-122220-115746

Li, S., & Zhou, B. (2022). A review of radiomics and genomics applications in cancers: The way towards precision medicine. *Radiation Oncology*, *17*(1), 217. https://doi.org/10.1186/s13014-022-02192-2

Lotter, W., Hassett, M. J., Schultz, N., Kehl, K. L., Van Allen, E. M., & Cerami, E. (2024). Artificial intelligence (AI) in oncology: Current landscape, challenges, and future directions. *Cancer Discovery*, *14*(5), 711–726. https://doi.org/10.1158/2159-8290.CD-23-1199

Marks, A. et al. (2021). Benevolent AI: Using AI to accelerate drug discovery. *Drug Discovery Today*, *26*(1), 180–185.

Mohammadian Rad, N., & Marchiori, E. (2021). Chapter 9—Machine learning for healthcare using wearable sensors. In A. Godfrey & S. Stuart (Eds.), *Digital Health* (pp. 137–149). Academic Press. https://doi.org/10.1016/B978-0-12-818914-6.00007-7

Muhammad, S. S., Shoaib, M., & Pervez, M. T. (2024). An integrated framework for analysis and prediction of impact of single nucleotide polymorphism associated with human diseases. *Evolutionary Bioinformatics Online*, *20*. https://doi.org/10.1177/11769343241249916

Odone, A., Barbati, C., Amadasi, S., Schultz, T., & Resnik, D. B. (2025). Artificial intelligence and infectious diseases: An evidence-driven conceptual framework for research, public health, and clinical practice. *The Lancet Infectious Diseases*, *0*(0). https://doi.org/10.1016/S1473-3099(25)00412-8

Ohia, D., & Daniel, C. (2024). *Advancements and Challenges in Machine Learning and Artificial Intelligence: Shaping the Future of Technology.*

Pillay, T. S., Khan, A. I., & Yenice, S. (2025). Artificial intelligence (AI) in point-of-care testing. *Clinica Chimica Acta*, *574*, 120341. https://doi.org/10.1016/j.cca.2025.120341

Popova, M. et al. (2021). De novo drug design for idiopathic pulmonary fibrosis using AI. *Nature Biotechnology*, *39*(6), 705–710.

Qahwaji, R., Ashankyty, I., Sannan, N. S., Hazzazi, M. S., Basabrain, A. A., & Mobashir, M. (2024). Pharmaco genomics: A genetic approach to drug development and therapy. *Pharmaceuticals*, *17*(7), 940. https://doi.org/10.3390/ph17070940

Rajpurkar, P. et al. (2022). Deep learning for medical image interpretation: Insights from an industry perspective. *Nature Communications*, *13*(1), 1–13.

Rao, X. et al. (2020). Graph neural networks for drug-drug interaction prediction. *Journal of Chemical Information and Modeling*, *60*(10), 4565–4573.

Rigby, E., Vidya, R., & Shaaban, A. M. (2025). Use of digital pathology and artificial intelligence (AI) in breast cancer diagnosis and management: Opportunities and challenges. *Diagnostic Histopathology*, *31*(3), 182–190. https://doi.org/10.1016/j.mpdhp.2024.12.005

Rojas López, A., & Barberis, M. (2024). Metabolic modeling for probiotic and prebiotic production to treat inflammatory disorders. *Chemical Engineering Journal*, *502*, 157852. https://doi.org/10.1016/j.cej.2024.157852

Rong, J., & Liu, Y. (2024). Advances in medical imaging techniques. *BMC Methods*, *1*(1), 10. https://doi.org/10.1186/s44330-024-00010-7

Roy, G., Prifti, E., Belda, E., & Zucker, J.-D. (2024). Deep learning methods in metagenomics: A review. *Microbial Genomics*, *10*(4), 001231. https://doi.org/10.1099/mgen.0.001231

Sanches, P. H. G., de Melo, N. C., Porcari, A. M., & de Carvalho, L. M. (2024). Integrating molecular perspectives: Strategies for comprehensive multi-omics integrative data analysis and machine learning applications in transcriptomics, proteomics, and metabolomics. *Biology*, *13*(11), 848. https://doi.org/10.3390/biology13110848

Schork, N. J. (2019). Artificial intelligence and machine learning in clinical development: A game changer. *Drug Discovery Today*, *24*(10), 1902–1909.

Sezgin, E., Hussain, S.-A., Rust, S., & Huang, Y. (2023). Extracting medical information from free-text and unstructured patient-generated health data using natural language processing methods: Feasibility study with real-world data. *JMIR Formative Research*, *7*, e43014. https://doi.org/10.2196/43014

Shang, M., Guo, J., & Guo, J. (2023). Point-of-care testing of infectious diseases: Recent advances. *Sensors & Diagnostics*, *2*(5), 1123–1144. https://doi.org/10.1039/D3SD00092C

Srivastava, R. (2025). Advancing precision oncology with AI-powered genomic analysis. *Frontiers in Pharmacology*, *16*, 1591696. https://doi.org/10.3389/fphar.2025.1591696

Struelens, M. J., Ludden, C., Werner, G., Sintchenko, V., Jokelainen, P., & Ip, M. (2024). Real-time genomic surveillance for enhanced control of infectious diseases and antimicrobial resistance. *Frontiers in Science*, *2*. https://doi.org/10.3389/fsci.2024.1298248

Swinckels, L., Bennis, F. C., Ziesemer, K. A., Scheerman, J. F. M., Bijwaard, H., de Keijzer, A., & Bruers, J. J. (2024). The use of deep learning and machine learning on longitudinal electronic health records for the early detection and prevention of diseases: Scoping review. *Journal of Medical Internet Research*, *26*, e48320. https://doi.org/10.2196/48320

Taherdoost, H., & Ghofrani, A. (2024). AI's role in revolutionizing personalized medicine by reshaping pharmacogenomics and drug therapy. *Intelligent Pharmacy*, *2*(5), 643–650. https://doi.org/10.1016/j.ipha.2024.08.005

Topol, E. J. (2019). High-performance medicine: The convergence of human and artificial intelligence. *Nature Medicine*, *25*(1), 44–56.

Unger, M., & Kather, J. N. (2024). Deep learning in cancer genomics and histopathology. *Genome Medicine*, *16*, 44. https://doi.org/10.1186/s13073-024-01315-6

Vayena, E. et al. (2018). Machine learning in medicine: Addressing ethical challenges. *PLoS Medicine*, *15*(11), e1002689.

Verma, K. K., Palmal, R., & Yadav, P. (2025). Unveiling the future: The impact of artificial intelligence in diagnostic pathology. *Surgical and Experimental Pathology*, *8*(1), 23. https://doi.org/10.1186/s42047-025-00197-1

Wang, J., Sourlos, N., Zheng, S., van der Velden, N., Pelgrim, G. J., Vliegenthart, R., & van Ooijen, P. (2023). Preparing CT imaging datasets for deep learning in lung nodule analysis: Insights from four well-known datasets. *Heliyon*, *9*(6), e17104. https://doi.org/10.1016/j.heliyon.2023.e17104

Wu, Y., & Xie, L. (2025). AI-driven multi-omics integration for multi-scale predictive modeling of genotype-environment-phenotype relationships. *Computational and Structural Biotechnology Journal*, *27*, 265–277. https://doi.org/10.1016/j.csbj.2024.12.030

Xu, Y. et al. (2022). AI-based toxicity prediction and drug design. *Advanced Drug Delivery Reviews*, *181*, 112217.

Yan, Y. et al. (2021). Deep learning in quantitative structure, activity relationship: Recent advances and challenges. *Journal of Chemical Information and Modeling*, *61*(2), 521–537.

Machine Learning and AI for Predicting Antimicrobial Resistance

11

Samuel Adeniyi Oyegbade, Mayowa Agbi, Iisominea Isozo Amos, Samson Olusola Adeboye, Gbotemi Jerry Oni, and Oluwafemi Adebayo Oyewole

11.1 INTRODUCTION

The mortality occurrences attributed to antimicrobial resistance (AMR) are more than 1.2 million, according to the World Health Organization (WHO), and impede food security and economic potential, particularly in low-income nations. By 2050, it is projected that, if not managed effectively, the number could rise to 10 million (WHO, 2023). It is present in various microbes such as bacteria, parasites, viruses, and fungi (Cedeño Muñoz et al., 2024; Prestinaci et al., 2015). An extreme resistance rate in key pathogens, such as 42% of *Escherichia coli* cases that resisted third-generation cephalosporins, has been reported by the WHO. Also, 20% of E. coli-linked urinary tract infections revealed lowered susceptibility to conventional antibiotics. *Klebsiella pneumoniae* is progressively resistant to high-priority antibiotics, demanding the utilization of drugs such as carbapenems, which are becoming less potent in nations that are low- or mid-income earning.

Microbes' capabilities to endure the impact of drugs and medications, which is known as AMR, result in ineffectiveness of the treatment, persistence of infections, and increased risk of mortality. A core 21st-century challenge ravaging the health of people all over the world is AMR (Cedeño Muñoz et al., 2024; Coque et al., 2023). AMR continues to threaten human existence and survival and instigates the quest for substitute techniques to curtail antibiotics in the production of animals (Oyedeji, 2025). AMR's sporadic rise is instigated by abuse of antibiotics in various sectors and spheres, including agriculture, human medicine, and livestock farming (Aslam et al., 2021). The slow pace of designing and developing novel antibiotics and the vast dissemination of resistant genes across borders blow

DOI: 10.1201/9781003727514-13

up the challenge even more. Also, the situation was aggravated by the COVID-19 pandemic, which skyrocketed the utilization of antibiotics for viral infections, which in turn influenced increased resistance patterns (Dhungel et al., 2023). Rising utilization of machine learning (ML) algorithms is shifting the paradigm in the medical sector and enhancing clinical practices. Deep and machine learning integration paves a peculiar path into antimicrobial identification (Ali et al., 2023).

11.2 FUNDAMENTALS OF AMR BIOLOGY

Through mechanisms such as impulsive mutations or horizontal gene transfer, bacteria gain resistance as vanA, blaNDM-1, and mcr-1 resistance genes spread globally. The prevalent rise in resistance in Gram-positive bacterial infections demonstrates a crucial issue for healthcare settings globally. Various means have been harnessed in tackling the growing threat by directing certain mechanisms, which influence the identification and susceptibility of antibiotics (Rajput et al., 2024).

11.2.1 Mechanisms of Resistance

Recently, AMR rose sporadically, and this necessitates a mechanistic understanding through which bacteria resist antibacterials (Zhu et al., 2013). Efflux pumps, enzymatic breakdown, alteration in antibacterials' structure, cell membrane permeability, and antibacterials' target site alterations are pivotal mechanisms that build up bacterial resistance (Crofts et al., 2017). Biofilm formation, a key feature of bacteria as well as the choice of antibacterial resistance genes, influences AMR spread (Røder et al., 2016).

A major mechanism that enables bacteria to lower antibacterials' potency through changing or defending their cellular targets, which in turn leads to disruption of antibacterials' binding (Blair et al., 2015). Bacteria also have the potential to change the targets critical for drug binding, fostering inefficient binding.

Efflux pumps are prevalent in numerous bacterial types and offer resistance to a vast range of antibiotics. They depend on energy in complex microbial systems found on the membrane of the cytoplast with the capacity to remove poisonous substances such as drugs from the cell, lowering the drug amounts and its potency within the cell region (Abdi et al., 2020). They also function pivotally in microbial physiology by fostering the regulation of their internal environment by microbes. The mechanism of the efflux pump enables the removal of antibacterial drugs within the cell (Peterson and Kaur, 2018).

Another key resistance mechanism of bacteria is the ability to generate enzymes that either degrade or change antibacterials, thereby making them ineffective (Egorov et al., 2018). β-lactamases, chloramphenicol acetyltransferases, and aminoglycoside-altering enzymes deactivate antibacterials, which are the three primary antibacterial deactivating enzymes (Peterson and Kaur, 2018; Reygaert, 2018).

11.2.2 Sources of Antimicrobial Resistance

The interwoven nature of human, microbial, and environmental factors edges AMR as a complex issue across the globe. The impact of AMR cuts across numerous regions and sectors, thereby demanding an intersectoral understanding. The following are sources and mechanisms responsible for its global spread and development. The predominant deployment of antibiotics in agriculture for disease prevention and yield boosting results in resistant bacteria emergence and water, food, and direct contact-based transmission to humans. Also, unsuitable prescription and utilization of antibiotics, particularly prominent in the treatment of viral infections such as flu and cold by using antibiotics for viral infections, incorrect dosages, and refusal to complete prescribed treatments, cause AMR. Healthcare facilities bereft of sanitation and infection protocols and proper hygiene are prone to resistant infections. Global trade and travel also facilitate the spread of resistant pathogens globally. Resistant bacteria can gain entry across borders through migrants, tourists, and imported food products. WHO (2022) put forward that antimicrobial-resistant strains, such as carbapenem-resistant Enterobacteriaceae, have been perceived globally within months of their emergence due to interconnected mobility across the globe.

11.3 PRINCIPLES OF MACHINE LEARNING AND AI

Artificial intelligence (AI) was first put forward by John McCarthy to expand design application systems, human intelligence technologies, and theoretical methods (Hamet and Tremblay, 2017; Haug and Drazen, 2023). A core aspect of AI referred to as machine learning (ML) encompasses the development of algorithms geared toward accurate prediction of outcome variables. Algorithms of ML are evaluated for their predictive capacity and accuracy on test datasets (Greener et al., 2022).

Generation of synergistic impacts in drug combinations, novel antibiotics design, and sequence-based AI analysis are notably applicable uses of ML technology in AMR research (Anahtar et al., 2021). Prediction of microorganisms likely to develop resistance to particular drugs occurs through analysis of patterns in resistance data and antimicrobial drug usage, helping professionals within the healthcare sector to make precise decisions (Sakagianni et al., 2023). In addition, ML models function in antibiotic resistance monitoring, analysis of enormous data for identification of novel resistance patterns, and potential hotspots.

On the other hand, deep learning (DL), an aspect of ML, entails a more in-depth expansion of AI in the simulation of multifaceted data processing trials. DL can rapidly identify chemicals in vast chemical libraries, thereby facilitating research into the search for appropriate antibiotic drugs (Talat and Khan, 2023).

11.3.1 Common Algorithms Used in AMR Prediction (SVM, Random Forests, and Neural Networks)

Convolutional and recurrent neural networks are DL models functioning primarily in the analysis of enormous microbiological datasets to sense resistance genes and predict antimicrobial susceptibility. CNNs utilize ML models that have been deployed for AMR prediction (Kuang et al., 2022). DL and ML models such as logistic regression, support vector machine (SVM), and random forest are engaged for post-data processing, identification, and categorization, as well as AMR utilization (Arango-Argoty et al., 2018).

Collection and preprocessing of data are key ML processes in AMR applications. The data applied chiefly covers single-nucleotide and whole-genome sequences with distinct phenotypes (Elsayed et al., 2023). Coding, data preprocessing, and feature extraction are implemented with ease using packages like NumPy and Pandas in Python (Harris et al., 2020).

11.4 MACHINE LEARNING AND DEEP LEARNING IN THE AMR CONTEXT

AI deployment in AMR covers an array of conventional DL and ML techniques, such as SVMs, random forest, decision trees, and logistic regression, which have showcased enviable performance and immense capacity in AMR prediction, with a pooled area under the curve of 0.82 from clinical data in predicting AMR (Sagar et al., 2025). Random forest and gradient-boosted decision trees, which are ensemble techniques in real-world systems and applications, have delivered better results than simpler models such as logistic regression (Syam and Kaul, 2021). These approaches however need structured data and often depend on predefined characteristics or known genetic marker sets, which can hamper the capacity to analyze raw and high-dimensional data.

DL tackles this drawback by utilizing multiple-layered deep neural networks for representation learning, that is, unraveling hierarchical feature representations from raw datasets directly. This ability is peculiarly essential where there is no room for manual feature engineering to function and analyze raw, unstructured genomic data sequences. DL models are known to be viable tools for mapping unstructured linkages between target labels and input features, functioning as potent interpretation and prediction tools (Prusa and Khoshgoftaar, 2017).

Evolution of simpler ML models and advanced DL ones is not predicated on chance; instead, it is a direct function of the rising available biological data pool and its corresponding growing complexity. Evolution of the field from analysis of structured clinical data and examination of whole genomes makes it more imperative to utilize specific DL architectures and transforming models to process high-dimensional data sequences via an automated process (Ogunjobi et al., 2024).

Antimicrobial resistance poses a public health risk, which escalates daily and weakens the potency of antibiotics in circulation, causing treatment efforts geared toward infectious diseases to become more complex and herculean (Salam et al., 2023). Traditional AMR prediction and detection approaches are often reactive, slow, and demand a lot of resources. These drawbacks give deep learning AI techniques an edge as a potent and revolutionary tool for AMR prediction and surveillance by harnessing the enormous available unstructured biological datasets. AI promises unlimited potential to combat AMR spread, as well as to discover novel therapeutic approaches (Sakagianni et al., 2024) (Figure 11.1).

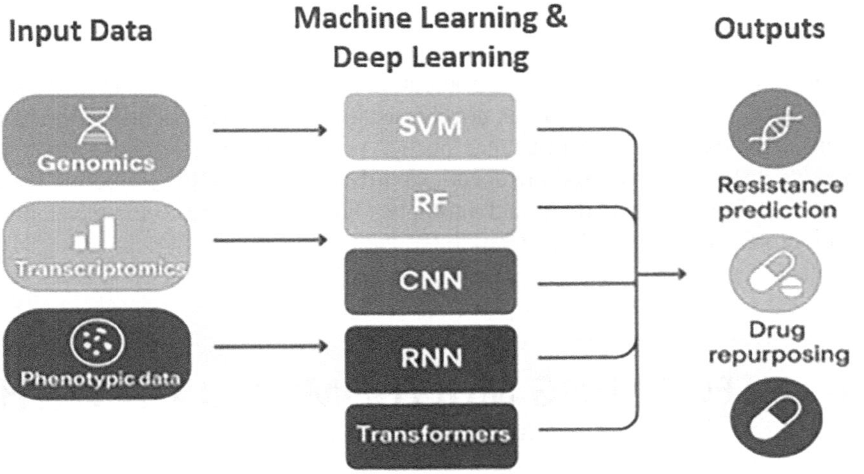

FIGURE 11.1 AI in antimicrobial resistance prediction and drug development.

11.5 DEEP LEARNING FOR AMR PREDICTION

DL is defined by man-made, diversely layered neural networks, which demonstrate effectiveness, and by handling high-dimensional biological data in predicting AMR.

11.5.1 CNNs for Raw Sequence Classification

CNNs are widely known for their image processing and analysis of genomic data sequences for categorization of raw sequences in the prediction of AMR. They can effectively identify understated signatures and motifs linked to resistance mechanisms and scan vast genomic sequences with no need for wide-ranging feature engineering. CNNs foster direct categorization of bacterial isolates on the basis of their whole-genome sequencing data, making it easier and faster to evaluate AMR profiles holistically (Green et al., 2022).

11.5.2 RNNs/LSTMs for Temporal AMR Data

The ever-evolving resistance patterns and geographical spread of AMR make it a dynamic phenomenon. Recurrent neural networks (RNNs) and long short-term memories (LSTMs) uniquely process sequential data for analysis of temporal AMR trends, modeling these temporal dependencies and picking from surveillance data sequences for prediction of future emergence and resistance trends, which in turn enable public health agencies to look forward to resistant strain outbreaks, effective resource allocation, and implementation of interventions at the appropriate time. For example, prediction of rise in carbapenem-resistant Enterobacteriaceae in certain regions by LSTMs through analysis of past surveillance data, and in turn enable healthcare centers and systems to prepare adequately for likely sporadic rise in resistant infections (Zhang et al., 2025).

11.5.3 Transformer-based Models for AMR Gene Discovery

Although it was designed initially for natural language processing, transformer networks have transformed sequence modeling as a result of their attention mechanisms, which enable weighing various sequence parts. These models can fast-track novel resistance mechanism identification by understanding complexities that exist between resistance phenotypes and gene sequences, which are essential for creating novel diagnostic tools and therapeutic approaches and also for predicting the influence of certain mutations on enzymatic activity and antibiotics binding, leading to the development of more effective antimicrobial substances (Ren et al., 2021).

11.6 INTEGRATING MULTI-OMICS DATA WITH AI

The intense complications surrounding AMR go far beyond pockets of sole genetic mutations. This further establishes intricate influences within microbial communities and host responses. The integration

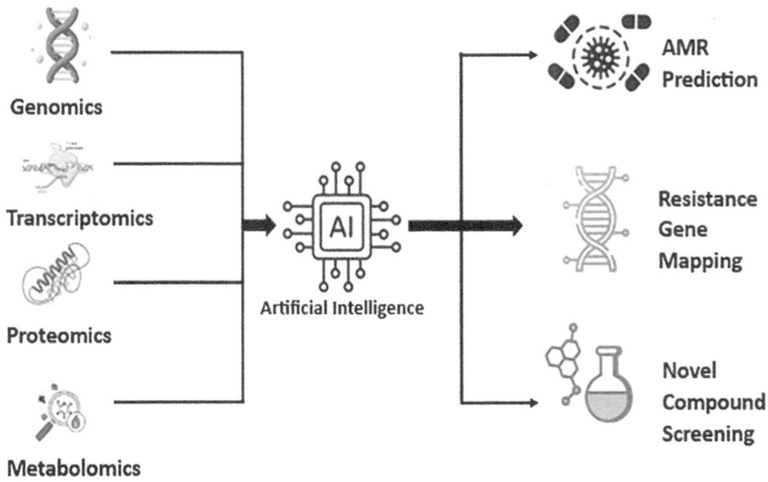

FIGURE 11.2 AI integration across omics layers for antimicrobial resistance (AMR) prediction. Multi-omics databases, such as genomics, transcriptomics, proteomics, and metabolomics, can be integrated with AI models to predict resistance patterns, map resistance genes, and screen novel antimicrobial compounds.

of numerous omics datasets with AI provides a more robust perspective, critical to make the most of this high-dimensional data (Figure 11.2).

11.6.1 Combining Genomics, Transcriptomics, and Metabolomics for AMR Prediction

Sole omics layers are limited based on the fact that they offer incomplete pictures. Hence, incorporation of metabolomics, transcriptomics, and genomics with AI will facilitate the creation of accurate and more comprehensive AMR predictive models. AI algorithms, such as multi-modal DL or Bayesian networks, foster the incorporation of disparate data forms to unravel complex linkages, offer predictions of highly accurate phenotypic resistance, and distinguish between acquired and inherent resistance mechanisms, thereby paving the way for a robust comprehension of the molecular underpinnings of resistance (Hasin et al., 2017).

11.6.2 Network-based AI Models for Microbial Interactions and Resistance Spread

AMR is more of interconnectedness within communities of microorganisms and horizontal resistance gene spread rather than single pathogens. A prominent example of network-based AI models specifically tailored for analysis of complicated interactions and modeling of microbial interaction networks are the graph neural networks (Subramanian et al., 2020).

Analysis of the networks fosters the prediction of resistant genes spread by AI within diverse host communities and environments and also enhances understanding of AMR epidemiology and development of strategies to disrupt the pathways. GNNs, for instance, can initiate prediction of resistance transfer risk from commensal to pathogenic bacteria within the microbiome of a patient (Wang et al., 2022).

11.7 AI IN DRUG DEVELOPMENT AGAINST AMR

AMR untamable rise coupled with the high cost of managing the impact of the menace drives the need to search out for new unexplored strategies and antimicrobial compounds continually.

11.7.1 Identifying Novel Antimicrobial Compounds

Conventional discovery of antimicrobials in time-exhausting and limiting as it entails high-throughput screening of vast compound libraries. AI can fit in to drastically ease the process seamlessly. Variational autoencoders and generative adversarial networks are generative AI models with predicted antimicrobial potential against certain resistant targets (Stokes et al., 2020) and can learn the chemical characteristics of numerous known antibiotics and bring about the generation of new chemical structures with enhanced or the same features. In addition, AI can predict the binding affinity of candidate compounds to efflux pumps and bacterial targets, focusing on the highest potential molecules to carry out further validation tests and experiments, which in turn raise the chances of discovering novel antibiotics while lowering the experimental burden.

11.7.2 AI for Repurposing Existing Drugs against Resistant Pathogens

Drug repurposing is a groundbreaking approach for tackling AMR since established safety profiles exist for these drugs. Potential repurposing candidates can be identified by AI via analysis of immense datasets, clinical outcome data, and drug-target interaction networks (Elalouf et al., 2025).

Identification of bacterial virulence factor-interfering drugs can be achieved through ML algorithms. These drugs can change efflux pumps or antibiotics to address resistance and pave the way for a cheaper and faster market path for new AMR treatments in comparison to the discovery of de novo drugs.

11.7.3 Predicting Resistance Emergence during Drug Development

A grave issue in developing new antimicrobials is forecasting whether bacteria will swiftly evolve resistance to them. AI can tackle this by building predictive models on the basis of the chemical structure of new drugs and already explored bacterial resistance mechanisms. By simulating molecular interactions and evolutionary pathways, AI can predict the chances and extent of resistance development against new compounds (Hughes and Andersson, 2023), thereby fostering early adjustment of compounds for drug developers to project more intense combination therapies, lowering resistance development. Such predictive proficiencies can implicitly extend the lifespan of novel antimicrobial drugs, corroborating their long-standing efficacy against growing threats.

11.8 DATA SOURCES FOR AMR STUDIES

New approaches for AMR surveillance, prediction, and detection gets more increasing demanded as the implications for public health aggravate (Arnold et al., 2025).

ML and AI offer efficient computational frameworks to tackle this issue; however, their effectiveness is largely influenced by the amount, diversity, and quality of available data. A robust comprehension and deployment of diverse data sources are essential for furthering the cause of vigorous ML models for AMR prediction. These sources are within an array of molecular-level biological data to all-inclusive epidemiological and phenotypic datasets (Al-Haboubi et al., 2021).

11.8.1 Genomic, Transcriptomic, and Proteomic Data

AMR research has been further advanced with the evolution of genomic data. A whole genome sequence provides a broad array of bacterial genomes, easing the process of recognizing chromosomal mutations, identifying of resistance-conferring genes, and mobile genetic elements that enable resistance spread (Baker and Holt, 2021). For ML models, genomic data can be denoted in various forms, which include k-mer frequencies, raw DNA sequences nonstop, presence/absence of specific resistance genes, and sole nucleotide polymorphisms (Nguyen et al., 2020). The rising accessibility and affordability of whole genome sequencing have resulted in an explosion of genomic data, establishing the basis for numerous sequence-grounded AMR prediction models.

Transcriptomic data obtained from RNA sequencing provide robust understanding into patterns of gene expression under various conditions, such as exposure to antibiotics. This data can bring about bacterial response to antimicrobial stress and identification of genes regulated in the presence of a drug, which may express resistance mechanisms and compensatory pathways (Stokes et al., 2020). An efflux pump gene, for instance, could be markedly upregulated in a resistant strain, even if the gene itself is not changed. These expression profiles can be analyzed by ML models to distinguish signatures predictive of resistance.

Proteomic data offer a straight evaluation of proteins present and their modifications within a bacterial cell and are typically derived through mass spectrometry techniques. Proteins are the direct machinery of the cell, and alterations in their structure, function, distribution, and resistance mediation are directly observed in changed drug targets and drug-inactivating enzymes (Zhu et al., 2021). Proteomic data can offer complementary insights, especially resistance mechanisms that cover complex protein-drug interactions and post-translational modifications than genomic data. Integrating these multi-omics datasets with ML holds weighty potential for a more robust AMR understanding and prediction (Rabbani et al., 2021).

11.8.2 Phenotypic Resistance Testing Datasets

Phenotypic resistance testing datasets are the gold standard for demonstrating the resistance or susceptibility of a bacterial isolate in response to a particular antimicrobial agent. These datasets characteristically encompass minimum inhibitory concentration values, which represent the lowermost dose of an antimicrobial that impedes the microbes' observable growth. Categorical groupings (susceptible, intermediate, and resistant) are derived from MIC values on the basis of clinical breakpoints (EUCAST and CLSI, 2022; Humphries and Hammerstrom, 2022).

For ML models, phenotypic data function as the central "ground truth" labels against which predictions from molecular or genomic data are trained and validated. Huge collections of MIC values, often associated with matching genomic sequences, are critical for training supervised ML models for resistance prediction (Nguyen et al., 2020). These datasets are characteristically generated in clinical microbiology laboratories, extensive surveillance programs, and reference laboratories. Phenotypic testing is accompanied by standardization and consistency, which are critical for the reliability of these datasets and the enhanced performance of ML models trained on them.

11.9 ML APPROACHES FOR PREDICTING RESISTANCE

ML has shifted the paradigm of AMR prediction to a more wide-ranging perspective forward and above sole gene presence. These approaches alter complicated patterns within biological data for novel resistance mechanisms' identification.

11.9.1 Predicting Minimum Inhibitory Concentrations (MICs)

Continuous MIC values' prediction, apart from susceptible/resistant classifications, paves the way for more informative and granular degrees of resistance, thereby depicting a regression task for ML models.

MIC prediction has been carried out using deep neural networks and ML algorithms like random forests, linear regression, and gradient boosting machines (Pati et al., 2021). Models are trained on datasets, and in this case, each bacterial isolate has both its genomic characteristics and experimentally determined MIC values. The aim is to predict the MIC value from the genomic data, although it is chiefly difficult because of the multifaceted genetic determinants of quantitative resistance and the constant nature of MICs.

Graph convolutional networks work on protein-protein interaction networks, while complex neural networks unswervingly process genomic sequences and have proved immense potential in enhancing the accuracy level in MIC prediction (Nguyen et al., 2020; Yang et al., 2022). Accurate MIC prediction can bring about informed dose modifications and more exact treatment decisions and enable novel antibiotics to advance by offering a developed quantitative understanding of resistance levels. Challenges remain in obtaining adequately diverse and standardized datasets for robust MIC prediction across numerous antibiotics and pathogens.

11.9.2 Cross-Species Resistance Prediction

AMR prediction across diverse bacterial species is predominantly challenging, though a vital application of ML. Resistance mechanisms are not always conserved or expressed correspondingly across species, even for the matching antibiotic. Conventional approaches often advance models for each species-drug combination autonomously.

ML has the capacity to build more generalizable models that can influence shared fundamental resistance mechanisms or recognize common genomic signatures across diverse bacterial taxa. Transfer learning method presents a model trained on one species and fine-tuned for another, while multi-task learning presents a solitary model that simultaneously learns to forecast resistance in numerous species that are being discovered (Gupta et al., 2021).

Core resistome identification and model buildup around these general characteristics, it may be possible to create more robust cross-species predictions. Furthermore, models that integrate bacterial pan-genomes or phylogenetic associations can account for shared ancestry and evolutionary distances, refining the transferability of resistance predictions (Zhao et al., 2020). Effective cross-species prediction would pointedly lower the need for robust species-specific training data and fast-track the identification of less explored species or resistance in emerging pathogens.

11.10 CASE STUDIES IN MACHINE LEARNING FOR AMR

The application of ML geared toward AMR prediction has yielded substantial improvements across numerous pathogens and antimicrobial classes, signifying the practical utilization of these computational approaches in both clinical and epidemiological settings.

11.10.1 ML Models for Tuberculosis Drug Resistance

Tuberculosis, which is triggered by *Mycobacterium tuberculosis*, with drug-resistant forms, poses a predominantly difficult challenge. Rapid detection of drug resistance is key to potent treatment and forestalling further dissemination. ML models have demonstrated effectiveness in predicting TB drug resistance from genomic data.

Studies have investigated whole genome sequence data for identification of mutations in certain genes that confer resistance, utilizing ML algorithms such as SVMs, neural networks, and random forests (Phelan et al., 2019; Yang et al., 2020), which are models trained on enormous datasets of isolates of *M. tuberculosis* with identified resistance phenotypes for prediction of resistance to first- and second-line anti-tuberculosis drugs. ML models advanced from numerous clinical isolates are extremely accurate and precise in predicting resistance to isoniazid and rifampicin, often outdoing rule-grounded gene-lookup tactics for intricate resistance profiles (Walker et al., 2021). Beyond sole drug resistance, ML models are also being advanced to predict multi-drug resistance by integrating manifold genetic markers (Zhou et al., 2022). ML's capacity to spot multifaceted epistatic interactions between subtle resistance-linked genomic variations and mutations has been key in fostering TB resistance prediction, leading to more tailored treatment regimens grounded on genomic data.

11.10.2 Prediction of β-Lactam and Fluoroquinolone Resistance

β-Lactams and fluoroquinolones are widely used antibiotic groups, though their resistance is rampant in several bacterial pathogens like *E. coli, Staphylococcus aureus*, and *K. pneumoniae*. Consequently, machine models are dynamically being advanced to predict resistance to this salient class of drug.

ML models regularly focus on detecting the presence of β-lactamase genes (e.g., blaCTX-M and blaNDM) from whole genome sequence data, efflux pump overexpression, and mutations in penicillin-binding proteins for β-lactam resistance (Gomez-Gonzalez et al., 2020). CNNs and GNNs can analyze genomic regions near known β-lactamase genes or whole genomes for resistance-linked complex pattern identification (Nguyen et al., 2020), and these models can predict resistance to specific β-lactam antibiotics accurately, supporting clinicians in choosing fitting empiric therapies.

For fluoroquinolone resistance, ML models target mutations in topoisomerase IV and gyrase (parC and parE) genes and the occurrence of plasmid-mediated quinolone resistance genes (for instance, qnr genes) (Klimova et al., 2021). By extracting these genetic characteristics from WGS data, ML algorithms can predict the resistance of an isolate to certain highly specific and sensitive fluoroquinolones. Recently, investigations have been carried out on the incorporation of phenotypic data from resistance screens with genomic data to train more capable ML models for fluoroquinolone resistance to bring about novel resistance mechanisms (Li et al., 2022).

11.10.3 ML-based Surveillance in Hospital and Environmental Settings

Asides from individual patient diagnosis, ML functions pivotally in more holistic AMR surveillance programs within environmental and clinical settings, where ML models can analyze routine microbiological data and whole genome sequence data from circulating pathogens for identification of outbreak prediction, emerging resistance trends, and drive efforts toward setting up infection control measures (Price et al., 2020). Also, ML identifies risk factors for acquiring resistant infections, detects resistant isolates' clusters, and predicts the future occurrence of certain resistance genes within a hospital unit in order to facilitate workable interventions to prevent further spread. AI-powered dashboards offer real-time insights into patterns of AMR, which in turn enable clinicians and hospital epidemiologists to offer data-driven decisions (Malhotra et al., 2022).

ML is utilized in tracking AMR spread in soil, wastewater, and animal populations in environmental settings. ML models also identify and quantify antibiotic resistance genes within diverse environmental niches through analysis of whole genome sequence data from metagenomic samples (Li et al., 2020) to enable monitoring of the environmental resistome, movement of resistance genes between reservoirs, and evaluate anthropogenic activities' impact on the spread of AMR. For instance, ML models can facilitate prediction of certain antibiotic-resistant genes in effluents of wastewater treatment plants based on upstream human activity data, offering an early warning system for potential public health risks (Yang et al., 2021). The efforts geared toward surveillance will greatly influence public health policies for robust comprehension of AMR's ecological dynamics.

11.11 CONCLUSION

In recent times, diagnostics, predictions, and treatment approaches for AMR are evolving immensely through the advent of ML and AI. Harnessing big data analytics and genome sequencing jointly will unveil substantial progress in early AMR detection and predictive modeling.

REFERENCES

Abdi, S. N., Ghotaslou, R., Ganbarov, K., Mobed, A., Tanomand, A., Yousefi, M., Asgharzadeh, M., & Samadi Kafil, H. (2020). Acinetobacter baumannii efflux pumps and antibiotic resistance. *Infection and Drug Resistance*, *13*, 423–434. https://doi.org/10.2147/IDR.S228089

Al-Haboubi, M., Glover, R. E., Eastmure, E., Petticrew, M., Black, N., & Mays, N. (2021). Quality and utility of information captured by surveillance systems relevant to antimicrobial resistance (AMR): A systematic review. *Antibiotics*, *10*(4), 431. https://doi.org/10.3390/antibiotics10040431

Ali, T., Ahmed, S., & Aslam, M. (2023). Artificial intelligence for antimicrobial resistance prediction: Challenges and opportunities towards practical implementation. *Antibiot. (Basel)*. *12*, 523. doi: 10.3390/antibiotics12030523

Anahtar, M. N., Yang, J. H., & Kanjilal, S. (2021). Applications of machine learning to the problem of antimicrobial resistance: An emerging model for translational research. *J. Clin. Microbiol*, *59*, e0126020. doi: 10.1128/JCM.01260-20

Arango-Argoty, G., Garner, E., Pruden, A., Heath, L. S., Vikesland, P., & Zhang, L. (2018). DeepARG: A deep learning approach for predicting antibiotic resistance genes from metagenomic data. *Microbiome, 6*, 23. doi: 10.1186/s40168-018-0401-z

Arnold, A., McLellan, S., & Stokes, J. M. (2025). *How AI can help us beat AMR*. npj Antimicrobials and Resistance, 3, Article 18. doi: https://doi.org/10.1038/s44259-025-00085-4

Aslam, B., Khurshid, M., Arshad, M. I., Muzammil, S., Rasool, M. H., Yasir, M., & Baloch, Z. (2021). Antibiotic resistance: A rundown of a global crisis. *Infection and Drug Resistance*, *14*, 1645–1658. https://doi.org/10.2147/IDR.S173867

Baker, S., & Holt, K. E. (2021). Whole-genome sequencing for surveillance of antimicrobial resistance. *Nature Reviews Microbiology*, *19*, 364–380. https://doi.org/10.1038/s41579-020-00443-1

Blair, J. M., Webber, M. A., Baylay, A. J., Ogbolu, D. O., & Piddock, L. J. (2015). Molecular mechanisms of antibiotic resistance. *Natural Reviews Microbiology*, *13*, 42–51. doi: 10.1038/nrmicro3380

Cedeño Muñoz, J. S., Aransiola, S. A., Reddy, K. V., Ranjit, P., Victor-Ekwebelem, M. O., Oyedele, O. J., Maddela, N. R., & Rodríguez Díaz, J. M. (2024). Antibiotic resistant bacteria and antibiotic resistance genes as contaminants of emerging concern: Occurrences, impacts, mitigations and future guidelines. *Science of the Total Environment*, *952*, 175906. https://doi.org/10.1016/j.scitotenv.2024.175906

Coque, T. M., Cantón, R., Pérez-Cobas, A. E., Fernández-de-Bobadilla, M. D., & Baquero, F. (2023). Antimicrobial resistance in the global health network: Known unknowns and challenges for efficient responses in the 21st century. *Microorganisms*, *11*(4):1050. doi: 10.3390/microorganisms11041050

Crofts, T. S., Gasparrini, A. J., & Dantas, G. (2017). Next-generation approaches to understand and combat the antibiotic resistome. *Natural Reviews Microbiology*, *15*, 422–434. doi: 10.1038/nrmicro.2017.28

Egorov, A. M., Ulyashova, M. M., & Rubtsova, M. Y. (2018). Bacterial enzymes and antibiotic resistance. *Acta Nat*. 10, 33–48. doi: 10.32607/20758251-2018-10-4-33-48

Elalouf, A., Elalouf, H., Rosenfeld, A., & Maoz, H. (2025). Artificial intelligence in drug resistance management. *3 Biotech*, *15*(5), 126. https://doi.org/10.1007/s13205-025-04282-w

Elsayed, H. A., Badawy, H., & Sharaf, D. K. (2023). Artificial intelligence for antimicrobial resistance prediction: Challenges and opportunities towards practical implementation. *Antibiotics*, *12*(3), 523. https://doi.org/10.3390/antibiotics12030523

EUCAST, CLSI. (2022). *The European Committee on Antimicrobial Susceptibility Testing (EUCAST) & Clinical and Laboratory Standards Institute (CLSI) guidelines for antimicrobial susceptibility testing*. (Note: Specific year may vary for latest versions, typically cited as the year of the specific guideline document used, e.g., CLSI M100 Ed. 32 2022).

Gomez-Gonzalez, M., García-Vázquez, E., & Delgado-Suárez, E. (2020). Prediction of beta-lactam resistance in Gram-negative bacteria using machine learning algorithms based on genomic data. *Frontiers in Microbiology*, *11*, 596135.

Green, A. G., Yoon, C. H., Chen, M. L., Ektefaie, Y., Fina, M., Freschi, L., & Farhat, M. R. (2022). A convolutional neural network highlights mutations relevant to antimicrobial resistance in mycobacterium tuberculosis. *Nature Communications*, *13*(1), 3805. https://doi.org/10.1038/s41467-022-31236-0

Greener, J. G., Kandathil, S. M., Moffat, L., & Jones, D. T. (2022). A guide to machine learning for biologists. *Nature Reviews Molecular Cell Biology*, *23*, 40–55. doi: 10.1038/s41580-021-00407-0

Gupta, R., Sharma, M., & Kumar, R. (2021). Cross-species antibiotic resistance prediction using transfer learning in machine learning. *Bioinformatics*, *37*(17), 2533–2540.

Hamet, P., & Tremblay, J. (2017). Artificial intelligence in medicine. *Metabolism: Clinical and Experimental*, *69S*, S36–S40. doi: 10.1016/j.artmed.2017.06.011

Harris, C. R., Millman, K. J., van der Walt, S. J., Gommers, R., Virtanen, P., Cournapeau, D., Wieser, E., Taylor, J., Berg, S., Smith, N. J., Kern, R., Picus, M., Hoyer, S., van Kerkwijk, M. H., Brett, M., Haldane, A., Del Río, J. F., Wiebe, M., Peterson, P., Gérard-Marchant, P., Sheppard, K., Reddy, T., Weckesser, W., Abbasi, H., Gohlke, C., & Oliphant, T. E. (2020). Array programming with numPy. *Nature*, 585, 357–362. doi: 10.1038/s41586-020-2649-2

Hasin, Y., Seldin, M., & Lusis, A. (2017). Multi-omics approaches to disease. *Genome Biology*, *18*, 83. https://doi.org/10.1186/s13059-017-1215-1

Haug, C. J., & Drazen, J. M. (2023). Artificial intelligence and machine learning in clinical medicine, 2023. *The New England Journal of Medicine*, *388*, 1201–1208. doi: 10.1056/NEJMra2302038 https://doi.org/10.1073/pnas.1222743110

Hughes, D., & Andersson, D. I. (2023). Forecasting antimicrobial resistance evolution. *Trends in Microbiology, 31*(2), 89–100. https://doi.org/10.1016/j.tim.2022.09.004

Humphries, R. M., & Hammerstrom, S. P. (2022). Challenges and opportunities in antimicrobial susceptibility testing for difficult-to-treat infections. *Clinical Infectious Diseases, 75*(Supplement_1), S19–S27.

Klimova, A., Kuligina, M., & Ivanov, S. (2021). Machine learning for predicting fluoroquinolone resistance in Gram-negative bacteria: A systematic review. *Antibiotics, 10*(7), 844.

Kuang, X., Wang, F., Hernandez, K. M., Zhang, Z., & Grossman, R. L. (2022). Accurate and rapid prediction of tuberculosis drug resistance from genome sequence data using traditional machine learning algorithms and CNN. *Scientific Reports, 12*, 2427. doi: 10.1038/s41598-022-06449-4

Li, J., Fang, W., & Wei, W. (2020). Metagenomics and machine learning for analyzing antibiotic resistance genes in environmental samples. *Environmental Science & Technology, 54*(22), 14758–14769.

Li, X., Wang, Y., & Chen, L. (2022). Integrating genomic and phenotypic data for improved fluoroquinolone resistance prediction using ensemble machine learning. *Journal of Antimicrobial Chemotherapy, 77*(8), 2139–2147.

Malhotra, S., Gupta, R., & Singh, A. (2022). AI-powered surveillance platforms for hospital-acquired infections and antimicrobial resistance. *Infection Control & Hospital Epidemiology, 43*(10), 1221–1227.

Nguyen, M., Chaudhari, S., & Chaudhari, R. (2020). Predicting antibiotic resistance using k-mer based genomic features and machine learning. *Scientific Reports, 10*(1), 10738.

Ogunjobi, T. T., Ohaeri, P. N., Akintola, O. T., Atanda, D. O., Orji, F. P., Adebayo, J. O., Abdul, S. O., Eji, C. A., Asebebe, A. B., Shodipe, O. O., & Adedeji, O. O. (2024). Bioinformatics applications in chronic diseases: A comprehensive review of genomic, transcriptomics, proteomic, metabolomics, and machine learning approaches. *Medinformatics, 52*, 371–381.

Oyedeji, O. A. (2025). Investigating the impact of antimicrobial resistance on global health and combating strategies. *International Journal of Medical Evaluation and Physical Report, 9*(3). DOI: 10.56201/ijmepr.v9.no3.2025.pg90.114

Pati, S., Choudhury, D., & Panda, R. (2021). Machine learning approaches for minimum inhibitory concentration prediction: A review. *Computers in Biology and Medicine, 138*, 104919.

Peterson, E., & Kaur, P. (2018). Antibiotic resistance mechanisms in bacteria: Relationships between resistance determinants of antibiotic producers, environmental bacteria, and clinical pathogens. *Frontiers in Microbiology, 9*, 2928. doi: 10.3389/fmicb.2018.02928

Phelan, J. E., O'Toole, R., & Warren, R. M. (2019). Whole-genome sequencing for tuberculosis drug resistance: The role of machine learning. *Clinical Microbiology and Infection, 25*(6), 661–666.

Prestinaci, F., Pezzotti, P., & Pantosti, A. (2015). Antimicrobial resistance: A global multifaceted phenomenon. *Pathogens and Globle Health, 109*(7), 309–318. doi: 10.1179/2047773215Y.0000000030

Price, J. R., Harris, R., & Hope, R. (2020). Using machine learning to predict antimicrobial resistance in hospital settings. *Clinical Infectious Diseases, 71*(Supplement_4), S333–S340.

Prusa, J. D., & Khoshgoftaar, T. M. (2017). Improving deep neural network design with new text data representations. *Journal of Big Data, 4*(1), 7.

Rabbani, M., Li, J., & Tang, J. (2021). Multi-omics integration for drug toxicity prediction using deep learning. *Bioinformatics, 37*(17), 2631–2638.

Rajput, P., Nahar, K. S., & Rahman, K. M. (2024). Evaluation of antibiotic resistance mechanisms in Gram-positive bacteria. *Antibiotics, 13*(12), 1197. https://doi.org/10.3390/antibiotics13121197

Ren, Y., Zhao, Y., Xu, J., Zeng, X., Wu, Y., Ma, Z., & Chen, X. (2021). Prediction of antimicrobial resistance based on whole-genome sequencing and machine learning. *Bioinformatics, 38*(2), 325–333. https://doi.org/10.1093/bioinformatics/btab681

Reygaert, W. C. (2018). An overview of the antimicrobial resistance mechanisms of bacteria. *AIMS Microbiology, 4*, 482–501. doi: 10.3934/microbiol.2018.3.482

Røder, H. L., Sørensen, S. J., & Burmølle, M. (2016). Studying bacterial multispecies biofilms: Where to start? *Trends in Microbiology, 24*, 503–513. doi: 10.1016/j.tim.2016.02.019

Sagar, A., Kolluru, V., Jaiswal, U., Kumavat, G., Hole, S. R., & Kumar, A. (2025). Machine Learning and Artificial Intelligence for Predictive Modeling in Antimicrobial Resistance Data Sets, Challenges, and Future Directions. In *2025 3rd International Conference on Smart Systems for applications in Electrical Sciences (ICSSES)* (pp. 1–6). IEEE.

Sakagianni, A., Koufopoulou, C., Koufopoulos, P., Kalantzi, S., Theodorakis, N., Nikolaou, M., Paxinou, E., Kalles, D., Verykios, V. S., Myrianthefs, P., & Feretzakis, G. (2024). Data-driven approaches in antimicrobial resistance: Machine learning solutions. *Antibiotics, 13*(11), 1052.

Sakagianni, A., Koufopoulou, C., Feretzakis, G., Kalles, D., Verykios, V. S., & Myrianthefs, P. (2023). Using machine learning to predict antimicrobial resistance: A literature review. *Antibiotics (Basel)*, *12*(3), 452. https://doi.org/10.3390/antibiotics12030452

Salam, M. A., Al-Amin, M. Y., Salam, M. T., Pawar, J. S., Akhter, N., Rabaan, A. A., & Alqumber, M. A. (2023). Antimicrobial resistance: A growing serious threat for global public health. *Healthcare*, *11*(13), 1946.

Stokes, J. M., Lopatkin, A. J., & Collins, J. J. (2020). Genome-wide insights into mechanisms of antibiotic resistance. *Nature Reviews Genetics*, *21*(9), 567–581.

Stokes, J. M., Yang, K., Swanson, K., Jin, W., Cubillos-Ruiz, A., Donghia, N. M., MacNair, C. R., French, S., Carfrae, L. A., Bloom-Ackermann, Z., Tran, V. M., Chiappino-Pepe, A., Badran, A. H., Andrews, I. W., Chory, E. J., Church, G. M., Brown, E. D., Jaakkola, T. S., Barzilay, R., & Collins, J. J. (2020). A deep learning approach to antibiotic discovery. *Cell*, *180*(4), 688–702.e13. https://doi.org/10.1016/j.cell.2020.01.021

Subramanian, I., Verma, S., Kumar, S., Jere, A., & Anamika, K. (2020). Multi-omics data integration, interpretation, and its application. *Bioinformatics and Biology Insights*, *14*, 1–24. https://doi.org/10.1177/1177932219899051

Syam, N., & Kaul, R. (2021). Random forest, bagging, and boosting of decision trees. In *Machine Learning and Artificial Intelligence in Marketing and Sales: Essential Reference for Practitioners and Data Scientists* (pp. 139–182). Emerald Publishing Limited.

Talat, A., & Khan, A. U. (2023). Artificial intelligence as a smart approach to develop antimicrobial drug molecules: A paradigm to combat drug-resistant infections. *Drug Discov. Today*, *28*, 103491. doi: 10.1016/j.drudis.2023.103491

Walker, T. M., Peto, T. E. A., & Crook, D. W. (2021). Genomic prediction of drug resistance in mycobacterium tuberculosis: A systematic review and meta-analysis. *Lancet Infectious Diseases*, *21*(4), 544–555.

Wang, X., Li, Z., Shen, Y., & Wu, J. (2022). Accurate and rapid prediction of tuberculosis drug resistance from genome sequence data using traditional machine learning algorithms and CNN. *Scientific Reports*, *12*, 6449. https://doi.org/10.1038/s41598-022-06449-4

World Health Organization WHO (2022). *Infection prevention and control.* Available at: https://www.who.int/teams/integrated-health-services/infection-prevention-World Health Organization control (accessed on 24 July 2024).

World Health Organization WHO (2023). *Food and Agriculture Organization of the United Nations.* Geneva: United Nations Environment Programme and World Organisation for Animal Health.

Yang, H., Liu, S., & Zhang, H. (2021). Machine learning for predicting antibiotic resistance genes in wastewater treatment plants. *Water Research*, *205*, 117709.

Yang, Y., Chen, J., & Xu, Y. (2022). Deep learning for predicting minimum inhibitory concentrations of antibiotics. *Bioinformatics*, *38*(5), 1184–1191.

Yang, Y., Zhou, X., & Chen, L. (2020). Machine learning models for predicting drug resistance in mycobacterium tuberculosis. *Emerging Microbes & Infections*, *9*(1), 1251–1259.

Zhang, D., Liu, Z., Wang, J., Xu, L., & Wang, H. (2025). Genomic insights into nontyphoidal salmonella: Prediction of antimicrobial resistance with whole genome-based machine learning. *Journal of Antimicrobial Chemotherapy*, *66*(5), 107575.

Zhao, Y., Xu, C., & Zhang, Y. (2020). Pan-genome based machine learning for cross-species antibiotic resistance prediction. *Bioinformatics*, *36*(24), 5585–5593.

Zhou, X., Li, X., & Yang, Y. (2022). Deep learning for predicting multi-drug and extensively drug-resistant tuberculosis from whole-genome sequencing data. *Frontiers in Microbiology*, *13*, 848987.

Zhu, Y., Johnson, T. A., Su, J., Qiao, M., Guo, G.-X., Stedtfeld, R. D., Hashsham, S. A., & Tiedje, J. M. (2013). Diverse and abundant antibiotic resistance genes in Chinese swine farms. *Proceedings of the National Academy of Sciences of the United States of America*, *110*, 3435–3440.

Zhu, Y., Li, J., & Wang, Q. (2021). Proteomic approaches for understanding and predicting antimicrobial resistance. *Journal of Proteome Research*, *20*(1), 324–335.

Dhungel, B., Thapa Shrestha, U., Adhikari, S., Adhikari, N., Bhattarai, A., Pokharel, S., Karkey, A., Limmathurotsakul, D., Ghimire, P., Rijal, K.R. and Cheah, P.Y. (2023). Use of antimicrobials during the COVID-19 pandemic: a qualitative study among stakeholders in Nepal. *PLOS Global Public Health*, *3*(11), p.e0002166

Applications of AI in Water and Wastewater Treatment

12

Sesan Abiodun Aransiola, Aishat Olorunkemi Salisu, Olamide Lasisi, Amos Musa Sumakada, Bolaji Oluwatosin Akanbi, Juliana Okwena Pondei, Olabisi Peter Abioye, and Naga Raju Maddela

12.1 INTRODUCTION

Water resources are essential for industrial processes and human life, and they are increasingly threatened by pollutants arising from natural events and human activities (Ray *et al.*, 2023). Every day, large quantities of wastewater are generated, most of which enclose harmful substances and are often discharged into the environment without adequate treatment. Typically, untreated sewage contains various nutrients, suspended solids (SSs), organic matters, organic micro-pollutants, and pathogenic and non-pathogenic microorganisms (Aransiola *et al.*, 2025; Chahal *et al.*, 2016). Wastewater can be grouped into six types: industrial, nuclear, agricultural, medical, domestic, and municipal. Among all these, domestic and municipal wastewaters are the most common, and the research on wastewater treatment (WWT) has primarily focused on these two categories (Jasim, 2020). Wastewater from different sources can vary greatly in their physical and chemical properties. Before selecting an appropriate treatment method, it is necessary to assess their characteristics. In order to safeguard human health, protect scarce water resources, preserve environmental quality, and address the rising demand for water. It is essential to explore WWT and promote its reuse as a resource. Physical, chemical, and biological methods of wastewater treatment remove pollutants from sewage (Abioye *et al.*, 2022; Lorena *et al.*, 2025; Zaibel *et al.*, 2022), and clean water is produced that can be safely discharged back into the environment. The main challenges in WWT are ways to effectively lower water pollution to a harmless level while minimizing environmental impacts and lowering energy consumption (Adelaja *et al.*, 2024). As a required process for water resources reuse and sustainable development, wastewater treatment plays a crucial role in protecting the environment and public health and has been extensively applied across agricultural and industrial sectors. Furthermore, to cut energy demand, enhance efficiency, and reduce costs, the development of effective and innovative technologies is required (Ma *et al.*, 2022).

DOI: 10.1201/9781003727514-14

Artificial intelligence (AI) is the capability of computer programs to achieve autonomous learning, judgment, reasoning, and decision-making by stimulating human intelligence. It has developed rapidly and found widespread applications in diverse fields, including computer vision (CV), autopilot, and natural language processing (NLP). Due to its elevated efficiency, AI can be applied to classification and regression analysis of vast amounts of data produced at any time and place. Thereby advancing the development of various sectors of society and driving industries forward. With the rapid progress in computer technology, machine language, which is a key branch of AI, leverages data, statistics, mathematical optimization, and algorithms to mimic human learning, steadily enhancing its accuracy and realizing intelligent performance. The rise of the big data era, combined with powerful supercomputing capabilities, has accelerated the popularity of machine language, enabling its successful application across industry, environmental protection, medicine, scientific research, agriculture, and many other areas (Malviya and Jaspal, 2021).

Water pollution is the tainting of water bodies (such as oceans, lakes, rivers, groundwater, and streams) with substances that make the water harmful to humans, animals, plants, or the environment. It can be caused by vital and essential human activities, both industrial and domestic; it mainly starts with alterations in energy level, radiation, and undesirable physical, chemical, and biological changes within the biosphere, which is the environment supporting all living organisms. These alterations can disrupt the ecological balance through water, food, air, and various agricultural products directly or indirectly. The pollutants in the environment are diverse in type and origin, with varying effects and meaning (Rashid, 2017). Water pollution can also occur when oxygen-depleting organic matter and solid wastes are discharged into water bodies that are majorly from industrial effluents and untreated urban sewage. Minute pollutants, including toxic chemicals and metals such as zinc, calcium, mercury, and lead, are exposed. Types of water pollutants include:

Physical pollution – It occurs when organic and inorganic materials are suspended in water. Such pollutants can alter the taste, odor, and color of the water. One example of physical pollution is increased temperature, which arises when factories and nuclear reactors release heat into water bodies. This increase in temperature reduces the level of dissolved oxygen and causes harm to aquatic life (Ewaid and Abed, 2017).

Chemical pollution – Occurrence of excessive amounts of acids, metals, dissolved salts, fertilizers, and pesticides can lead to this type of pollution. The metals are soluble in water, including toxic ones like mercury, lead, barium, and cadmium. Non-toxic metals such as magnesium, calcium, sodium, copper, and iron can also cause health issues when present in high concentrations; for example, excess sodium makes water unpleasant to drink, increases the risk of heart and kidney problems, and harms plants. Most organic substances dissolve in water and can be either dissolve by microorganisms or remain as non-degradable pollutants such as detergents or pesticides. Fertilizers, which contain phosphorus and nitrogen, encourage uncontrolled growth of aquatic plants. This process can lead to early maturation of lakes, which may eventually change into dry land or swamps (Ewaid and Abed, 2017).

Biological pollution – Biological contamination involves harmful pollutants, which include viruses, pathogenic bacteria, and parasites. These mainly come from human and animal excreta, which enter water through agricultural runoff or sewage, causing infection in humans, including diseases like cholera. To ensure safe drinking water, disinfectants such as chlorine can be used to abrogate these biological pollutants (Ewaid and Abed, 2017).

Radioactive pollution – This type of pollution occurs primarily from the disposal of nuclear waste and nuclear activities. Radioactive substances can seep into water bodies and be taken in by living organisms and pass directly to humans causing various effects on genetics (Ewaid and Abed, 2017).

The WWT process involves water quality monitoring, laboratory-scale studies, and process design. In recent years, AI models have gained significant attention in wastewater-related research, where they have been introduced to predict and optimize treatment processes (Safeer *et al.*, 2022). Some previous articles have demonstrated the strong predictive and optimization capabilities of AI models (Bhardwaj *et al.*, 2022), showing successful applications in process design, wastewater treatment process parameters optimization, and water quality monitoring (Wang *et al.*, 2022). This chapter highlights AI as a powerful tool with substantial achievements in WWT applications. Moreover, most existing AI reviews in WWT concentrate on specific techniques or process aspects, such as membrane processes, wastewater treatment plants (WWTPs), membrane bioreactors, and adsorption.

12.2 OVERVIEW OF ARTIFICIAL INTELLIGENCE

AI involves the ability of computational systems to execute tasks strictly associated with human intelligence, such as learning, reasoning, problem-solving, perception, and decision-making. The key drivers in AI are machine learning (ML) and deep learning (DL), as shown in Figure 12.1 (Capodaglio and Callegari, 2025).

According to Lowe *et al.* (2022), AI refers to the notion of furnishing algorithms with the ability to execute tasks and make conclusions an intelligent human would make in the same position.

12.3 HISTORY

The use of AI dates back as far as 1940s, when AI "precursors", which were the artificial neural networks (ANNs) and expert systems (ES), were experimented with, and the limitations were excessive cost and deficient availability of computing capacity. Backpropagation neural networks (BNN) were

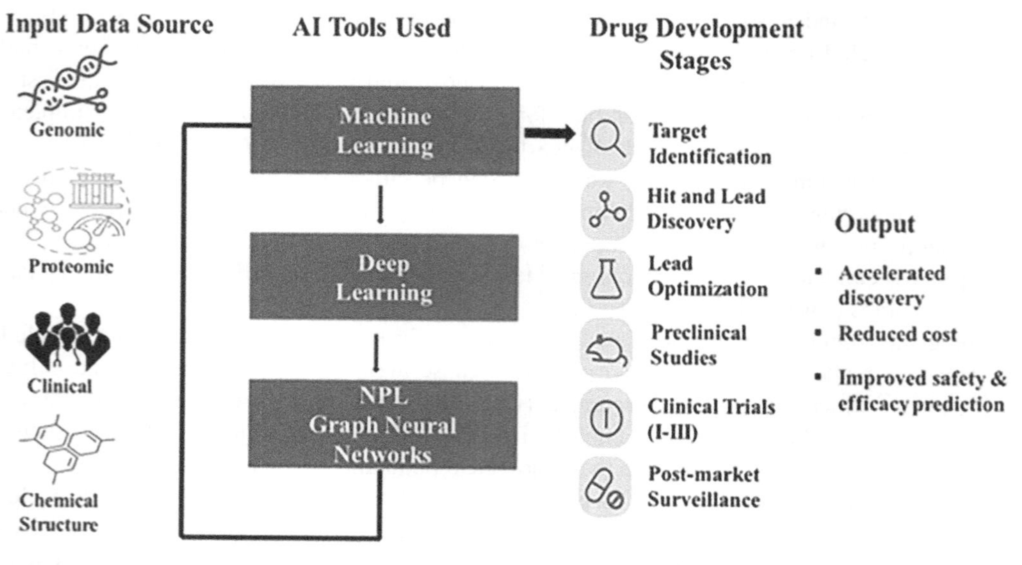

FIGURE 12.1 Representation of artificial intelligence drivers.

developed in the 1980s while there was a rapid improvement in the computing hardware's capabilities. DL was developed in the early 2000s, and this elevated the AI application scenario. The AI models mentioned, such as ANN, ES, genetic algorithms (GA), and fuzzy logic (FL), are capable of forecasting results, configuring complex data mapping, and solving difficult problems. The applications of AI are ideal in the cases of complex or several interconnected operations with highly variable factors influencing the overall systems.

There are two foundations of AI: graphic processing units and big data.

- Graphic processing units (GPUs), first produced in the 1990s, are electronic circuits (known as chips) that are able to use massive parallel processing to perform mathematical calculations at high speed. They are able to elaborate on a day's result that would take a CPU days, weeks, or months to elaborate. It is invariably one of the essential foundations of AI. A setback faced by this massive processing is its enormous demand for resources. Presently, it consumes 1–2% of the total generated power.
- Big data is also a foundation or requirement for AI applications. The concept of big data encompasses the technology of storing, processing, and analyzing intricate datasets, which does not allow canonical processing. The characteristics of these data are large volume of information and high input speed, greater than the established processing speed. Big data is associated with Internet of Things (IoT) in the production processes, and this is defined as a network of physical objects embedded with sensors, network connectivity, and software that allow the accumulation, storage, and sharing of a continuous flow of data. IoT devices allow the collection of real-life data from WWTPs in the water sector (Capodaglio and Callegari, 2025).

12.4 DRIVERS OF ARTIFICIAL INTELLIGENCE

12.4.1 Machine Learning (ML)

An intelligent system that can adapt its behavior to newly provided information during a system-training stage can be related to ML (Lowe *et al.*, 2022). This process also enables computing machines to make decisions about real event and also learn and analyze big data. ML algorithms analyze enormous datasets generated by IoT devices to recognize and differentiate patterns, make predictions, and transmute raw data into effective insights (Capodaglio and Callegari, 2025). In addition, machine-learning algorithms with a specific target expand the learning process of computing machines in an exponential manner (Krishnan *et al.*, 2022).

12.4.2 Deep Learning (DL)

This uses multilayered deep neural networks to replicate the decision-making power of the human brain and is a subset of ML. It is also a part of the powering factors for AI applications (Capodaglio and Callegari, 2025).

The use of traditional computer simulations enables researchers to recapitulate analytical methods and physical laws for the benefit of computers, whereas in AI, the approach is inspired by human babies, in which no one teaches them how to recognize objects; they learn on their own. In addition, discovering the statistical structure of data and establishing rules for automating a task are achieved using AI rather than being programmed explicitly by humans, which is the traditional way (Ojadi *et al.*, 2025).

The use of an AI model has enabled the computer to learn characteristics of objects, which can be text, sound, data, or images, through appropriate training models (ANNs) and can accurately recognize them in other circumstances. The use of the models (ANNs and DNNs) is declined in many variations depending on the structure and algorithms used (Capodaglio and Callegari, 2025). AI-driven frameworks, in order to oversee and enhance water consumption patterns, detect leakages, and improve WWT processes, integrate ML algorithms, real-time data analytics, and IoT sensors (Ojadi *et al.*, 2025).

12.5 WATER AND WASTEWATER TREATMENT (WWT)

Access to fresh and clean drinking water is one of the greatest challenges of our time, forming the central pillar of the United Nations Sustainable Development Goals (SDGs) (Alenezi, 2025; Chowdhury, 2024). One of the most important environmental problems arises from the pollution of rivers and streams with chemical contaminants. This causes a great deal of destruction to the water bodies (flora and fauna), living things dependent on such water. There are water pollution that are from natural causes, that is, human activities. This is caused by the daily use of water. The used water returns to the source after usage (groundwater). The used water is termed wastewater. If discharged into waterways, it can cause serious pollution (Sonune and Ghate, 2004). Howbeit, water pollution continues to increase due to rapid urbanization, industrialization, and population growth, leading to unprecedented crisis in water resources (Alenezi, 2025; Chowdhury, 2024). Each year, about 300–400 million tons of pollution and contaminants are released into water ecosystems, threatening agricultural productivity, biodiversity, and human health (Mousazadehgavan, 2025). There are other sources of water contamination, which are industry, mines, and infiltration. Water is categorized into four types, which are rainwater, domestic wastewater, agricultural water, and industrial wastewater (Crini and Lichtfouse, 2019). According to Sonune and Ghate (2004), wastewater can be defined as the fusion of liquid waste removed from industries, institutions, and residences together with groundwater, surface water, and runoffs. It contains pathogenic agents, organic materials, minerals, and sediments that may be toxic to living organisms. Industrial wastewater can be further subdivided into cooling water, washing effluent, and manufacturing or process water. Industrial water is the most dangerous as it is toxic. Wastewater differs from drinking water due to the volume of contaminants in wastewater and the composition of wastewater. This chapter depicts how water and wastewater can be treated to be consumable and less harmful to health with AI.

There are conventional ways of WWT, namely

- Preliminary treatment: This involves the removal of solid and large materials, which consist of wood, paper, plastics, etc.
- Primary treatment: This involves the removal of organic and inorganic solids/materials, which consist of oil, grease, heavy metals, biological oxygen demand (BOD), and SSs by sedimentation and flotation.
- Secondary treatment: This involves further treatment of the wastewater by introducing biological agents (microorganisms) in a controlled environment (WWTPs).

The method to be used for treatment is determined by the type of effluent or waste. Each treatment has its advantages and disadvantages (Crini and Lichtfouse, 2019).

12.6 DEVELOPMENTS IN ADVANCED WASTEWATER TREATMENT METHODS

BOD and SSs are majorly removed by the primary and secondary treatment in wastewater. However, this technique has proven to be insufficient in providing reusable water for industrial and domestic recycling, hence the need for advanced techniques. The types of advanced WWT are:

- Tertiary treatment: Here operation units are incorporated into the flow scheme. This involves simple (filter for solids removal) or complex (unit processes for nitrogen or phosphorus removal) additions to the secondary treatment to enhance the process.
- Physiochemical treatment: This is a mixture of biological and physicochemical treatment processes to get suitable outflow.
- Combined biological-physical treatment: This is when biological and physiochemical treatments are combined (Sonune and Ghate, 2004).

The use of AI is another way to enhance WWT. It is a promising tool for the optimization of WWT systems. The WWT processes are enhanced using sophisticated algorithms, ML, and data analytics. The applications consist of various benefits, which are control optimization, real-time monitoring, predictive maintenance, decision support tools, and fault detection. Inasmuch as the applications are diverse, there are setbacks or challenges associated with the utilization of such techniques (Dalibard *et al.*, 2025).

12.7 AI MODELS USED IN WATER AND WASTEWATER TREATMENT (WWT)

According to Alprol *et al.* (2024), there are five different models of AI as depicted in Figure 12.2, which are:

- Recurrent Neural Network (RNN): This operates by repeating a sequence evolution and taking the sequence data as input. It has memory, parameter sharing, and turning results that make it advantageous. The most commonly used RNN for solving gradient disappearance is long short-term memory (LSTM).
- Artificial Neural Network (ANN): Mathematical models that replicate the conduct of biological neural networks to process data are termed ANN. It comprises an input layer, an output layer, and multiple hidden layers between them. They perform complex nonlinear computations.
- Fuzzy Neural Network (FNN): The combination of FL and ANNs produces a hybrid neural network model that handles ambiguous or uncertain situations, termed FNN. FL reasoning is utilized to develop input data, and an ANN is employed to train and produce results by FNN.

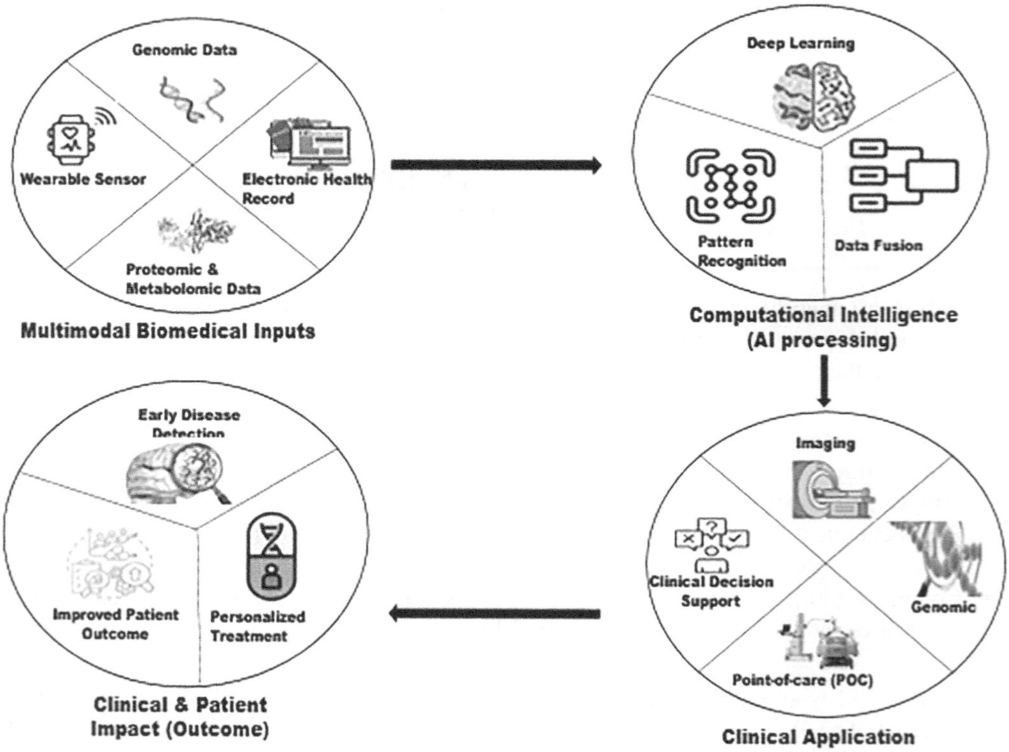

FIGURE 12.2 Artificial intelligence models used in water and wastewater treatment.

- Deep Neural Network (DNN): This comprises multiple layers hidden between the output and input layers. The architectural feature of DNN allows it to learn level data representations by combining successive layers that characterize lower-level features to learn higher-level features.
- Convolutional Neural Network (CNN): Convolutional computations and a deep structure feedforward neural network are utilized under CNN. CNN uses are employed in computer vision, natural language development, and other fields. It is a DL technique.

12.8 WASTEWATER TREATMENT PLANTS (WWTPs)

Nourani *et al.* (2018) defined WWTPs as equipment used to treat wastewater, which allows anthropogenic activities and industrial outflow to be disposed of without danger to human health or harmful damage to the natural environment. The proper operation and control of this highly dynamic system is paramount in order to ensure the safeguarding of the public from environmental health issues. The operations of WWTP have to be optimized, for example, the measurement of dominant variables such as biological oxygen demand (BOD), chemical oxygen demand (COD), total SSs, etc., and this can be made possible through the use of AI. The WWTP systems are dependent on environmental and operational conditions; they can also be cyclic in nature (recycling system) (Seshan *et al.*, 2023). In addition, the generation of useful resources of water from wastewater can be recovered through AI (Hernández-Chover *et al.*, 2024).

12.9 APPLICATIONS OF ARTIFICIAL INTELLIGENCE IN WATER AND WASTEWATER TREATMENT

As stated above, WWTPs are dynamic and highly sophisticated systems that require critical optimization. The use of AI in water and WWTPs enhances the performance of the equipment in some areas, such as:

- System Simulation: ANN-particle swarm optimization (ANN-PSO) and ANN-genetic algorithm (ANN-GA) are two hybrid approaches that have been developed to enhance predictive accuracy in adsorption processes. These processes involve nonlinear interactions between multiple input variables, including adsorbent dose, pH, and contaminant concentration (Liu *et al.*, 2025). These models show flexibility and adaptability across a range of treatment scenarios and have been successfully used to remove dyes, heavy metals, medications, and pesticides (Chowdhury, 2024). ANNs are able to reproduce human brain interactions to execute parallel complex computations. The replication of WWTP unit/process behavior and evaluation of the overall performance are made possible by using ANN. Some predictive modeling, compared to the mechanistic modeling, uses ANN models, which have depicted better forecasting abilities, like N_2O production in a nitrification plant, where ANNs were effective in predicting effluent membranes' transmembrane pressure (MTP) and COD pressure. ML algorithms are used to enhance the powerful performance of the WWTPs (Capodaglio and Callegari, 2025). In addition, AI and ML play a significant role in optimizing the water and WWT stages and processes, which leads to efficient resource utilization (Rane *et al.*, 2023).
- Real-time Monitoring and Control: When teamed with IoT sensors, AI is then manually enabled to constantly monitor water kindness elements such as dissolved oxygen, turbidity, microbial contamination, and pH. These systems may also include automatic adjustment of water treatment parameters (i.e., coagulant dosage), anomaly detection, membrane cleaning frequency, and anticipation of failures (Alenezi, 2025) in order to remain in compliance with required regulations. These AI models have been in use in the treatment plants of consumable water for coagulation, optimization of chlorination dosing, and chemical demand reduction, which ultimately reduces DBPs according to Narayanan *et al.*, (2024). Along these lines, CIB-based AI systems, such as the supervisory control and data acquisition (SCADA), amplify reliance even further by autonomously responding to minor changes in system quality. AI and ML are applied in WWT for real-time monitoring and control. Periodic sampling and laboratory activities, being traditional methods of platform devices such as Raspberry Pi to continuously monitor various parameters such as dissolved oxygen, nutrient level, and pH in real time (Capodaglio and Callegari, 2025; Rane *et al.*, 2023), have outperformed monitoring the process. The use of bioelectrochemical sensors has made it possible to detect the development of microbial community, product, or substrate production using biological materials such as enzymes, antibodies, or bacterial cells as receptors (Capodaglio and Callegari, 2025).

- Anomaly and Fault Detection in WWTPs: Increased operational costs and suboptimal performance are the results of anomalies in WWT processes (Rane *et al.*, 2023). Opportunities for fault diagnosis, management, and decision-making in urban WWTP operations could be provided through advances in AI and data analytics. SCADA, DNN, and wireless sensor networks (WSN) are AI models used to detect faults and anomalies in WWTPs (Capodaglio and Callegari, 2025).
- Digital Twins and Predictive Maintenance: Among the most promising solutions of digital twins and predictive maintenance is the area of AI-driven digital twins, virtual models of physical water treatment systems that mimic system performance in real time. With these tools, it is possible for operators to experience operational practice, predict equipment performance, and optimize the use of energy and resources before any alteration to physical plant (Wang *et, al.*, 2024). Case studies indicate that AI-enabled digital twins save up to 50% energy use in water and WWTPs to secure compliance with good quality, as Mousazadehgavan (2025) noted.
- Water Quality Forecasting and Environmental Applications: AI has been used to forecast the quality of water in rivers and lakes to spot any sign of contamination before they spread. XGBoost models have been able to nail predictions of key water quality measures like total nitrogen and COD, even in the tangled flow of complex treatment systems (Chowdhury, 2024). Even outside of water treatment plants, AI now forecasts harmful algal blooms, tracks river systems, and gauges how farming activities affect water quality, which helps guide the wider ecosystem management, from clear mountain streams to coastal wetlands (Smart Water Magazine, 2025).

In a study by Dikmen *et al.* (2025), the research carried out on AI-driven wastewater management through comparative analysis of feature selection techniques and predictive models revealed that effluent volatile suspended solids (VSS) held the highest predictive significance across all feature selection methods, and the presence of VSS indicates microbial activity in the treatment system. Other parameters were predicted in this research, such as BOD, total nitrogen, and total phosphate present in the wastewater. The findings in this research indicate the importance of AI-driven approaches to WWT processes.

12.10 ADVANTAGES OF AI IN WATER AND WASTEWATER TREATMENT

Using AI in the treatment of water has numerous advantages, as depicted in Table 12.1, which include detecting any tiny changes in the quality of drinking water and then adjusting it in real time. First, it boosts efficiency by allowing real-time monitoring, predictive control, and automated decisions like adjusting a conveyor belt the moment a sensor detects a jam cutting down the need for human intervention (Alenezi, 2025). Second, AI models can more precisely capture complex, nonlinear processes than traditional mathematical methods, which make them ideal for tasks like adsorption, membrane filtration, and disinfection, much like tracing the swirl of cream in coffee. Third, AI-powered digital twins make it easier to fine-tune designs, reduce costs, and help its resilience, according to Wang *et al.*, (2024). Fourth, AI and ML models enhance the energy efficacy of water and WWTPs due to the energy-intensive nature of water treatment processes. Innovative solutions are offered through these

TABLE 12.1 Advantages and disadvantages of artificial intelligence applications in water and wastewater treatment

	ADVANTAGES	*DISADVANTAGES OR LIMITATIONS*
Artificial Intelligence Applications in Water and Wastewater Treatment	Boosts treatment plants' efficiency (Alenezi, 2025) Capture complex processes Fine-tuning of plants' designs (Wang *et al.*, 2024) Energy efficacy Minimizing operational cost (Rane *et al.*, 2023) Detects contamination in water instantly (Liu *et al.*, 2025)	Expensive to manage and install (Mousazadehgavan, 2025) Unpredictable processes (ArXiv, 2025a) Omissions of processes when providing solutions (Narayanan *et al.*, 2024) Inability of the machines to accept the technologies (Rane *et al.*, 2023) Staff training is expensive (Chowdhury, 2024) Environmental impact of power sources (Wikipedia, 2025)

technologies to optimize the process, minimize operational cost, and reduce environmental impact of the treatment process (Rane *et al.*, 2023).

Finally, AI helps to preserve and safeguard public health by detecting and spotting contamination instantly, for example, detecting a spike in harmful bacteria that is in the drinking water supply, and it will send a real-time alert immediately (Liu *et al.*, 2025).

12.11 DISADVANTAGES AND LIMITATIONS OF AI IN WATER AND WASTEWATER TREATMENT

AI is a promising tool for the treatment of water and wastewater, but putting it to work is a big challenge. The hurdles include complex regulations, aging pipes, and tight budgets. AI's biggest problem is its total dependence on data, which demands huge sum of money for training so as to get reliable, clean, and verified information and data to work with (Mousazadehgavan, 2025). Second, some often criticize AI models, especially DL networks, describing them as "black boxes". They claim that they hide information on how they work, making them unpredictable and, as such, unsafe, which in turn weakens accountability, slows regulatory approval, and erodes trust (ArXiv, 2025a). Also harmonizing and standardizing the data generated from various sensors and devices in wastewater systems is a challenge in the implementation of AI (Rane *et al.*, 2023). Third, neural networks can be tricky to reproduce because they start with random initialization, sometimes landing on a solution that works well in one small corner but misses the best answer altogether (Narayanan *et al.*, 2024). Also, existing WWT facilities with fixed capacities are sometimes unable to accept scalability requirements of the advanced technologies (Rane *et al.*, 2023). Fourth, the price of putting the system in place, along with the need for strong computing power, sensors, and trained staff, still puts it out of reach for many small utilities (Chowdhury, 2024). AI systems are facing growing cybersecurity risks, because adversarial attacks can slip in altered inputs like a single corrupted sensor reading and throw off digital twin predictions, disrupting operations (ArXiv, 2025b). Finally, AI's environmental toll from the heavy power it burns to the discarded circuit boards piling up raises serious concerns about long-term sustainability (Wikipedia, 2025).

12.12 CHALLENGES IN AI APPLICATIONS IN WATER AND WASTEWATER TREATMENT PLANTS

Technologies such as AI, ML, IoT can transmogrify the WWT systems, hence leading to ameliorated efficiency, better optimization, and above all, reduction in environmental impact. However, the assimilation of these technologies comes with its setbacks.

- Data Limitations – Quantity, Quality, and Representativeness: One of the biggest problems in using AI for the treatment of water and wastewater is the lack of solid and representative data; sometimes, it is just a handful of data available. Many water treatment plants operate with only a handful of sensors, which collect data at uneven intervals and keep scanty records of past contamination events. These challenges model accuracy, and they also weaken its ability to generalize and raise the risk of overfitting operating conditions (Pérez-Beltrán, 2024). These gaps make it hard to build ML models in this field. This is in agreement with broader findings of other researchers that ML models for water quality often falter when the training data is narrowly focused, like testing a river's health from just a few drops as sample; this greatly reduces how well they can predict outcomes (Pérez-Beltrán, 2024).
- Heterogeneity and Transferability of Models across Sites: AI models built in one plant often fail when used elsewhere because they are tripped up by different shifting process setups, sensors, mismatched data formats, and the quirks of each facility's proprietary SCADA. Thin documentation and sparse metadata make it even harder to transfer the model (UNESCO, 2024). Working in silos slows the process down, making it hard to roll out AI solutions across the board just like trying to push water through a narrow pipe. Many reviews support this limitation: a model trained on local hydrology or water quality data often stumbles when it's dropped into a different climate or run under new operating systems, like shifting from a humid delta to an arid plateau (Gacu *et al.*, 2025).
- Nonstationarity and the Need for Continuous Adaptation: Water sources are constantly running; this sometimes changes their source quality, which weakens the behavioral control of the water and consequently gives it new bad behavior (Galindo and Wertheimer, 2025).
- Detecting Rare, High-Impact Events: Supervised AI can miss rare but critical events like a hidden fault in a machine or a sudden chemical spike in the air because there are not enough labeled examples to learn from. What make matters worse is the strict privacy rules and regulations that keep utilities from sharing sensitive event data, like blocking the creation of broad outage logs and reliable models (Pérez-Beltrán, 2024; UNESCO, 2024). Worries about trustworthiness like interpretability, robustness, and fairness have not been fully addressed, especially in models built to predict the quality of drinking water for large communities, where a single wrong reading can mislead an entire region (Xia *et al.*, 2025).
- Institutional, Ethical, and Governance Barriers: Across the globe, utility crews do not have the trusted systems they need to bring AI tools on board, according to UNESCO (2024). Multidisciplinary review sheds light on these socio-technical dimensions, which encourage the use of fair algorithms, explainable AI, and solid governance frameworks to guide the adoption of AI (Gacu *et al.*, 2025).

- Regulatory Compliance: Strict environmental regulations govern WWT, and implementing new technologies will require meticulous consideration of compliance issues and adhering to existing standards (Rane *et al.*, 2023).

12.13 FUTURE ADVANCEMENTS IN ARTIFICIAL INTELLIGENCE IN WATER AND WASTEWATER TREATMENT

- Human-Machine Synergy: The connection of WWT operators to the AI and ML systems is a user-friendly process, as it will build a relationship between the machine and the operators for easy operation.
- Academic and Industry Collaboration: Skills gap is a challenge in the utilization of AI and ML systems. Bridging this gap and promoting further technology adoption will enhance the optimization of water and WWTPs.
- Renewable Energy sources Fusion: The fusion of renewable energy sources into the treatment plants is a future development. AI algorithms can be used to kick-start this process.
- Edge Computing: Positioning of data processing to a plant sensor reduces the real-time decision-making in WWT. The infusion of AI and ML algorithms into the edge device will heighten data processing efficiency.
- Smart Water Grids synergy: Incorporation of WWT systems with smart water grids influenced by IoT provides a comprehensive approach to water management. This fusion will promote minimized water loss, more water distribution, and boost water resource sustainability (Capodaglio and Callegari, 2025).

12.14 CONCLUSION

The introduction of AI and ML technologies has emerged as a driving force enhancing the WWT process. This chapter examines the applications of AI and ML, including their potential to strengthen the sustainability, efficiency, effectiveness, and reliability of water and WWT processes. The impact of these technologies is significant in optimizing the treatment operations using AI models and ML algorithms. AI algorithms and models identify patterns, detect anomalies and faults, facilitate predictive maintenance, and aid decision-making. Similarly, ML models enable continuous learning of the system and improve their performance, resulting in more efficient and effective operations of the WWTPs. The applications of AI-driven WWT processes enable significant optimization of the operations and come with some setbacks. Some of the challenges experienced in the application of AI and ML in water and WWTs are data integration and compatibility, skill gap and training, cost, and infrastructure constraints, to mention a few. Overcoming these hurdles will take a united effort, pouring huge resources into advanced sensor networks that can detect the faintest signal, building AI that is both transparent and dependable, and creating good governance frameworks that will support and keep innovation alive while holding decision-makers accountable. The optimization of WWT processes through AI-driven technologies is a huge progress in health and environmental sustainability; however, future advancements need to be integrated to revolutionize the process.

REFERENCES

Abioye, O.M., Adeniran, K.A., & Abadunmi, T. (2022). Poultry wastes effect on water quality of shallow wells of two farms in two locations of Kwara State, Nigeria. *Nature Environment and Pollution Technology*. Vol 21 (1), 1–6.

Adelaja, O.A., Babaniyi, B.R., Aransiola, S.A., Rodriguez-Diaz, J.M., Maddela, N.R., & Prasad, R. (2024). Biosorption of dyes in wastewater using Chitosan/polyethylene nanoparticles as adsorbent. *Discover catalysis*, 1 (1), 3.

Alenezi, A. (2025). Artificial intelligence applications in water treatment and desalination. *Water, 17*(8), 1169. https://doi.org/10.3390/w17081169

Alprol, A. E., Mansour, A. T., Ibrahim, M. E. E.-D., & Ashour, M. (2024). Artificial intelligence technologies revolutionizing wastewater treatment: Current trends and future prospective. *Water, 16*(314), 1–26. https://doi.org/10.3390/w16020314

Aransiola, S.A., Bhatnagar, S., & Maddela, N.R. (Eds.). (2025). *Marine Bioresources: Prospects and Obstacles. Elsevier.*

ArXiv. (2025a). Identifying trustworthiness challenges in deep learning models for continental-scale water quality prediction. *arXiv Preprint*. https://arxiv.org/abs/2503.09947

ArXiv. (2025b). The dark side of digital twins: Adversarial attacks on AI-driven water forecasting. *arXiv Preprint*. https://arxiv:2504.20295

Bhardwaj, A., Kishore, S., & Pandey, D. K. (2022). Artificial intelligence in biological sciences. *Life, 12*, 1430.

Capodaglio, A. G., & Callegari, A. (2025). Use, potential, needs, and limits of AI in wastewater treatment applications. *Water, 17*, 170. https://doi.org/10.3390/w17020170

Chahal, C., Van Den Akker, B., Young, F., Franco, C., Blackbeard, J., & Monis, P. (2016). Pathogen and particle associations in wastewater: Significance and implications for treatment and disinfection processes. *Advances in Applied. Microbiology, 97*, 63–119.

Chowdhury, S. (2024). Applications of artificial intelligence (AI) in drinking water treatment processes. *Chemosphere, 354*, 139912. https://doi.org/10.1016/j.chemosphere.2024.141958

Crini, G., & Lichtfouse, E. (2019). Advantages and disadvantages of techniques used for wastewater treatment. *Environmental Chemistry Letters, 17*, 145–155. https://doi.org/10.1007/s10311-018-0785-9

Dalibard, A., Kriem, S. L., Beckett, M., Scherle, S., Yeh, Y., & Schließmann, U. (2025). Challenges and requirements of AI-based waste water treatment systems. *At-Automatisierungstechnik, 73*(1), 40–49.

Dikmen, F., Demir, A., Özkaya, B., Raza, M. O., Rasheed, J., Asuroglu, T., & Alsubai, S. (2025). AI-driven wastewater management through comparative analysis of feature selection techniques and predictive models. *Scientific Reports, 15*, 25347. https://doi.org/10.1038/s41598-025-07124-0

Ewaid, S. H., & Abed, S. A. (2017). Water quality index for Al-Gharraf River, Southern Iraq. *Egyptian Journal of Aquatic Resolution, 43*(2), 117–122.

Gacu, J. G., Monjardin, C. E. F., Mangulabnan, R. G. T., Pugat, G. C. E., & Solmerin, J. G. (2025). Artificial intelligence (AI) in surface water management: A comprehensive review of methods, applications, and challenges. *Water, 17*(11), 1707. https://doi.org/10.3390/w17111707

Galindo, J. I., & Wertheimer, N. (2025, July 24). How AI is rewriting and enhancing water risk management. *World Economic Forum*. https://www.weforum.org/stories/2025/07/ai-water-risk-management/

Hernández-Chover, V., Bellver-Domingo, Á, Castellet Viciano, L., & Hernández-Sancho, F. (2024). AI applied to the circular economy: An approach in the wastewater sector. *Sustainability, 16*, 1365. https://doi.org/10.3390/su16041365

Jasim, N. A. (2020). The design for wastewater treatment plant (WWTP) with GPS modeling. *Cogent Engineering, 7*, 1723782.

Krishnan, S. R., Nallakaruppan, M. K., Chengoden, R., Koppu, S., Iyapparaja, M., Sadhasivam, J., & Sethuraman, S. (2022). Smart water resource management using artificial intelligence—A review. *Sustainability, 14*, 13384. https://doi.org/10.3390/su142013384

Liu, Y., Mancuso, G., Petrotto, L., Lavrnić, S., Dong, Z., Tian, Y., Zhang, J., & Toscano, A. (2025). AI-driven solutions in wastewater treatment and agricultural systems. *Journal of Environmental Management, 393*, 127008. https://doi.org/10.1016/j.jenvman.2025.127008

Lowe, M., Qin, R., & Mao, X. (2022). A review on machine learning, artificial intelligence, and smart technology in water treatment and monitoring. *Water, 14*, 1384. https://doi.org/10.3390/w14091384

Ma, Z., Cheah, W. Y., N.G, I. S., Chang, J. S., Zhao, M., & Show, P. L. (2022). Microalgae-based biotechnological sequestration of carbondioxide for net zero emissions. *Trends Biotechnology, 40*, 1439–1453.

Malviya, A., & Jaspal, D. (2021). Artificial intelligence as an upcoming technology in wastewater treatment: A comprehensive review. *Environmental Technology Reviews, 10*, 177–187.

Mousazadehgavan, M. (2025). The critical role of artificial intelligence in optimizing water treatment. *ACS ES&T Water.* Advance online publication. https://doi.org/10.1021/acsestwater.5c00238

Narayanan, D., Bhat, M., Paul, N.R.S., Khatri, N., & Saroliya, A. (2024). Artificial intelligence driven advances in wastewater treatment: Evaluating techniques for sustainability and efficacy in global facilities. *Desalination and Water Treatment, 320*, 100618. https://doi.org/10.1016/j.dwt.2024.100618

Nourani, V., Elkiran, G., & Abba, S. I. (2018). Wastewater treatment plant performance analysis using artificial intelligence – An ensemble approach. *Water Science & Technology, 78*(10), 2064–2076.

Ojadi, J. O., Owulade, O. A., Odionu, S. C., & Onukwulu, C. E. (2025). AI-driven optimization of water usage and waste management in smart cities for environmental sustainability. *Engineering and Technology Journal, 10*(3), 4284–4306. doi: 10.47191/etj/v10i03.36, I.F.– 8.482

Pérez-Beltrán, C. H. (2024). Artificial intelligence and water quality: From drinking water to wastewater. *ScienceDirect/Bohrium*. doi: 10.1016/j.trac.2024.117597

Rane, N., Choudhary, S., & Rane, J. (2023). Leading-edge Artificial Intelligence (AI), Machine Learning (ML), Blockchain, and Internet of Things (IoT) technologies for enhanced wastewater treatment systems. Pp. 1–24. http://dx.doi.org/10.2139/ssrn.4641557

Rashid, S. (2017). The effects of economic problems of environmental pollution in Iraq and ways to address them. *Al-Mustansiriya Journal for Arab and International Studies, 14*, 103.

Ray, S. S., Verma, R. K., Singh, A., Ganesapillai, M., & Kwon, Y. N. (2023). A holistic review on how artificial intelligence has redefined water treatment and seawater desalination processes. *Desalination, 546*, 116221.

Safeer, S., Pandey, R. P., Rehman, B., Safdar, T., Ahmad, I., Hasan, S. W., & Ullah, A. (2022). A review of artificial intelligence in water purification and wastewater treatment. Recent advancements. *Journal of Water Process Engineering, 20*, 1089–1109.

Seshan, S., Vries, D., van Duren, M., van der Helm, A., & Poinapen, J. (2023). AI-based validation of wastewater treatment plant sensor data using an open data exchange architecture. *IOP Conf. Series: Earth and Environmental Science, 1136*, 012055. doi: 10.1088/1755-1315/1136/1/012055

Smart Water Magazine. (2025, July 31). Engineering water systems with AI and digital twins. *Smart Water Magazine*. https://smartwatermagazine.com/news/engineering-water-systems-ai-and-digital-twins

Sonune, A., & Ghate, R. (2004). Developments in wastewater treatment methods. *Desalination, 167*, 55–63.

UNESCO. (2024–2025). *Applications of artificial intelligence for water management.* UNESCO Publications. https://unesco.org/en/articles/applications-artificial-intelligence-water-management

Wang, A. J, Li, H., He, Z., Tao, Y., Wang, H., Yang, M., Savic, D., Daigger, G.T., & Ren, N. (2024). Digital twins for wastewater treatment: A technical review. *Engineering, 36*, 21-35. DOI: 10.1016/j.eng.2024.04.012

Wang, G., Su, W., Hu, B., Al-Huqail, A., Majdi, H. S., Algethami, J. S., Jiang, Y., & Ali, H. E. (2022). Assessment in carbon-based layered double hydroxides for water and wastewater. Application of artificial intelligence and recent progress. *Chemosphere, 308*, 136303.

Wikipedia. (2025). Environmental impact of artificial intelligence. In *Wikipedia*. https://en.wikipedia.org/wiki/Environmental_impact_of_artificial_intelligence

Xia, X., Liu, X., Liu, J., Fang, K., Lu, L., Oymak, S., Currie, W. S., & Liu, T. (2025). Identifying trustworthiness challenges in deep learning models for continental-scale water quality prediction. *arXiv Preprint*. https:cell.com/nexus/pdfExtended/S2950-1601(25)00051-8

Zaibel, I., Arnon, S., & Zilberg, D. (2022). Treated municipal wastewater as a water source for sustainable aquaculture. A review. *Review of Aquaculture, 14*, 362–377.

Production of Biofuels from Biomass 13

Using Artificial Intelligence Modeling to Forecast Bioenergy

Abdulrazaq Izuafa, Josephine Nathaniel, Chiedu Epiphany Edoamodu, Tajudeen Olanrewaju Yahaya, Abdurrashid Usman Gulumbe, and Oluwafemi Adebayo Oyewole

13.1 INTRODUCTION

The modern global energy scene is under a severe crisis, with climate change, resource exhaustion, and the increased interest in issues of security, respectively, increasing the urgency of feasible alternatives to fossil fuels (Maddela *et al.*, 2024a). In answer, countries with climate goals and aiming at energy security are becoming inclined toward biomass as a tactical renewable resource that can be used to ease the shift toward a low-carbon future. Among the most versatile and ubiquitous sources of renewable energy, biomass can be sourced using agricultural residue, forest products, algae, energy crops, and so forth (Goel *et al.*, 2024; Varghese *et al.*, 2024). However, the biofuel industry faces a lot of uncertainty that hinders expansion and investment. The variability in the quality of feedstock, seasonality, process effectiveness, and market dynamics poses problems that the conventional forecasting tools are not adequately prepared to address. The barriers therefore result in the creation of economic threats, ineffectiveness in the use of resources, and a loss of the opportunity to develop the potential of bioenergy (Maddela *et al.*, 2024a). This is further complicated by the fact that producers have to balance a variety of variables, such as the type of feedstock, process conditions, and environmental and economic limitations (Luo and Ierapetritou, 2023).

Artificial intelligence (AI) has become a disruptive technology in the bioenergy sector, with advanced functionalities to manipulate large, complex, multidimensional data and identify emerging patterns that can be used to predict and optimize. The forecasting systems based on AI combine a wide array of data (feedstock properties, weather, market dynamics, and process variables) to provide

 DOI: 10.1201/9781003727514-15

real-time predictions and optimization programs that are superior to traditional solutions. Such systems complement predictive accuracy and support adaptive decision-making in dynamic operating environments (Madanchian, 2024; Sharma *et al.*, 2024). There are three main goals of this chapter. It discusses first the different methods of transforming biomass into biofuels based on traditional thermochemical methods and the newer biochemical and mixed methods based on biochemical methods. Second, it examines new AI modeling methods, which have a high level of prediction of bioenergy outputs. Third, it points out the existing uses of AI in biofuel production and assesses the future trends in a field that is quickly evolving (Santoso & Surya, 2024).

13.2 BIOMASS AS A RENEWABLE RESOURCE

Biomass is one of the richest and most varied renewable resources on Earth, which includes organic substances produced by plants, animals, and microorganisms, which can be converted into forms of utilizable energy. Biomass feedstocks can be classified into a number of different categories, each with its own specific characteristics and conversion capabilities. The most abundant terrestrial resource is lignocellulosic biomass, which consists of agricultural residues (corn stover, wheat straw, rice husks, and others). It has compound polymers of cellulose, hemicellulose, and lignin that require special processing methods. Algal biomass has inherent benefits such as faster growth rates, high lipid levels that can be used to make biodiesel, and the ability to grow in non-arable soils and in salty areas, making it specifically appealing to the generation of sustainable biofuels. The sources of biomass in waste, including municipal solid waste, sewage sludge, and industrial organic waste, offer two benefits, where waste management is not only a possible solution but also energy production in light of the circular economy (Phogat, 2025; Silva *et al.*, 2024; Varghese *et al.*, 2024).

The world has access to immense potential in the use of biomass to meet future energy needs, and it is estimated that 20–50% of the primary energy in 2050 world energy needs may comprise biomass. Recent production trends show that there is a consistent rise in using biomass to generate energy, and the global primary energy consumption is approaching 55 exajoules per year, or about 10% biomass energy use. These differences are strong regionally: developing economies are more dependent on the use of traditional biomass that is used in heating and cooking; the developed ones are turning toward modern biomass technologies that produce electricity and transportation fuel (Aransiola et al., 2024; Sertolli *et al.*, 2022; Sliper, 2024).

The benefits of biomass compared to fossil fuels are strong and multifarious. The most notable one is perhaps carbon neutrality since biomass releases only CO_2 that was recently captured by the atmosphere and placed into the plant during the process of photosynthesis, thus creating a closed carbon cycle that could achieve net-zero, or even negative, emissions when combined with carbon capture technologies. The domestic biomass resources also increase the energy security by decreasing the reliance on unstable imports of fossil fuels and ensuring the stability of prices for the energy consumers. The resulting benefits of rural economic development include biomass production, creating employment and income-generating opportunities in agricultural and forestry industries, which serve to support local communities and economies (Ahmad *et al.*, 2022; Rabbi *et al.*, 2022; Rocha-Meneses *et al.*, 2023).

Although biomass utilization has all these benefits, there are major challenges that hinder the extensive use of biomass. The seasonal changes in the supply of biomass lead to complicated supply-chain dynamics, with high harvesting periods requiring large storage facilities and preservation technologies to assure the quality of feedstock all year round. Logistical concerns or issues are due to the fact that biomass is bulky and has lower energy density, which leads to a higher cost of transportation and the need to locate processing facilities distributed over the area to make it economically viable. Fluctuations in moisture content have significant impacts on the conversion of energy: high moisture reduces heating values and increases processing costs, whereas low moisture content increases the risk of fires and dust forming. Also, because of the heterogeneity of biomass feedstocks, quality-control challenges also exist and need to be addressed with sophisticated characterization and standardization methods to guarantee similar processing performance (Adeleke *et al.*, 2021; Hess and Tumuluru, 2024; Jayarathna *et al.*, 2020; Zhang *et al.*, 2023).

13.3 BIOFUEL PRODUCTION PATHWAYS FROM BIOMASS

Biomass-to-biofuel conversion follows two major routes, i.e., thermochemical and biochemical transformation, each of which has its own benefits and limitations depending on the properties of the feedstock and the end product required. These routes have continued to develop significantly through technological progress to produce greater efficiency, reduced costs, and increased feedstock compatibility (Callegari *et al.*, 2020).

Thermochemical conversion involves using thermal energy, together with chemical reactions, to break up the complex organic molecules that occur in biomass into more simple components that can be refined into biofuels. Pyrolysis is the most important of the thermochemical procedures that involve heating the biomass to temperatures between 400 and 600°C in the absence of oxygen to produce bio-oil, biochar, and syngas in varying ratios depending on temperature and residence time. Optimized to produce bio-oils, fast pyrolysis has an approximate yield of liquid of 60–75%, which is further converted to transportation fuels through a catalytic process. Gasification takes place at a high temperature (700–1000°C) with a measured input of oxygen or steam to produce synthesis gas (syngas) that contains high hydrogen and carbon monoxide contents; this mainly can be converted to liquid fuels using Fischer-Tropsch synthesis or fermentative processes. The simplest thermochemical option is combustion, but in this mode, biomass is directly oxidized to generate heat and electricity; nevertheless, it does not produce liquid fuels and is mostly used to produce stationary power generation (Chidhambaram *et al.*, 2023; Do and Nguyen, 2024; Reza *et al.*, 2023). Feedstock properties, especially moisture content, ash composition, and volatile matter content, also play a significant role in determining the efficiency of thermochemical processes. Other recent developments include torrefaction, which is a light pyrolysis that is performed at 200–300°C, creating a fuel that is similar to coal but has better handling properties and energy density, making it easier to co-fire with coal in existing power plants (Adeoye *et al.*, 2023; Mpungu *et al.*, 2024).

Biochemical conversion takes advantage of biological agents, such as bacteria, yeast, and enzymes, to depolymerize biomass to biofuels by metabolic pathways. The process of anaerobic digestion uses consortia of microorganisms to break down organic matter in the absence of oxygen to create biogas that mainly comprises methane and CO_2; the biogas can be converted to renewable natural gas or used to generate electricity. It is particularly efficient with wet biomass feedstocks, including animal manure, sewage sludge, and food waste, and yields are up to 200–400 m^3 of methane per ton of volatile solids (da Silva Fernandes *et al.*, 2022; Kabeyi *et al.*, 2024; Soni and Soni, 2025). Biomass sugars and starches are converted into ethanol by fermentation, which is the oldest commercially developed biochemical pathway. The global production of ethanol is over 100 billion liters each year. In enzymatic conversion, cellulose and hemicellulose in lignocellulosic biomass are hydrolyzed using specialized enzymes into fermentable sugars to produce ethanol using agricultural residues and energy crops that do not have effects on food production. The main problem of enzymatic conversion is related to the pretreatment procedures that break cellulose walls and make them vulnerable to the enzymes; different chemical, physical, and biological pretreatment techniques are being developed to maximize the production of sugars (Wu *et al.*, 2023).

New hybrid technologies combine both thermochemical and biochemical reactions to maximize their conversion efficiency and diversify product lines. Microbial electrofuels represent an innovative technology where electroactive bacteria directly convert CO_2 and organic food biomass into liquid fuel under the impact of electrical energy, representing a prospect of creating carbon-negative fuels. Hydrothermal liquefaction (HTL) enables the treatment of wet biomass at moderate temperatures (280–370°C) and under high pressures, resulting in bio-crude oil with similar energy densities to petroleum crude and without requiring the use of drying processes, which engage high-energy usage (Shukla *et al.*, 2025; Tyagi *et al.*, 2025).

Comparisons of performance show that thermochemical reactions often achieve high energy conversion efficiencies (50–80%) compared to biochemical reactions (30–50%), but biochemical reactions often yield higher-quality products that need less downstream upgrading. Combining several conversion pathways into process integration strategies can achieve overall efficiencies higher than 85% using all the biomass components and using by-products as energy. Economic modeling proves that the choice of optimal pathways depends on the availability of feedstock, operation scale, product specifications, and local market conditions, and artificial-intelligence-based optimization is being more and more utilized to find the most efficient conversion strategy under specific conditions (Durán-Valle and López-Coca, 2024; Hidalgo *et al.*, 2025; Ro *et al.*, 2023). Figure 13.1 compares thermochemical (pyrolysis and gasification) and biochemical (fermentation and enzymatic) pathways for biofuel production.

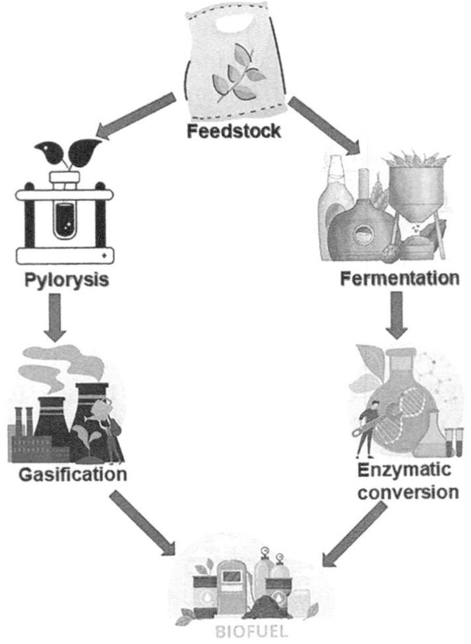

FIGURE 13.1 Thermochemical and biochemical biofuel production pathways.

13.4 ROLE OF ARTIFICIAL INTELLIGENCE IN BIOFUEL RESEARCH

AI has transformed the bioenergy research area by offering solutions to underlying problems that have, in the past, limited the performance and scale of biofuel generation systems. The disruptive potential of AI is due to its exclusive ability to handle the inherent complexity of biological and chemical reactions in biomass transformation and, thus, offer previously unknown insights and optimization opportunities that are inaccessible to traditional methods (Ananda *et al.*, 2025; Saju *et al.*, 2025).

The bioenergy sector produces enormous amounts of multi-dimensional, multi-faceted data, including feedstock characterization, process control, environmental measurements, and product quality measurements. Traditional methods of analysis do not detect trends and links in this diverse data environment, especially with the nonlinear relationships between several variables that determine conversion efficiency (Butean *et al.*, 2025). AI can be useful at multivariate optimization, taking dozens or even hundreds of variables into consideration that have a direct impact on biofuel production, such as feedstock composition and moisture content, reactor temperature, pressure, residence time, and catalyst activity. It requires such ability to optimize any complex systems where minimal changes in some parameters can significantly change overall results (Ahmad, 2023; Ajao *et al.*, 2024; Shah *et al.*, 2025). One more notable benefit of AI in the sphere of bioenergy is the ability to make predictions. By using machine learning (ML) technologies, prediction of biofuel productivity, product quality, equipment behavior, and possible failures of processes can be done with amazing precision using both historical trends and real-time conditions of the processes. This proactive management feature helps in preemptive operational control, hence less wastage, less downtime, and efficient utilization of resources (Patrício *et al.*, 2025; Qudus, 2025).

The artificial neural network (ANN) is one of the most commonly used AI methods in modern biofuel studies and is widely used to estimate fuel properties and yields of conversion. ANNs are also particularly strong at nonlinear prediction of input versus target variables, which, in the case of biodiesel property prediction, have correlation coefficients (R2) higher than 0.95 in many cases. In recent works, the prediction of cetane number, flash point, and kinematic viscosity with a model based on ANN represents 96.69%, 99.07%, and 95.80%, based on the fatty-acid composition values only (Ganeshmoorthy *et al.*, 2019; Zhu *et al.*, 2025). Support vector machines (SVMs) have been shown to be useful in regression and classification problems in the bioenergy arena, especially when dealing with small- to medium-sized datasets and high dimensionality issues. Compared to traditional statistical models, SVM models of wind and solar energy prediction exhibit better performance and better generalization to various environmental conditions. The accurate prediction of biochemical-parameter-based yield predictions and optimization of biochemical transformations are also dependent on SVM algorithms, which help to accomplish the ultimate goal of achieving maximum conversion efficiency of biofuel applications (Benti *et al.*, 2023; Dong *et al.*, 2024; Hassan and Baghban, 2025). RFs are very interpretable and strong in bioenergy, able to handle missing data, and give insights into the rank of variable significance to inform the prioritization of crucial process parameters. These ensemble methods make use of a large number of decision trees to achieve high predictive precision and still be computationally feasible enough to be deployed in a real-time environment (Abbasi *et al.*, 2025; Sakib *et al.*, 2025). Deep learning-based models such as convolutional neural networks (CNNs) and long short-term memory (LSTM) network models are the latest trend of AI applications in bioenergy. CNNs suit well at the task of image-based analysis to characterize biomass and evaluate its quality, whereas LSTMs learn temporal relationships in time-series data of continuous biofuel-manufacturing processes. ELG ensemble model that combines empirical mode decomposition, LSTM, and Gradient Boosting machine gained impressive performance with

RMSE = 0.089 and MAPE = 2.02% in the prediction of microalgae biofuel production efficiency (Shrivastava *et al.*, 2025; Simsek *et al.*, 2025).

In the bioenergy sector, modern AI applications may be applied using the advanced process-simulation computing platforms, such as Aspen Plus, MATLAB, and Python-based frameworks, to build end-to-end modeling environments. Coupling it with Aspen Plus allows AI algorithms to use high-fidelity thermodynamic models and process-simulation abilities, leading to physics-informed ML, which integrates mechanistic and data-driven optimization. Those who study AI/ML models have been able to create models that are based on CPFD and Aspen-based simulations, thus resulting in hybrid methods that utilize both basic understanding of the core processes and trends in experimental data (Pereira *et al.*, 2025; Shash *et al.*, 2025). Python-based systems, including scikit-learn, TensorFlow, and PyTorch, offer comprehensive ML systems that can be easily combined with process-modeling systems to allow the rapid prototyping and deployment of AI-based solutions to bioenergy problems. MATLAB environments include specialized statistics-, ML-, and deep learning-toolboxes, allowing the development and implementation of algorithms in bioenergy studies, and include built-in functions to perform time-series, optimization, and neural-network design (Marella and Palakurti, 2025). The interaction of these tools results in new hybrid modeling environments where AI algorithms are able to view real-time process data, handle more complex calculations, and make optimization suggestions that consider both data-driven understanding and fundamental engineering concepts and thus reflect the state-of-the-art in building bioenergy systems nowadays (Khazaleh *et al.*, 2025; Lin *et al.*, 2025). Figure 13.2 shows the

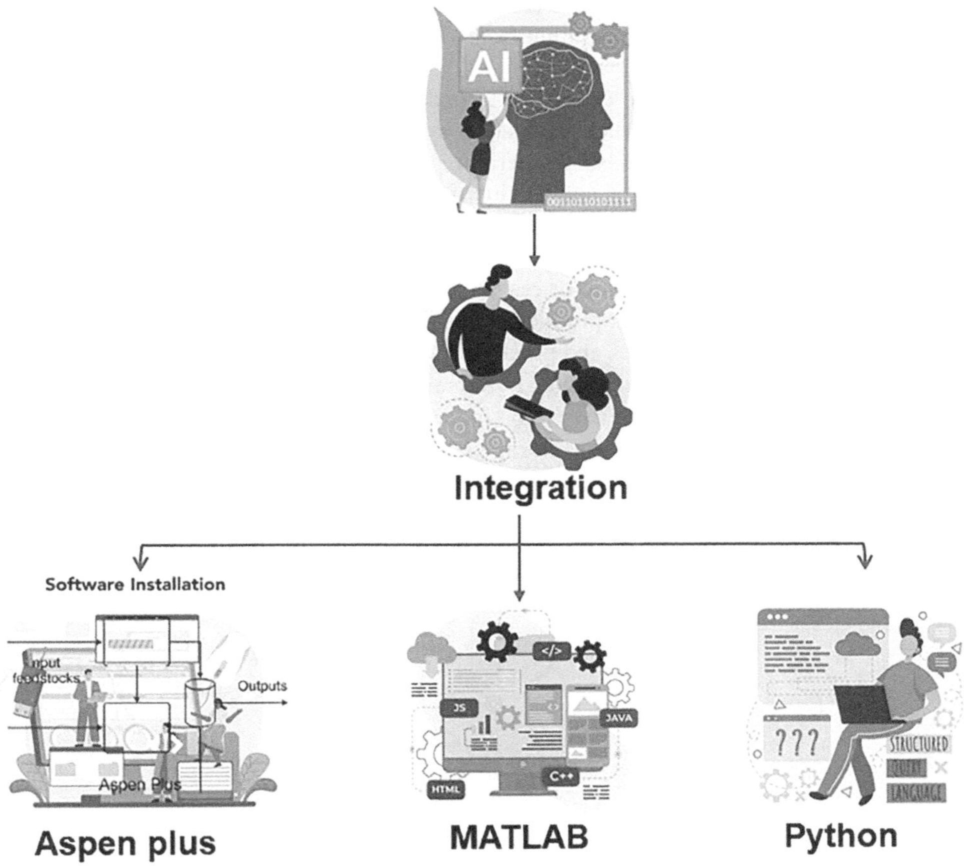

FIGURE 13.2 Integration of AI with process simulation software.

integration of AI with process simulation software, linking Aspen Plus, MATLAB, and Python for enhanced modeling and optimization.

13.5 AI MODELING FOR BIOENERGY FORECASTING

The advanced AI models to predict bioenergy require careful attention to the input variables, data preprocessing methods, model structures, and the validation methods to achieve accurate and practical predictions that can be applied in the industrial context (Dong *et al.*, 2023; Saju *et al.*, 2025).

Feedstock characteristics are the basis of bioenergy prediction models, necessitating comprehensive characterization of the chemical constituents, physical qualities, and quality parameters. Compositional analysis involves the measurement of the cellulose, hemicellulose, and lignin processes in the lignocellulosic biomass since these factors determine conversion yields and processing demands directly. The moisture content is one of the variables that are crucial in determining heating value, storage stability, and processing energy requirements; therefore, AI models use both the absolute moisture content and the temporal patterns of moisture variation. Measures of calorific value give direct information on the energy potential, and AI algorithms are trained to estimate heating value based on proximate and ultimate analysis data instead of requiring a bomb calorimetry reaction to estimate heating value per feedstock sample (Hoque, 2024; Singh *et al.*, 2023). Figure 13.3 depicts AI models that optimize bioenergy by integrating feedstock traits, environmental conditions, and process variables for efficiency.

Process parameters refer to the operational conditions, which define efficiency of conversion and quality of products in the biofuel production systems. The temperature profiles do not just contain set-point temperatures but also temperature gradients, heating rates, and thermal history effects, which affect reaction kinetics and product formation. Gas–liquid equilibria, reaction rates, and mass transfer effects are also influenced by pressure conditions, especially in thermochemical conversion processes where pressure optimization can be of great significance in yield enhancement. Parameters such as reaction time or residence time can be used to determine the degree of conversion and selectivity of different products, and AI models are capable of learning the best time profiles with different feedstock types and processing conditions (Khalid *et al.*, 2024; Webster *et al.*, 2025). The environment has significant roles in biomass production, storage, and processing, which requires meteorological and seasonal data to be incorporated into prediction models. The effects of humidity on the rate of feedstock-drying, storage stability, and the energy requirements of the process are factored in, and in AI-based systems, it is trained to predict seasonal rotation and its effect

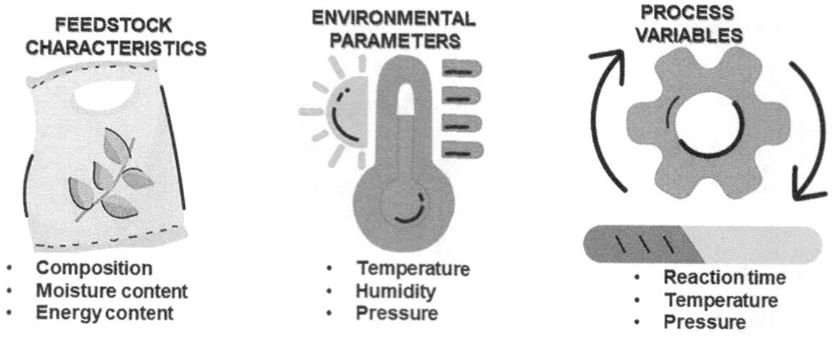

FIGURE 13.3 Key input variables for AI models.

on the overall performance of the system. Patterns in seasonal variation cause changes in the supply of feedstock and its quality and costs associated with logistics, which allows AI models to optimize the production schedules and inventory management tactics (Et-Taibi *et al.*, 2024; Sajja *et al.*, 2025; Simonič and Klančnik, 2025).

The effective AI modeling presupposes the existence of powerful data acquisition systems combining various types of sensors, analyzing tools, and external data. The Internet of Things (IoT) sensor networks installed all over bioenergy facilities deliver real-time data about the temperature, pressure, flow rates, pH, dissolved oxygen, and other vital process variables. Contemporary installations are fitted with intelligent sensors that are connected wirelessly and have the ability of edge computing and predictive maintenance, which guarantee the quality of the data and reliability of the system (Dębowski *et al.*, 2025; Ficili *et al.*, 2025; Ortiz *et al.*, 2024). Before ML algorithms can be used, raw sensor data must be preprocessed into formats comprehensible to ML algorithms. The processes of data cleaning isolate and rectify sensor drift, outliers, gaps in measurements, and measurement errors that may undermine model performance. Normalization and scaling ensure that variables with different units and ranges make their contribution to the model training, whereas feature engineering makes derived variables and reflects the necessary relationships and patterns that are not so obvious in raw data (Bhagat and Bakariya, 2025; Gul, 2025; Sharma *et al.*, 2024).

The sophisticated training methodologies have since evolved past the use of simple train-test splits to include the use of sophisticated validation techniques that will guarantee model robustness and generalization. Cross-validation methods, such as k-fold cross-validation and time series cross-validation, give an unbiased estimate of how well a model works and identify overfitting. Ensemble techniques are used to identify the use of multiple individual models in order to enhance prediction accuracy and minimize variance: bagging methods, boosting methods, and stacking methods have demonstrated better results compared to the use of single-model methods (Rimal *et al.*, 2024; Seyam *et al.*, 2025). Hyperparameter optimization uses automated methods such as grid search, random search, and Bayesian optimization to find an optimal model configuration. More recent investigations of Bayesian optimization for CNN-BiLSTM models showed that the models achieved a massive improvement in performance, and the R2 values showed to be more than 0.81 when used to predict winter wheat yield, which is significantly higher than traditional LSTM methodologies (Szczepaniak *et al.*, 2025).

One of the most established uses of AI in biofuel manufacturing is ethanol yield prediction, and neural-network models have achieved prediction appraisals of greater than 95% in not just cellulosic manufacturing but also in corn-based production. New applications allow predicting the yields of ethanol 25 days before harvest precisely, and they provide the opportunity to plan the supply chain in advance and manage stocks (Ajao *et al.*, 2024; Jain *et al.*, 2025). The prediction of biodiesel quality using ANN models has demonstrated impressive accuracy, with literature showing an accuracy of 99.40% in density prediction, 99.07% in flash point, and 96.69% in cetane number only using fatty acid methyl ester composition data. These models can quickly assess quality without the need for expensive laboratory testing and can greatly lower the cost of production and permit quality control in real time (Saju *et al.*, 2025).

Ensemble-based methods in biogas volume forecasting applications have recorded improvement in volume forecasting using ELG model, which is based on empirical mode decomposition, LSTM, and gradient boosted machines, with the model representing an RMSE of 0.089 and an MAPE of 2.029% on standardized data. These levels of accuracy make accurate planning of production and the resource allocation optimal in the anaerobic digestion facilities (Ahmad *et al.*, 2024). Coefficient of determination (R2) values exceeding 0.95, root-mean-square error (RMSE) less than 5% of average values of production, mean absolute error (MAE) less than 3% on key parameters, and mean absolute percentage error (MAPE) less than 5% are typical performance targets of bioenergy AI models. These high accuracy standards are based on the economic significance of high accuracy predictions in industrial bioenergy practices, where the slight improvement of the accuracy of the prediction can immediately translate into large savings and efficiency improvements (Román *et al.*, 2024; Saju *et al.*, 2025).

13.6 REAL-WORLD APPLICATIONS AND SUCCESS STORIES

The application of AI to bioenergy systems in practice has already produced striking outcomes in a variety of applications and proves the possibilities of the transformative nature of intelligent systems in the real environment of production (Ejiyi *et al.*, 2025). The use of precision agriculture has transformed predicting and optimizing energy production using agricultural residues, where AI technology collects satellite images, weather information, soil conditions, and crop management activities to predict the availability and quality of biomass. Combining remote sensing with machine learning AI solutions allows farmers and bioenergy producers to estimate the energy content of crop residues with almost unbelievable precision and accuracy, with R2 values of 0.8702 between normalized difference vegetation index (NDVI) measurements and biomass volume estimations (Aarif et al., 2025; Akande, 2024).

Multisource data fusion techniques, which integrate conventional remote sensor variables like the enhanced vegetation index (EVI) and leaf area index (LAI) with the solar-induced chlorophyll fluorescence (SIF) data, have shown to be more effective in yield estimation. The model that combines all these different data streams, BO-CNN-BiLSTM, had an R2 of 0.81, RMSE of 616.99 kg/ha-1, and MRE of 7.14% when predicting winter wheat yield, significantly higher than that of traditional forecasting models. These systems offer essential information to bioenergy supply-chain planning, which allows producers to ensure the optimal supplies of feedstock and design processing capacities far ahead of harvesting seasons (Alemde, 2025; Priya and Yadav, 2025; Xiao *et al.*, 2025; Zhang *et al.*, 2024).

One of the most effective uses of AI in the biofuel production sphere is the idea of real-time process optimization, where smart systems are permanently changing the parameters of the operations to achieve the highest rate of conversion efficiency and product quality. IoT-based real-time monitoring systems have been used at Lootah Biofuels in the United Arab Emirates (UAE) to pilot biofuel conversion of DIC factory with success, producing fully automated, end-to-end production of zero waste during processing processes. This application indicates that AI-based systems could be used to ensure that environmental sustainability and economic optimization are both achieved (Hussain *et al.*, 2022; Ojadi *et al.*, 2025).

AI applications in biogas production have been applied when optimizing the feeding strategies of the methane-producing organisms and the maximum yields of methane by fine-tuning the sensor data (temperature, pH, and gas makeup) in real time using ML algorithms. Research in Uganda showed that AI-controlled biogas production systems based on genetic algorithms and ANNs with backpropagation training were highly efficient at producing biomethane whilst keeping the fermentation processes stable. The AI-based controlled feeding approach also resulted in better substrate physicochemical conditions, increased fermentation processes, and more predictable and stable biogas production with a higher control of the process (Paavani *et al.*, 2025; Zhang *et al.*, 2025). Optimization of biodiesel production that uses AI solutions to tune the parameters has shown amazing results in the yield and quality indicators. Producers have been able to optimize the selection of a catalyst, reaction temperature, pressure, and residence time jointly by ML models examining the relationships between feedstock properties, reaction conditions, and product characteristics, with yield improvements of 1525% over the conventional methods of optimization (Mekuye *et al.*, 2024; Swami *et al.*, 2025).

This has become a highly important field of application: bioenergy market prediction, where AI models have the capability of incorporating climate data, policy signals, and economic variables to predict market trends and make investment decisions. The NRBO-CNN-BiLSTM-Attention model has been shown to be more effective in predicting global bioenergy returns, using inputs such as the Global Climate Policy Uncertainty Index, the Global Green Bond Index Returns, the Arctic Oscillation Index, and even the NASDAQ BioClean Fuel Index. This model is also applicable and consistently predicts better over five, ten, and fifteen days and gives useful information to both the market participants and policymakers (Fang *et al.*, 2025; Guo *et al.*, 2025). Policy impact assessment is a method using AI algorithms that evaluates the usefulness of renewable energy policies and the impact they have on the rates of biofuels adoption. The use of ML algorithms to analyze information provided by Energy Information Administration (EIA) and National Renewable Energy Laboratory (NREL) datasets has shown that predicting trends in renewable energy is highly accurate, and Gradient Boosting models have shown the best results in identifying the rates of energy production. These bring to the fore the evidence-based policy formulations and minimize regulatory structures to speed up bioenergy implementation (Shovon *et al.*, 2025; Soni and Soni, 2025).

Supply-chain optimization also uses AI to improve the level of transparency and efficiency across the bioenergy value chain. Recent applications have shown that predictive maintenance using AI can save 20–30% of the downtime in the bioenergy plants and enhance the efficiency of the biomass conversion by 15%. They combine a variety of data, such as the quality of feedstock measurements, performance measurements of equipment, and the environmental conditions, to optimize the logistics, the inventory control, and the production scheduling of whole supply networks (Calvin *et al.*, 2021; Pasupuleti *et al.*, 2024).

These practical applications show that AI is no longer considered an experimental technology in the bioenergy industry but a time-tested mechanism that makes it possible to attain considerable progress in efficiency, sustainability, and economic performance in a variety of operational settings. Since these implementations are in the process of maturing and scaling, they act as blueprints for wider implementations across the entire bioenergy sector in the world (Priya and Yadav, 2025).

13.7 CHALLENGES AND LIMITATIONS

Despite the impressive achievements reflected in AI applications to bioenergy, there are still several large issues that hinder the widespread use of intelligent systems and their maximum efficiency in the sector (Adewoyin *et al.*, 2025; Saju *et al.*, 2025). One of the most basic issues of AI implementation in bioenergy systems is data availability constraints. Most bioenergy operations are poorly equipped with large historical datasets to train powerful ML models (unlike more established industries that have decades of digital data collection). This difficulty is further compounded by the heterogeneous nature of biomass feedstocks, with each type of feedstock needing to be characterized, and the seasonal and geographical variability adding another layer of data needs, which are either not available or incomplete (Akande, 2024; Ananda *et al.*, 2025; Inayat *et al.*, 2025).

The issues of data quality have a great influence on model performance and reliability. Breakdowns in the quality of data lead to the frequent retraining of the models because of ineffective performance, thus costing more resources to compute and decreasing the performance reliability of the systems. Measurement errors due to sensor drift and calibration in adverse bioenergy processing conditions spread through AI models, which may cause suboptimal decisions or failures in the process. The non-standardization of measurement procedures in various facilities and research institutes brings in discrepancies, which restrict the cross-applicability of AI models between operations (Huang and

Khabusi, 2025; Li *et al.*, 2024). The challenge of data privacy and security occurs when bioenergy companies become more worried about the possibility of sharing their proprietary process data needed to make joint developments of AI models. This is a setback in terms of creating industry-wide AI curative measures and compels all separate companies to build models using smaller datasets, which may not be as representative (Pesapane *et al.*, 2025; Wagner *et al.*, 2025).

Issues with overfitting are especially problematic in bioenergy applications where datasets can be small, and models can learn to follow certain patterns that are not generalized to new operating conditions and feedstock types. The high variability of biological and chemical mechanisms in the production of biofuels gives many opportunities to the models to define spurious correlations instead of the underlying relationships and, therefore, perform poorly in response to changing conditions (Malashin *et al.*, 2025; Wang *et al.*, 2024). Generalization problems appear when AI models that are trained on certain types of feedstock, regions, or times of the year cannot work sufficiently in other situations. The extreme susceptibility of certain AI algorithms, in particular the Random Forest models, to outliers may result in erratic predictions when facing some strange yet valid operating conditions. This is especially worrisome in the case of bioenergy, as the quality of the feedstock and the conditions of the process can change dramatically on a daily basis (Kumar *et al.*, 2025; Schroer and Just, 2023).

The challenges of model validation are explained by the fact that there are few independent test datasets available that actually represent the full scope of operating conditions that are encountered in the commercial bioenergy operations. Laboratory-level measurements do not necessarily include the complications and variation inherent to full-scale industrial operations, thus causing overly optimistic performance projections when developing a model (Schroer and Just, 2023).

Barriers to technology integration lead to major barriers in trying to adopt AI solutions in existing bioenergy plants. Outdated control systems, vendor-specific software platforms, and non-standard data formats can be expensive to upgrade their infrastructure to allow them to use AI. The shortage of qualified people who are trained in both bioenergy processes and AI technologies makes it difficult to implement, since the successful implementation of AI requires a profound knowledge of both fields (Arora, 2025; Shah *et al.*, 2025). The computational demands of advanced AI models and the necessity to make decisions in real-time in industrial processes lead to the appearance of real-time implementation constraints. In fast processes, latency problems may be severe, as prediction lag will no longer be useful in process control. Industrial-specific reliability needs also impose a requirement of AI systems that have fail-safe features and graceful degradation that most research-driven models do not have (Mahmood and Szabolcsi, 2025; Rai *et al.*, 2024). The question of scalability is evident when the challenge arises of deploying AI solutions designed to run in particular facilities to other work settings. The physical remoteness between the renewable energy production and processing locations poses challenges to the adoption of integrated AI systems to optimize the whole value chain (Panda, 2025; Ukoba *et al.*, 2024).

Regulation issues render uncertainty to AI implementation in bioenergy systems, especially as AI-specific regulations are introduced in various jurisdictions with dissimilar needs. The absence of globally accepted standards in the use of AI energy has caused problems with scaling solutions in international markets and evaluating their real effectiveness. The building efficiency and other renewable energy practices are extremely different in different regions, making AI systems' design and implementation difficult (Soltani *et al.*, 2025). The issues of environmental impact emerge due to the amount of energy needed to train and run AI models that will counterbalance certain environmental advantages of bioenergy systems. The process of training advanced deep learning models also produces large carbon dioxide emissions, as one study reported that the training of a single AI model produced more than 626,000 pounds of CO_2. This leads to a paradox of the idea of AI systems that are supposed to make the world more sustainable, causing environmental issues themselves when not addressed carefully (Frimpong, 2025; Hasan *et al.*, 2025).

These biases and fairness problems may arise when AI models developed on data in a particular geographic area, feedstock, or production process systemically do not perform well elsewhere, thus potentially causing some producers or communities economic disadvantages. The need to be more transparent and explainable is also on the rise, with the regulatory bodies and stakeholders requiring knowledge of the decision-making mechanism of AI systems that influence the production of energy and the environment (Chen *et al.*, 2024; Talaat *et al.*, 2025).

The reliance on prominent actors in the development and implementation of AI poses a threat to the accessibility of new technology and the concentration of the market. Centralization of R&D to a small number of large technology firms could restrict the supply of specialized bioenergy AI products, to the disadvantage of smaller operators, which could further worsen the industry inequalities that are already present (Ramesh, 2025).

13.8 FUTURE PERSPECTIVES

The intersection of AI and bioenergy systems will gain exponential growth in the next decade, as new technological advances, scaling opportunities, and sustainability demands on a global level continue to transform the energy landscape (Cara and Cara, 2025; Yin *et al.*, 2025). The autonomous biorefineries with AI will be the frontier of bioenergy automation that will go past the current process optimization to fully independent plants that will also be able to self-regulate complete production cycles. These systems combine highly developed robotics with AI decision-making algorithms to manage the feedstock reception, preprocessing, conversion, and finishing of the products with the minimum human involvement. Recent advances in independent laboratory systems indicate the viability of closed-loop experimentation and optimization, with systems like the Autonomous Lab (ANL) able to optimize medium conditions to recombinant strains of E. coli and, at the same time, enhance cell growth rate and optimal cell density. Such prototype applications present roadmaps toward autonomous operation at industrial bioenergy plants (Bidyalakshmi *et al.*, 2025; Liu *et al.*, 2025; Tom *et al.*, 2024).

The IoT–AI convergence toward real-time monitoring is turning bioenergy facilities into intelligent and self-optimizing systems. Recent sensor networks incorporate edge computing functionality, which allows local AI processing, which results in less latency and faster response times to critical control decisions. Ultra-low latency communication in distributed systems. Integration of 5G connectivity with IoT sensors can connect the components of a distributed system to coordinate optimization across the full bioenergy value chains. According to recent applications, sensor accuracy of soil moisture in real-time transmission of the data wirelessly with a precision of approximately 3% can be used, which allows to regulate irrigation and optimize the use of resources (Majumdar *et al.*, 2025; Seyar and Ahamed, 2024).

Hybrid AI-blockchain systems are currently becoming potent solutions to supply chain visibility and automated contract execution in bioenergy markets. Ledger technology, with AI-based quality evaluation and prediction, is brought together in these systems to establish transparent automated trading platforms of biomass feedstocks and biofuels. Integration of blockchain can also save 30–50% compliance costs in industries that need high levels of regulatory controls and shall also lower the risks of regulatory breaches by 25%. AI-based smart contracts allow autonomous execution of supply contracts using real-time quality metrics and yield projections and simplify business dealings and mitigate counterparty risks in bioenergy chains of supply (Abikoye *et al.*, 2024; Ebrahimi *et al.*, 2025; Shukla, 2025).

The integration of quantum computing should be seen as a game changer in the application of bioenergy AI programs, with a billionfold increase in computational ability for intricate optimization tasks. Molecular dynamics simulations by quantum algorithms would transform the design of catalysts, allowing the identification of new catalysts with previously unseen selectivity and activity in bioreactions to produce biofuels. The huge dimensionality of integrated biorefinery optimization could be dealt with by quantum ML methods, and the process conditions, feedstock selection, and product distribution across the whole facilities are concurrently optimized (Malashin *et al.*, 2024; Mishra *et al.*, 2025).

AI-enhanced digital twin technology is developing holistic virtual models of bioenergy plants that can be used for predictive maintenance, simulation of scenarios, and optimization without interfering with the real production. Physics-informed AI models integrate mechanistic knowledge with data-driven learning to produce one of the most accurate predictive models capable of simulating the interactions of processes and predicting the optimal operating strategies. The initial applications demonstrate the chance of cost reduction of maintenance by 20–30% and the overall equipment performance enhancement through predictive maintenance methods (Al-Shetwi *et al.*, 2025; Quarteroni *et al.*, 2025; West *et al.*, 2024).

Joint AI systems are coming to overcome data scarcity issues through secure collaboration between bioenergy producers that is privacy-preserving. Federated learning methods can empower different facilities to jointly train AI models without exposing sensitive proprietary information, building industry-wide AI capabilities that benefit all the players without compromising competitive advantages. These platforms might also quicken the adoption of AI amongst smaller operators that do not have the resources to develop a model independently but can contribute to network effects that enhance model performance as additional facilities join (Mehmood *et al.*, 2025; Rane *et al.*, 2025).

AI-based bioenergy systems will be vital in optimizing carbon emissions as a way of meeting the global net-zero objectives. Entire bioenergy chains can be optimized by AI systems to yield maximal carbon abatement by taking into account lifecycle emissions of feedstock production to final fuel combustion. The more sophisticated AI solutions combine carbon accounting and production optimization to find strategies that maximize economic benefit and carbon emission reduction to support policy frameworks that price carbon emissions accordingly (Manikandan *et al.*, 2025; Saju *et al.*, 2025).

Connecting with renewable energy grids uses AI to optimize the nexus between variable renewable electricity generation and bioenergy production to form hybrid renewable systems that are more reliable and efficient. Demand-response systems powered by AI can reposition energy-intensive biofuel production processes to times of surplus renewable electricity production to enhance grid stability and lower the costs of producing biofuels. Retrained ML models that consume real-time grid information of regional transmission organizations are highly accurate in predicting the level of renewable energy production, with Gradient Boosting models showing the best results in the classification of the energy production level (Cavus, 2025; Miller *et al.*, 2025; Suanpang and Jamjuntr, 2024).

To build effective incentive regimes around bioenergy deployment, policy support mechanisms also become based on AI-driven analysis to design effective incentive schemes. Predictive policy impact models are machine-driven models that predict the effectiveness of various regulatory methods to enable evidence-based policy evolution that optimizes the use of bioenergy without causing economic discontinuity. The AI solutions, which trace the uncertainty scores of global climate policy and renewable energy markets, are also valuable tools to policymakers working on the development of long-term energy transition policies (Adewoyin *et al.*, 2025; Ekeh *et al.*, 2025).

Integration of AI in the renewable energy plans has been explicitly incorporated in the schematic outlines of national biofuel policies that states like the UAE are currently developing. It is accepted in these policies that all energy systems of the future will be both inherently data-driven and AI-enabled, thus making it necessary to develop regulatory frameworks that promote innovation and, at the same time, put safety, security, and environmental stewardship first. Specifically, the introduction of a biofuel blend requirement and the encouragement of AI and IoT technologies' use in production sites offered by the UAE can serve as an example to other countries that are currently developing comprehensive bioenergy policies according to net-zero goals (Arun *et al.*, 2025; Dsilva *et al.*, 2024; Sreekanth, 2024).

These trends have come together to present the bioenergy industry with a transformation that has never been witnessed before. AI is no longer an optimization tool but a fundamental enabler, which makes the economics of large-scale sustainable bioenergy economically rational and beneficial to the environment. Early implementation results are also empirically supportive of the scalable nature of the realization of these opportunities, thus leading to the development of a bioenergy sector significantly more efficient, sustainable, and responsive to global climate demands than it had ever been imagined (Khan *et al.*, 2025).

13.9 CONCLUSION

The use of AI in the production chain of biomass-based biofuels is driving the paradigm shift in the global transition to renewable energy. Preliminary pilot projects have developed into operational solutions that improve feedstock selection, optimize processes, logistics within the supply chain and market predictions. The empirical data suggest that AI always has a higher performance in comparison with the traditional approaches, as ML algorithms deliver accuracy greater than 95% when predicting fuel yield, quality, and performance indicators. Field trials have reported yield gains of up to 25%, minimized downtime, and massive savings in costs due to a smarter use of resources.

AI is versatile in thermochemical as well as biochemical conversion routes, enabling the use of AI in innovations, including predictive maintenance, autonomous biorefineries, and zero-waste systems. The combination of IoT sensors, edge computing, and cloud analytics is making smart platforms, which learn and optimize operations in real-time. Digital twins, blockchain-based supply chains, and quantum-enhanced computing are emerging technologies with the potential to unlock additional opportunities. Moreover, AI is incorporated into the world's climate goals because it helps to reduce carbon emissions and guide effective strategies for integrating renewable energy. Although the issues associated with the quality of data and system integration are still present, they are not core barriers but technical barriers. In general, AI is becoming a strategic guide to bioenergy innovation, which would help to create a more resilient and cleaner energy future that could be economically viable.

REFERENCES

Aarif KO, M., Alam, A., & Hotak, Y. (2025). Smart sensor technologies shaping the future of precision agriculture: Recent advances and future outlooks. *Journal of Sensors*, *2025*(1), 2460098.

Abbasi, M., Váz, P., Silva, J., & Martins, P. (2025). Machine learning approaches for predicting maize biomass yield: Leveraging feature engineering and comprehensive data integration. *Sustainability*, *17*(1), 256.

Abikoye, B. E., Umeorah, S. C., Adelaja, A. O., Ayodele, O., & Ogunsuji, Y. M. (2024). Regulatory compliance and efficiency in financial technologies: Challenges and innovations. *World Journal of Advanced Research and Reviews*, *23*(1), 1830–1844.

Adeleke, A. A., Odusote, J. K., Ikubanni, P. P., Lasode, O. A., Malathi, M., & Paswan, D. (2021). Essential basics on biomass torrefaction, densification and utilization. *International Journal of Energy Research*, *45*(2), 1375–1395.

Adeoye, A. O., Lawal, O. S., Quadri, R. O., Malomo, D., Aliyu, M. T., Dang, G. E., Emojevu, E. O., Maikato, M. J., Yahaya, M. G., Omonije, O. O., & Edidem, V. G. (2023). Sustainable energy via thermochemical and biochemical conversion of biomass wastes for biofuel production. In Verma, K. K., Song, X.-P., Xu, L., Bhatt, R., & Li, Y.-R. (Eds.) *Transportation Energy and Dynamics* (245–306). Apple Academic Press, Taylor & Francis Group, London, UK.

Adewoyin, M. A., Adediwin, O., & Audu, A. J. (2025). Artificial intelligence and sustainable energy development: A review of applications, challenges, and future directions. *International Journal of Multidisciplinary Research and Growth Evaluation*, *6*(2), 196–203.

Ahmad, A., Yadav, A. K., Singh, A., & Singh, D. K. (2024). A comprehensive machine learning-coupled response surface methodology approach for predictive modeling and optimization of biogas potential in anaerobic co-digestion of organic waste. *Biomass and Bioenergy*, *180*, 106995.

Ahmad, I. (2023). Advances in Machine Learning for Monitoring, Control, and Optimization of Temperature of Reactors.

Ahmad, R., Ilyas, H. N., Li, B., Sultan, M., Amjad, M., Aleem, M., Abbas, A., Imran, M. A., & Riaz, F. (2022). Current challenges and future prospect of biomass cooking and heating stoves in Asian countries. *Frontiers in Energy Research*, *10*, 880064.

Ajao, J., Akanbi, O., Akinwole, I., Adedokun, A., Ogunsakin, O., & Ige, A. (2024). Research advances on machine learning technologies for enhanced biodiesel production: A comprehensive review. *Petroleum & Coal*, *66*(1), 1–29.

Akande, O. (2024). Challenges and Opportunities in Machine Learning for Bioenergy Crop Yield Prediction: A Review. *Available at SSRN 4898518*.

Alemde, V. O. (2025). Innovative process technologies: Advancing efficiency and sustainability through optimization and control. *International Journal of Research Publication and Reviews*, *6*(2), 1941–55.

Al-Shetwi, A. Q., Atawi, I. E., El-Hameed, M. A., & Abuelrub, A. (2025). Digital twin technology for renewable energy, smart grids, energy storage and vehicle-to-grid integration: Advancements, applications, key players, challenges and future perspectives in modernising sustainable grids. *IET Smart Grid*, *8*(1), e70026.

Ananda, A., Sujeeth, R. K., & Archana, S. (2025). Integrating automation in biomass transformation: Opportunities, challenges, and future directions. *BioEnergy Research*, *18*(1), 1–25.

Aransiola, S. A., Oyewole, O. A., Maddela, N. R., Ijah, U. J. J., & Manjunatha, B. (Eds.). (2024). *Marine Greens: Environmental, Agricultural, Industrial and Biomedical Applications* (1st ed.). CRC Press, Taylor and Francis Group, Boca Raton, FL, USA. https://doi.org/10.1201/9781003369738

Arora, A. (2025). Challenges of Integrating Artificial Intelligence in Legacy Systems and Potential Solutions for Seamless Integration. *Available at SSRN 5268176*.

Arun, M., Barik, D., Chandran, S. S., Praveenkumar, S., & Tudu, K. (2025). Economic, policy, social, and regulatory aspects of AI-driven smart buildings. *Journal of Building Engineering*, *99*, 111666.

Benti, N. E., Chaka, M. D., & Semie, A. G. (2023). Forecasting renewable energy generation with machine learning and deep learning: Current advances and future prospects. *Sustainability*, *15*(9), 7087.

Bhagat, M., & Bakariya, B. (2025). A comprehensive review of cross-validation techniques in machine learning. *IJSAT-International Journal on Science and Technology*, *16*(1), 1–4.

Bidyalakshmi, T., Jyoti, B., Mansuri, S. M., Srivastava, A., Mohapatra, D., Kalnar, Y. B., Narsaiah, K., & Indore, N. (2025). Application of artificial intelligence in food processing: Current status and future prospects. *Food Engineering Reviews*, *17*(1), 27–54.

Butean, A., Cutean, I., Barbero, R., Enriquez, J., & Matei, A. (2025). A review of artificial intelligence applications for biorefineries and bioprocessing: From data-driven processes to optimization strategies and real-time control. *Processes*, *13*(8), 2544.

Callegari, A., Bolognesi, S., Cecconet, D., & Capodaglio, A. G. (2020). Production technologies, current role, and future prospects of biofuels feedstocks: A state-of-the-art review. *Critical Reviews in Environmental Science and Technology*, *50*(4), 384–436.

Calvin, K., Cowie, A., Berndes, G., Arneth, A., Cherubini, F., Portugal-Pereira, J., Grassi, G., House, J., Johnson, F. X., Popp, A., & Rounsevell, M. (2021). Bioenergy for climate change mitigation: Scale and sustainability. *GCB Bioenergy*, *13*(9), 1346–1371.

Cara, R., & Cara, F. (2025). Transforming the energy landscape: The role of emerging firms in shaping a sustainable future. *Trans Motauto World*, *10*(2), 60–65.

Cavus, M. (2025). Advancing power systems with renewable energy and intelligent technologies: A comprehensive review on grid transformation and integration. *Electronics*, *14*(6), 1159.

Chen, C. F., Napolitano, R., Hu, Y., Kar, B., & Yao, B. (2024). Addressing machine learning bias to foster energy justice. *Energy Research & Social Science*, *116*, 103653.

Chidhambaram, N., Kay, S. J. J., Priyadharshini, S., Meenakshi, R., Sakthivel, P., Dhanbalan, S., Shanavas, S., Kamaraj, S. K., & Thirumurugan, A. (2023). Magnetic nanomaterials as catalysts for syngas production and conversion. *Catalysts*, *13*(2), 440.

da Silva Fernandes, F., de Souza, ÉS., Carneiro, L. M., Alves Silva, J. P., de Souza, J. V. B., & da Silva Batista, J. (2022). Current ethanol production requirements for the yeast saccharomyces cerevisiae. *International Journal of Microbiology*, *2022*(1), 7878830.

Dębowski, M., Kazimierowicz, J., & Zieliński, M. (2025). Multi-sensing monitoring of the microalgae biomass cultivation systems for biofuels and added value products synthesis—Challenges and opportunities. *Applied Sciences*, *15*(13), 7324.

Do, P. T. M., & Nguyen, L. X. (2024). A review of thermochemical decomposition techniques for biochar production. *Environment, Development and Sustainability*, 1–57. doi: 10.1007/s10668-024-05841-6

Dong, L., Sharma, P., Abass, R. R., Al-shuwaili, S. J., Kumar, A., Al-Abdeen, S. H. Z., Qassem, L. Y., & Alam, M. M. (2024). Optimization of biofuel production from biomass using montmorillonite catalyst by development of predictive models. *Journal of Molecular Liquids*, *403*, 124768.

Dong, Z., Bai, X., Xu, D., & Li, W. (2023). Machine learning prediction of pyrolytic products of lignocellulosic biomass based on physicochemical characteristics and pyrolysis conditions. *Bioresource Technology*, *367*, 128182.

Dsilva, J., Shaikh, M. I., & Usmani, F. Z. A. (2024). Net zero policies and climate resilience: A focus on emerging technologies and implementation strategies in the Middle East. In Singh, R., Crowther, D. (Eds.), *Transition towards a Sustainable Future: Net Zero Policies and Environmental Sustainability* (119–138). Springer Nature, Singapore.

Durán-Valle, C. J., & López-Coca, I. M. (2024). Biochemical and thermochemical conversion technologies for agriculture waste transformation. In Arora, J., Joshi, A., Ray, R.C. (Eds.), *Transforming Agriculture Residues for Sustainable Development: From Waste to Wealth* (47–84). Springer, Cham, Switzerland.

Ebrahimi, S., Chen, J., Bridgelall, R., Szmerekovsky, J., & Motwani, J. (2025). Unlocking the commercialization of SAF through integration of Industry 4.0: A technological perspective. *Sustainability*, *17*(16), 7325.

Ejiyi, C. J., Cai, D., Thomas, D., Obiora, S., Osei-Mensah, E., Acen, C., Eze, F. O., Sam, F., Zhang, Q., & Bamisile, O. O. (2025). Comprehensive review of artificial intelligence applications in renewable energy systems: Current implementations and emerging trends. *Journal of Big Data*, *12*(1), 169.

Ekeh, A. H., Apeh, C. E., Odionu, C. S., & Austin-Gabriel, B. (2025). Leveraging machine learning for environmental policy innovation: Advances in data analytics to address urban and ecological challenges. *Gulf Journal of Advance Business Research*, *3*(2), 456–482.

Et-Taibi, B., Abid, M. R., Boufounas, E. M., Morchid, A., Bourhnane, S., Hamed, T. A., & Benhaddou, D. (2024). Enhancing water management in smart agriculture: A cloud and IoT-based smart irrigation system. *Results in Engineering*, *22*, 102283.

Fang, Y., Lee, C. C., & Li, X. (2025). Assessing the impact of artificial intelligence on the transition to renewable energy? Analysis of US states under policy uncertainty. *Renewable Energy*, *246*, 122969.

Ficili, I., Giacobbe, M., Tricomi, G., & Puliafito, A. (2025). From sensors to data intelligence: Leveraging IoT, cloud, and edge computing with AI. *Sensors*, *25*(6), 1763.

Frimpong, V. (2025). The Sustainability Paradox of Artificial Intelligence: How AI Both Saves and Challenges Resource Management Efforts. *Available at SSRN 5176930.*

Ganeshmoorthy, V., Muthukannan, M., & Thirugnanasambandam, M. (2019). Comparison of linear regression and ANN of fish oil biodiesel properties prediction. *Journal of Physics: Conference Series*, *1276*(1), 012074.

Goel, S., Thapliyal, D., & Arya, R. K. (2024). Renewable energy production using crop waste. In Arya, R.K., Verros, G.D., Verma, O.P., Hussain, C.M. (Eds.), *From Waste to Wealth* (281–311). Springer Nature, Singapore.

Gul, A. (2025). A novel hybrid ensemble framework for stock price prediction: Combining bagging, boosting, dagging, and stacking. *Computational Economics*, 1–31. doi: 10.1007/s10614-025-10979-0

Guo, L., Cheng, Q., He, X., Su, M., & Li, H. (2025). Bioenergy market predictions using AI: Integrating climate change and green finance. *Renewable Energy*, *250*(4665), 123328.

Hasan, S. M., Islam, T., Saifuzzaman, M., Ahmed, K. R., Huang, C. H., & Shahid, A. R. (2025). Carbon emission quantification of machine learning: A review. *IEEE Transactions on Sustainable Computing*. doi: 10.1109/TSUSC.2025.3578834

Hassan, R., & Baghban, A. (2025). Predicting CO_2 adsorption in KOH-activated biochar using advanced machine learning techniques. *Scientific Reports*, *15*(1), 24410.

Hess, J. R., & Tumuluru, J. S. (2024). Biomass supply chain logistics: Challenges and technological advancements. In Bisaria, V. (Ed.), *Handbook of Biorefinery Research and Technology: Biomass Logistics to Saccharification* (3–24). Springer Nature, Dordrecht, The Netherlands.

Hidalgo, D., Urueña, A., Martín-Marroquín, J. M., & Díez, D. (2025). Integrated approach for biomass conversion using thermochemical routes with anaerobic digestion and syngas fermentation. *Sustainability, 17*(8), 3615.

Hoque, A. (2024). Artificial intelligence in post-harvest drying technologies: A comprehensive review on optimization, quality enhancement, and energy efficiency. *International Journal of Scientific Research, 13*(11), 493–502.

Huang, Y. P., & Khabusi, S. P. (2025). Artificial intelligence of things (AIoT) advances in aquaculture: A review. *Processes, 13*(1), 73.

Hussain, R. F., Mokhtari, A., Ghalambor, A., & Salehi, M. A. (2022). IoT for Smart Operations in the Oil and Gas Industry: From Upstream to Downstream.

Inayat, A., Dafalla, M., Asaad, S., Jamil, F., Al-Haj, L., Shah, F. M., Ghenai, C., & Shanableh, A. (2025). Sustainable energy production from waste: A review of hybrid approaches combining anaerobic digestion and gasification. *International Journal of Energy Research, 2025*(1), 6644084.

Jain, N., Rohatgi, A., Suransh, J., & Kumar, D. (2025). Artificial intelligence in biofuel applications. In Kuila, A., & Kumar, D. (Eds.), *Optimizing Biofuel Production with Artificial Intelligence* (1–32). Scrivener Publishing LLC, Beverly, MA, USA.

Jayarathna, L., Kent, G., O'Hara, I., & Hobson, P. (2020). A geographical information system based framework to identify optimal location and size of biomass energy plants using single or multiple biomass types. *Applied Energy, 275*, 115398.

Kabeyi, M. J. B., Olanrewaju, O. A., & Akpan, J. (2024). Biogas Production and Process Control Improvements. In *From Biomass to Biobased Products*.

Khalid, H., Amin, F. R., Gao, L., Chen, L., Chen, W., Javed, S., & Li, D. (2024). Syngas conversion to biofuels and biochemicals: A review of process engineering and mechanisms. *Sustainable Energy & Fuels, 8*(1), 9–28.

Khan, T., Emon, M. M. H., Rahman, M. A., Hamid, A. B. A., & Yaakub, N. I. (2025). Bridging the gap: Realizing GreenTech potential. In R. Kumar, A. Abdul Hamid, N. Binti Ya'akub, T. Nyamasvisva, & R. Tiwari (Eds.), *AI and Green Technology Applications in Society* (91–122). IGI Global Scientific Publishing, USA.

Khazaleh, O. R., Alawneh, F. R., & Alkhuzaie, A. R. (2025). Machine learning-based prediction of wind power using MATLAB. *International Journal of Computing and Related Technologies, 5*(1), 49–60.

Kumar, R. S., Singh, A. R., Narayana, P. L., Chandrika, V. S., Bajaj, M., & Zaitsev, I. (2025). Hybrid machine learning framework for predictive maintenance and anomaly detection in lithium-ion batteries using enhanced random forest. *Scientific Reports, 15*(1), 6243.

Li, N., Qi, Y., Li, C., & Zhao, Z. (2024). Active learning for data quality control: A survey. *ACM Journal of Data and Information Quality, 16*(2), 1–45.

Lin, Y., Tang, J., Guo, J., Wu, S., & Li, Z. (2025). Advancing AI-enabled techniques in energy system modeling: A review of data-driven, mechanism-driven, and hybrid modeling approaches. *Energies, 18*(4), 845.

Liu, F., Wang, T., Liu, X. H., Xu, N., & Pan, X. L. (2025). Efficient biodegradation and upcycling of polyethylene terephthalate mediated by cell-factories. *Frontiers in Microbiology, 16*, 1599470.

Luo, Y., & Ierapetritou, M. (2023). Multifeedstock and multiproduct process design using neural network surrogate flexibility constraints. *Industrial & Engineering Chemistry Research, 62*(5), 2067–2079.

Madanchian, M. (2024). The role of complex systems in predictive analytics for e-commerce innovations in business management. *Systems, 12*(10), 415.

Maddela, N. R., Aransiola, S. A., Ezugwu, C. I., Winkelstroter Eller, L. K., Scalvensi, L., & Meng, F. (Eds.). (2024a). Microbial Biotechnology for Bioenergy (1st ed.). ISBN-9780443141126. https://doi.org/10.1016/C2022-0-02979-2

Mahmood, A., & Szabolcsi, R. (2025). A systematic review on risk management and enhancing reliability in autonomous vehicles. *Machines, 13*(8), 646.

Majumdar, P., Mitra, S., Bhattacharya, D., & Bhushan, B. (2025). Enhancing sustainable 5G powered agriculture 4.0: Summary of low power connectivity, internet of UAV things, AI solutions and research trends. *Multimedia Tools and Applications, 84*(17), 17389–17433.

Malashin, I., Martysyuk, D., Tynchenko, V., Gantimurov, A., Semikolenov, A., Nelyub, V., & Borodulin, A. (2024). Machine learning-based process optimization in biopolymer manufacturing: A review. *Polymers, 16*(23), 3368.

Malashin, I. P., Martysyuk, D. A., Tynchenko, V. S., Gantimurov, A. P., Nelyub, V. A., & Borodulin, A. S. (2025). Integrating machine learning and multi-objective optimization in biofuel systems: A review. *IEEE Access, PP*(99), 1–1.

Manikandan, S., Kaviya, R. S., Shreeharan, D. H., Subbaiya, R., Vickram, S., Karmegam, N., Kim, W., & Govarthanan, M. (2025). Artificial intelligence-driven sustainability: Enhancing carbon capture for sustainable development goals—A review. *Sustainable Development, 33*(2), 2004–2029.

Marella, B. C. C., & Palakurti, A. (2025). Harnessing python for AI and machine learning: Techniques, tools, and green solutions. In P. William & S. Kulkarni (Eds.), *Advancing Social Equity through Accessible Green Innovation* (237–250). IGI Global Scientific Publishing, USA.

Mehmood, K., Kiani, A., Rehman, H., Alshibani, S. M., & Piccardi, P. (2025). Can platform leadership drive twin transitions in greening SMEs? Exploring the nexus between AI infrastructure flexibility, information effects, and green sustainable practices. *Business Ethics, the Environment & Responsibility*, *34*(4), 2274–2292.

Mekuye, B., Mebratie, G., Abera, B., Yibeltal, A., Lake, A., & Tefera, A. (2024). Energy: An overview of type, form, storage, advantages, efficiency, and their impact. *Energy Science & Engineering*, *12*(12), 5678–5707.

Miller, T., Durlik, I., Kostecka, E., Kozlovska, P., Staude, M., & Sokołowska, S. (2025). The role of lightweight AI models in supporting a sustainable transition to renewable energy: A systematic review. *Energies*, *18*(5), 1192.

Mishra, A. S., Lade, V. G., Barmar, J. R., Bindwal, A. B., & Birmod, R. P. (2025). A mini review on revolutionizing hydrogenation catalysis: Unleashing transformative power of artificial intelligence. *Journal of Molecular Modeling*, *31*(5), 1–18.

Mpungu, I. L., Maube, O., Nziu, P., Mwasiagi, J. I., Dulo, B., & Bongomin, O. (2024). Optimizing waste for energy: Exploring municipal solid waste variations on torrefaction and biochar production. *International Journal of Energy Research*, *2024*(1), 4311062.

Ojadi, J. O., Owulade, O. A., Odionu, C. S., & Onukwulu, E. C. (2025). AI-driven optimization of water usage and waste management in smart cities for environmental sustainability. *Engineering and Technology Journal*, *10*(3), 4284–4306.

Ortiz, B. L., Gupta, V., Kumar, R., Jalin, A., Cao, X., Ziegenbein, C., Singhal, A., Tewari, M., & Choi, S. W. (2024). Data preprocessing techniques for AI and machine learning readiness: Scoping review of wearable sensor data in cancer care. *JMIR mHealth and uHealth*, *12*(1), e59587.

Paavani, K., Agarwal, K., Alam, S. S., Dinda, S., & Abrar, I. (2025). Advances in plastic to fuel conversion: Reactor design, operational optimization, and machine learning integration. *Sustainable Energy & Fuels*, *9*(1), 54–71.

Panda, S. (2025). Scalable artificial intelligence architectures: Cloud-native, edge-AI, and hybrid models. In: *Scalable Artificial Intelligence Systems: Cloud-Native, Edge-AI, MLOps, and Governance for Real-World Deployment* (1–26). Deep Science Publishing. New Delhi, India.

Pasupuleti, V., Thuraka, B., Kodete, C. S., & Malisetty, S. (2024). Enhancing supply chain agility and sustainability through machine learning: Optimization techniques for logistics and inventory management. *Logistics*, *8*(3), 73.

Patrício, L., Varela, L., & Silveira, Z. (2025). Proposal for a sustainable model for integrating robotic process automation and machine learning in failure prediction and operational efficiency in predictive maintenance. *Applied Sciences*, *15*(2), 854.

Pereira, G. R., Bispo, H., Costa, T. V. D., & Neiro, S. M. D. S. (2025). From process simulation to digital twin: Custom gasifier modelling, complex kinetics and optimisation in AVEVA process simulation. *Digital Twins and Applications*, *2*(1), e70009.

Pesapane, F., Hauglid, M. K., Fumagalli, M., Petersson, L., Parkar, A. P., Cassano, E., & Horgan, D. (2025). The translation of in-house imaging AI research into a medical device ensuring ethical and regulatory integrity. *European Journal of Radiology*, *182*, 111852.

Phogat, P. (2025). Waste-and refuse-derived fuels: Circular energy solutions from waste streams. *Journal of Material Science and Technology Research*, *12*, 22–44.

Priya, R., & Yadav, P. (2025). Cost Analysis of Artificial Intelligent-Based Biofuels Industry. *Optimizing Biofuel Production with Artificial Intelligence*, 285–310.

Quarteroni, A., Gervasio, P., & Regazzoni, F. (2025). Combining Physics-Based and Data-Driven Models: Advancing the Frontiers of Research with Scientific Machine Learning. *arXiv preprint arXiv:2501.18708*.

Qudus, L. (2025). Leveraging artificial intelligence to enhance process control and improve efficiency in manufacturing industries. *International Journal of Computer Applications Technology and Research*, *14*(02), 18–38.

Rabbi, M. F., Popp, J., Máté, D., & Kovács, S. (2022). Energy security and energy transition to achieve carbon neutrality. *Energies*, *15*(21), 8126.

Rai, P. R., Nanjundan, P., & George, J. P. (2024). Enhancing industrial operations through AI-driven decision-making in the era of Industry 4.0. In *AI-Driven IoT Systems for Industry 4.0* (42–55). CRC Press Taylor & Francis Group, London UK.

Ramesh, S. (2025). Renewable energy technologies. In *The Political Economy of Contemporary Human Civilisation, Volume II: From Quantum Computing and Nuclear Fusion to War and Conflict* (183–218). Palgrave Macmillan, Cham, Switzerland.

Rane, J., Chaudhari, R. A., & Rane, N. L. (2025). AI, and federated learning. In *Enhancing Sustainable Supply Chain Resilience through Artificial Intelligence and Machine Learning: Industry 4.0 and Industry 5.0 in Manufacturing* (163). Deep Science Publishing, New Delhi, India.

Reza, M. S., Iskakova, Z. B., Afroze, S., Kuterbekov, K., Kabyshev, A., Bekmyrza, K. Z., Kubenova, M. M., Bakar, M. S. A., Azad, A. K., Roy, H., & Islam, M. S. (2023). Influence of catalyst on the yield and quality of bio-oil for the catalytic pyrolysis of biomass: A comprehensive review. *Energies*, *16*(14), 5547.

Rimal, Y., Sharma, N., & Alsadoon, A. (2024). The accuracy of machine learning models relies on hyperparameter tuning: Student result classification using random forest, randomized search, grid search, bayesian, genetic, and optuna algorithms. *Multimedia Tools and Applications*, *83*(30), 74349–74364.

Ro, J. W., Zhang, Y., & Kendall, A. (2023). Developing guidelines for waste designation of biofuel feedstocks in carbon footprints and life cycle assessment. *Sustainable Production and Consumption*, *37*, 320–330.

Rocha-Meneses, L., Luna-delRisco, M., González, C. A., Moncada, S. V., Moreno, A., Sierra-Del Rio, J., & Castillo-Meza, L. E. (2023). An overview of the socio-economic, technological, and environmental opportunities and challenges for renewable energy generation from residual biomass: A case study of biogas production in Colombia. *Energies*, *16*(16), 5901.

Román, S. A. R. L., Collazos, S. Z., & Chavez, J. A. C. (2024). Model for reducing mean absolute percentage error through smoothing and time series forecasting in a tourism SME: A case study. *Journal of Machine Intelligence and Data Science (JMIDS)*, *5*, 109–116.

Sajja, G. S., Addula, S. R., Meesala, M. K., & Ravipati, P. (2025, June). Optimizing Inventory Management through AI-Driven Demand Forecasting for Improved Supply Chain Responsiveness and Accuracy. In *AIP Conference Proceedings* (Vol. 3306, No. 1, 050003).

Saju, L., Selvaraj, D., & Vairaperumal, T. (2025). Artificial intelligence and machine intelligence: Modeling and optimization of bioenergy production. In Dubey, A. K., Srivastav, A. L., Kumar, A., Pati, U. C., García Márquez, F. P., & García-Díaz, V. (Eds.), *Computer Vision and Machine Intelligence for Renewable Energy Systems* (163–176). *Elsevier, Amsterdam, The Netherlands.*

Sakib, M., Mustajab, S., & Alam, M. (2025). Ensemble deep learning techniques for time series analysis: A comprehensive review, applications, open issues, challenges, and future directions. *Cluster Computing*, *28*(1), 73.

Santoso, A., & Surya, Y. (2024). Maximizing decision efficiency with edge-based AI systems: Advanced strategies for real-time processing, scalability, and autonomous intelligence in distributed environments. *Quarterly Journal of Emerging Technologies and Innovations*, *9*(2), 104–132.

Schroer, H. W., & Just, C. L. (2023). Feature engineering and supervised machine learning to forecast biogas production during municipal anaerobic co-digestion. *ACS Es&t Engineering*, *4*(3), 660–672.

Sertolli, A., Gabnai, Z., Lengyel, P., & Bai, A. (2022). Biomass potential and utilization in worldwide research trends—A bibliometric analysis. *Sustainability*, *14*(9), 5515.

Seyam, A., Mathew, S. S., Du, B., El Barachi, M., & Shen, J. (2025). A stacking ensemble model for food demand forecasting: A preventative approach to food waste reduction. *Cleaner Logistics and Supply Chain*, *15*, 100225.

Seyar, M. H., & Ahamed, T. (2024). Optimization of soil-based irrigation scheduling through the integration of machine learning, remote sensing, and soil moisture sensor technology. In Ahamed, T. (Eds.), *IoT and AI in Agriculture: Smart Automation Systems for Increasing Agricultural Productivity to Achieve SDGs and Society 5.0* (275–299). Springer Nature, Singapore.

Shah, M., Wever, M., & Espig, M. (2025). A framework for assessing the potential of artificial intelligence in the circular bioeconomy. *Sustainability*, *17*(8), 3535.

Sharma, P., Gounder, M., & Kumar, K. (2024, July). Mastering Feature Engineering: Unlocking the Art of Data Transformation for Enhanced Predictive Modeling with Neural Networks. In *International Conference on Data Science and Applications* (451–464).

Sharma, S., Sharma, K., & Grover, S. (2024). Real-time data analysis with smart sensors. In Gulati, S. (Ed.), *Application of Artificial Intelligence in Wastewater Treatment* (127–153). Springer, Cham, Switzerland.

Shash, A. Y., Abdeltawab, N. M., Hassan, D. M., Darweesh, M., & Hegazy, Y. G. (2025). Computational methods, artificial intelligence, modeling, and simulation applications in green hydrogen production through water electrolysis: A review. *Hydrogen*, *6*(2), 21.

Shovon, M. S. S., Gomes, C. A., Reza, S. A., Bhowmik, P. K., Gomes, C. A. H., Jakir, T., Amjad, M. H. H., Hossain, A., & Hasan, M. S. (2025). Forecasting renewable energy trends in the USA: An AI-driven analysis of electricity production by source. *Journal of Ecohumanism*, *4*(3), 322–345.

Shrivastava, R., Rajak, R., Kumar, S., & Saxena, R. (2025). Bioenergy prediction using computer vision and machine intelligence: Modeling and optimization of bioenergy production. In Dubey, A. K., Srivastav, A. L., Kumar, A., Pati, U. C., García Márquez, F. P., & García-Díaz, V. (Eds.), *Computer Vision and Machine Intelligence for Renewable Energy Systems* (141–162). Elsevier, Amsterdam, The Netherlands..

Shukla, C., Lakhmani, D., & Rajasekhar Reddy, B. (2025). Hydrothermal liquefaction of solid residues and organic waste to bio-crude oil. In Aslam, M., Mishra, S., & Aburto Anell, J. A. (Eds.), *Production and Biorefining of Biocrude Oil: Current Status and Future Developments* (87–108). Springer Nature Singapore.

Shukla, R. P. (2025). Blockchain for enhanced transparency and traceability in circular supply chains. In E. Ozen, A. Singh, S. Taneja, R. Rajaram, & J. Davim (Eds.), *Innovating Sustainability through Digital Circular Economy* (203–228). IGI Global Scientific Publishing, Hershey, PA, USA.

Silva, T. A. L., Varão, L. H. R., & Pasquini, D. (2024). Lignocellulosic biomass. In Thomas, S., Hosur, M., Pasquini, D., & Jose Chirayil, C. (Eds.), *Handbook of Biomass* (105–142). Springer Nature, Singapore.

Simonič, M., & Klančnik, S. (2025). Temporal and statistical insights into multivariate time series forecasting of corn outlet moisture in industrial continuous-flow drying systems. *Applied Sciences*, *15*(16), 9187.

Simsek, A. I., Özer, C., & Tasar, İ. (2025). A novel hybrid machine learning model proposal for biodiesel consumption: A feature engineering based predictive framework. *Energy*, *333*, 137407.

Singh, S. S., Lim, L. T., & Manickavasagan, A. (2023). Imaging and spectroscopic techniques for microstructural and compositional analysis of lignocellulosic materials: A review. *Biomass Conversion and Biorefinery*, *13*(1), 499–517.

Sliper, N. J. (2024). Energy utilization and conversion in modern biomass conversion technologies. *Journal of Computational Intelligence in Materials Science*, *2*, 001–011.

Soltani, S., Abbasnejad, B., Gu, N., Yu, R., & Maxwell, D. (2025). A multi-faceted analysis of enablers and barriers of industrialised building: Global insights for the Australian context. *Buildings*, *15*(2), 214.

Soni, S. K., & Soni, R. (2025). Biochemical and biotechnological approaches: Harnessing Nature's toolbox for sustainable resource recovery. In *Green Biorefinery Solutions: Transforming Biodegradable Waste into Resources* (119–160). Springer Nature Singapore.

Soni, S. K., & Soni, R. (2025). Policy and regulatory frameworks for green biorefinery: Driving sustainable waste valorization. In *Green Biorefinery Solutions: Transforming Biodegradable Waste into Resources* (315–349). Springer Nature Singapore.

Sreekanth, K. J. (2024). Energy policy analyses, energy transition and sustainability. In *Energy Efficiency Improvements with Emission Abatement for Energy Sustainability: Energy Statistics* (135–167). Springer Nature Singapore, Singapore.

Suanpang, P., & Jamjuntr, P. (2024). Machine learning models for solar power generation forecasting in microgrid application implications for smart cities. *Sustainability*, *16*(14), 6087.

Swami, S., Suthar, S., Singh, R., Thakur, A. K., Gupta, L. R., & Sikarwar, V. S. (2025). Integration of anaerobic digestion with artificial intelligence to optimise biogas plant operation. *Environment, Development and Sustainability*, *27*(5), 9773–9803.

Szczepaniak, I., Olech, I., & Szymańska, E. J. (2025). The use of canola for biofuel production in the context of energy security—A systematic literature review. *Energies*, *18*(10), 2410.

Talaat, F. M., Kabeel, A. E., & Shaban, W. M. (2025). Towards sustainable energy management: Leveraging explainable artificial intelligence for transparent and efficient decision-making. *Sustainable Energy Technologies and Assessments*, *78*, 104348.

Tom, G., Schmid, S. P., Baird, S. G., Cao, Y., Darvish, K., Hao, H., Lo, S., Pablo-García, S., Rajaonson, E. M., Skreta, M., & Yoshikawa, N. (2024). Self-driving laboratories for chemistry and materials science. *Chemical Reviews*, *124*(16), 9633–9732.

Tyagi, U., Aslam, M., & Sarma, A. K. (2025). Current patents and innovation in biocrude production. In Aslam, M., Mishra, S., Aburto, J. A., Anell, A., & Sarma, A. K. (Eds.), *Biocrude Oil Biorefinery: An Emerging Biorefining Approach* (505–529). Springer Nature, Switzerland AG.

Ukoba, K., Olatunji, K. O., Adeoye, E., Jen, T. C., & Madyira, D. M. (2024). Optimizing renewable energy systems through artificial intelligence: Review nd future prospects. *Energy & Environment*, *35*(7), 3833–3879.

Varghese, R. T., Antony, T., & Chirayil, C. J. (2024). Biomass: Abundance, classification, energy potential. In Thomas, S., Chirayil, C.J., Varghese, R.T. (Eds.), *Handbook of Advanced Biomass Materials for Environmental Remediation.* Materials Horizons: From Nature to Nanomaterials. (1–12). Springer Nature, Singapore.

Wagner, M., Song, Q., Borg, M., Engström, E., & Lysek, M. (2025). AI Act High-Risk AI Compliance Challenge and Industry Impact: A Multiple Case Study. *Available at SSRN 5221279.*

Wang, R., He, Z., Chen, H., Guo, S., Zhang, S., Wang, K., & Ho, S. H. (2024). Enhancing biomass conversion to bioenergy with machine learning: Gains and problems. *Science of The Total Environment*, *927*, 172310.

Webster, S., Jeong, J., Mogonye, J. E., Zuback, J., Liao, S., Rocher, J., Ehmann, K., & Cao, J. (2025). Effect of closed-loop coaxial melt pool temperature control on thermal history and microstructure of nickel alloy 718 in directed energy deposition. *Journal of Materials Processing Technology*, *337*, 118725.

West, J., Siddhpura, M., Evangelista, A., & Haddad, A. (2024). Improving equipment maintenance—Switching from corrective to preventative maintenance strategies. *Buildings*, *14*(11), 3581.

Wu, W., Li, P., Huang, L., Wei, Y., Li, J., Zhang, L., & Jin, Y. (2023). The role of lignin structure on cellulase adsorption and enzymatic hydrolysis. *Biomass*, *3*(1), 96–107.

Xiao, G., Huang, J., Zhuo, W., Huang, H., Song, J., Du, K., Wang, J., Yuan, W., Sun, L., Zeng, Y., & Su, W. (2025). Progress and Perspectives of Crop Yield Forecasting with Remote Sensing: A Review. *IEEE Geoscience and Remote Sensing Magazine.*

Yin, S., Gao, Z., & Mahmood, T. (2025). Artificial intelligence-driven bioenergy system: Digital green innovation partner selection of bioenergy enterprises based on interval fuzzy field model. *Kybernetes*, *54*(3), 1344–1372.

Zhang, B., Biswal, B. K., Zhang, J., & Balasubramanian, R. (2023). Hydrothermal treatment of biomass feedstocks for sustainable production of chemicals, fuels, and materials: Progress and perspectives. *Chemical Reviews*, *123*(11), 7193–7294.

Zhang, L., Li, C., Wu, X., Xiang, H., Jiao, Y., & Chai, H. (2024). BO-CNN-BiLSTM deep learning model integrating multisource remote sensing data for improving winter wheat yield estimation. *Frontiers in Plant Science*, *15*, 1500499.

Zhang, Z., Liu, J., Yi, W., & Wang, S. (2025). Automated machine learning-assisted analysis of biomass catalytic pyrolysis for selective production of benzene, toluene, and xylene. *Energy*, *320*, 135389.

Zhu, C. Z., Samuel, O. D., Taheri-Garavand, A., Elboughdiri, N., Paramasivam, P., Hussain, F., Enweremadu, C. C., & Ayanie, A. G. (2025). ANN-ANFIS model for optimising methylic composite biodiesel from neem and castor oil and predicting emissions of the biodiesel blend. *Scientific Reports*, *15*(1), 5638.

14 Applications of Artificial Intelligence and Machine Learning in Food Quality Control and Safety Assessment

David Oyeyemi Aboyeji, Helen Shnada Auta, Isiaka Adedayo Adelere, Ifeanyi Famous Ossamulu, Taofeek Olajide Hammed, Hussaini Anthony Makun, and Hadiza Lami Muhammad

14.1 INTRODUCTION

Ensuring food quality and safety remains a global priority, especially as international food supply chains expand and food systems become more complex. The global food system is undergoing rapid transformation due to population growth, urbanization, the expansion of international supply chains, and the increasing complexity of food processing and distribution networks (Babaniyi *et al.*, 2026). Global food systems have grown in complexity in terms of production, processing, distribution, and retail and now operate at an unprecedented scale across longer supply chains and under higher throughput than ever before (Aransiola *et al.*, 2025; Savary *et al.*, 2020). This complexity raises new challenges for ensuring consistent food quality and safety, because the points at which contamination, adulteration, or degradation can occur have multiplied across multiple nodes of the value chain. Food-borne diseases continue to impose significant public health and economic burdens, with the World Health Organization (WHO) estimated over 600 million cases of food-borne illness annually (Mallapaty, 2022; Almaary, 2023). Hence, there is a great demand for reliable, rapid, and cost-effective methods of ensuring food quality and safety.

Traditional laboratory-based testing, focused on physical inspection, chemical assays, and microbiological analyses, has served as the backbone of food quality control for decades. However, these methods are often labor-intensive, time-consuming, and limited in their ability to meet the demands of

DOI: 10.1201/9781003727514-16

real-time quality assurance (Ain *et al.,* 2024; Aransiola et al., 2024; Mandal *et al.,* 2025). The methods cannot meet the speed, scale, and continuous monitoring needs of modern supply chains. Consequently, there is a rapidly growing interest in computational, automated methods that integrate sensor technologies, data analytics, and machine learning (ML) to provide continuous, proactive monitoring for real-time food quality control and predictive safety assessment (Chhetri, 2024; Yu *et al.,* 2025).

Artificial intelligence (AI) and ML have emerged as transformative technologies in modern food quality control capable of addressing these challenges. AI-driven systems can process complex datasets, provide tools to extract patterns from high-dimensional, multimodal sensor data (e.g., spectral, visual, and time-series), detect patterns invisible to human inspectors, and predict future states such as spoilage risk or microbial contamination at machine speeds that exceed the speed and accuracy of traditional manual techniques (Kazanskiy *et al.,* 2025; Mandal *et al.,* 2025). AI methods, from classical ML algorithms to advanced deep learning (DL) architectures, now support a wide range of tasks such as food grading, contamination detection, shelf-life prediction, microbial risk profiling, adulteration analysis, and real-time traceability (Chhetri, 2024; Dhal and Kar, 2025). In doing so, these models can automate decision-making on production lines, reduce inspection bottlenecks, lower human error, and trigger timely interventions before food quality deteriorates. Furthermore, by integrating AI with Internet of Things (IoT) sensors and cloud/edge computing, food businesses can move from reactive testing to proactive management, improving both safety outcomes and economic efficiency (Ficili *et al.,* 2025).

AI and ML are increasingly critical in achieving safe, traceable, efficient, and sustainable food supply chains. The food sector generates enormous volumes of data from IoT sensors, industrial cameras, hyperspectral imaging systems, supply-chain logs, and laboratory instruments that traditional statistical approaches cannot fully leverage. ML models can extract value from these datasets, offering continuous monitoring, early warning systems for hazards, and automated decision-support tools (Mu *et al.,* 2024). Liu *et al.* (2023) conducted a bibliometric analysis of AI applications in food safety over a decade, showing steep growth in research and identifying emerging hotspots such as predictive analytics, computer vision, and traceability. These advances highlight an important inflection point that AI is not just a research novelty but a practical tool reshaping food safety frameworks.

This chapter surveys the conceptual foundations of AI and ML applicability in food quality control and safety assessment. It reviews the relevant technologies and data architectures of AI and ML in food safety and presents major applications (e.g., vision-based sorting, shelf-life prediction, pathogen risk modeling, and adulteration detection) and their benefits across dairy, meat, produce, and beverage sectors. It also discusses practical challenges (data quality, deployment, and regulatory and ethical issues) and highlights future directions and opportunities of AI and ML for global food systems.

14.2 OVERVIEW OF ARTIFICIAL INTELLIGENCE AND MACHINE LEARNING

AI refers to computational systems capable of performing tasks that typically require human intelligence, including classification, prediction, reasoning, and pattern recognition. ML, a subset of AI, focuses on algorithms that improve performance with data. It enables systems to learn from data and improve their performance without explicit programming. ML is generally categorized into supervised learning, unsupervised learning, semi-supervised learning, and reinforcement learning, each with unique applications within the food sector (Table 14.1) (Jamali *et al.,* 2025).

Each approach differs in its methodology and analytical power and aligns differently with stages of the food value chain (production, storage, sorting, and distribution) and supports different operational goals (quality classification, predictive maintenance, and anomaly detection) (Chhetri, 2024).

TABLE 14.1 Common ML approaches

ML APPROACH	*EXPLANATION*
Supervised learning	Models are trained on labeled data (e.g., "defect/no defect" and "high/low contamination"). Examples include: CNNs – widely used for image-based inspection of produce, packaging, or defects; SVMs – useful for classification tasks such as defect detection; and Random Forests – robust for handling high-dimensional data in chemical analysis.
Unsupervised learning	Discovering structure in unlabeled data, e.g., anomaly detection in sensor streams. When labels are not available, clustering (e.g., k-means) or dimensionality reduction (PCA) helps discover structure in the data (e.g., grouping similar spoilage profiles).
Semi-supervised learning	Combines a small labeled dataset with a larger unlabeled dataset very useful in food contexts where obtaining labeled contamination examples is expensive or difficult.
Reinforcement learning	Agents learn sequential decision-making policies, such as how a robotic sorter should pick items to minimize damage or maximize yield.

DL, an ML subfield based on artificial neural networks, has rapidly gained attention due to its ability to automatically extract complex, hierarchical features from raw data. DL models such as convolutional neural networks (CNNs), recurrent neural networks (RNNs), vision transformers (ViTs), and Generative Models are increasingly used for image-based inspection, infrared spectroscopy analysis, contamination detection, and process optimization in food manufacturing (Trigka and Dritsas, 2025). In food applications, ML tasks typically include classification (e.g., defective vs. acceptable), regression/prediction (e.g., forecasting shelf-life), clustering (e.g., grouping samples with similar contamination profiles), and anomaly detection (e.g., identifying rare adulteration events or unexpected spoilage) (Liu *et al.*, 2023).

DL architectures enable more sophisticated modeling in food inspection. It has enabled breakthroughs in non-destructive testing, hyperspectral imaging, automated grading, and biomarker detection. CNNs outperform classical image-processing methods in defect detection and are widely used for feature extraction from imagery (RGB and hyperspectral), while recurrent networks (RNNs) and temporal architectures (e.g., LSTM) model time-series data (temperature logs and sensor drift) (Yang *et al.*, 2023). Transformer architectures and multimodal models have shown promise in multimodal food quality analysis and are emerging for fusing text (traceability logs), images, and spectra into unified predictions (Rashed *et al.*, 2025). The major AI techniques in food quality applications are summarized in Table 14.2.

TABLE 14.2 Major AI techniques in food quality applications

AI TECHNIQUE	*COMMON APPLICATION*	*ADVANTAGES*	*REFERENCES*
CNN	Visual defect detection	High accuracy in image tasks	Jha and Babiceanu (2023)
Random Forest	Shelf-life prediction	High robustness	Li *et al.* (2024)
SVM	Texture & sensory modeling	Performs well on small datasets	Liu *et al.* (2023)
ANN	Spectral data interpretation	Good nonlinear modeling	Jiang *et al.* (2023)
Transformer models	Multimodal data analysis	Handles complex, high-dimensional data	Rajender and Gopalachari (2024); Rashed *et al.* (2025)

The integration of AI into food systems is driven by three main factors, including growing availability of high-dimensional data (spectral data, imaging, IoT sensors, and microbial sequencing data), advances in computing power, enabling real-time inference, and demand for automation in food processing, driven by labor shortages and regulatory pressures (Wu *et al.,* 2025).

14.2.1 Emerging AI Technologies Relevant to Food Quality and Safety

Recent advances including Explainable AI (XAI) techniques like SHAP (SHapley Additive exPlanations) or LIME allow us to understand which features (e.g., spectral bands and pixels) drive a model's decision (Vimbi *et al.,* 2024). This is critical for regulatory transparency and trust. Generative AI (models) like generative adversarial networks (GANs) can simulate adulteration scenarios, sensory attributes, or generate synthetic data for rare events, aiding training when real data are scarce (Gupta *et al.,* 2025). Federated learning models can be trained across multiple sites (factories and labs) without centralizing sensitive data, improving privacy and collaboration (e.g., in global food supply networks). Digital twins simulate production lines that mirror real-world factories, enabling "what-if" safety analyses and safety scenario testing without disrupting operations, as well as for virtual simulations of food processing plants (Vidyalakshmi *et al.,* 2025). These technologies mitigate data sharing barriers and improve trust in automated decisions.

14.3 TRADITIONAL APPROACHES TO FOOD QUALITY CONTROL AND SAFETY ASSESSMENT

Food quality and safety assessment has historically relied on a combination of sensory evaluation, chemical analysis, microbiological testing, and visual inspection. Traditional methods form the foundation of food quality control. Physical methods include visual inspection, color measurement, and textural analysis. Classical quality-control tools include sensory panels, plate counts, and other microbiological assays, chromatography and mass spectrometry for chemical contaminants, and physical measurements (e.g., firmness and colorimetry). Chemical assays assess attributes such as acidity, moisture, fat content, and chemical residues, while microbiological tests, including culture-based techniques and biochemical identification, detect pathogens and spoilage organisms (Gillani *et al.,* 2024). Although these methods remain the regulatory gold standard across global food control systems, they face well-documented limitations related to speed, cost, scalability, and human subjectivity. A good understanding of these legacy methods provides essential context for appreciating why AI and ML have gained such unprecedented attention in the last decade.

14.3.1 Sensory and Manual Visual Inspection

For centuries, frontline quality assessment in food processing has depended on human inspectors who rely on sight, smell, and touch to judge attributes such as color uniformity, texture, shape, bruising, ripeness, and the presence of visible defects. In several industries such as fruit grading, fish filleting, and bakery product sorting, manual inspection is still dominant. However, sensory inspection is inherently subjective; judgments vary between inspectors based on experience, fatigue, lighting conditions, and cultural expectations (Ogidi *et al.,* 2025).

Multiple studies highlight the low interobserver reliability associated with manual grading, especially during high-throughput operations where speed compromises precision. For example, fruit packing facilities commonly report misclassification rates of 10–30% when relying solely on human sorting, with quality drift occurring during long shifts (Bollen and Prussia, 2022; Garg *et al.*, 2022; Walsh, 2018). These shortcomings have motivated the adoption of optical sorting and computer vision systems as more consistent alternatives.

14.3.2 Microbiological Testing

Microbiological analysis is central to food safety evaluation. Regulatory bodies typically require routine testing for pathogens such as *Salmonella, Listeria monocytogenes, E. coli* O157:H7, *Campylobacter*, and various spoilage organisms. Traditional microbiological methods include culture-based assays requiring selective media incubation (24–72 hours), colony counting used to estimate microbial load, biochemical tests for species differentiation, and PCR-based detection, which shortens diagnostic time but still requires laboratory infrastructure (Ferone *et al.,* 2020).

Although culture methods are accurate and standardized, their slow turnaround times make them poorly suited for real-time decision-making. A batch may sit in storage for days while awaiting microbiological clearance, increasing energy costs and risking unnoticed spoilage. Molecular techniques (e.g., qPCR and multiplex PCR) offer faster detection but remain resource-intensive and require skilled personnel (Kaushik *et al.,* 2025). Moreover, all laboratory sampling methods suffer from a fundamental limitation; only a small portion of a batch is tested, leaving room for heterogeneity and false negatives. These operational constraints are major drivers behind the growing interest in predictive microbiology and AI-based microbial risk modeling.

14.3.3 Chemical and Spectroscopic Analysis

Chemical assays are widely used to evaluate food composition, nutritional attributes, shelf-life indicators, and contaminants. Common analytical methods include: Chromatography (HPLC, GC-MS) for pesticide residues, adulterants, and chemical contaminants; Spectroscopy (FTIR, NIR, Raman, and UV-Vis) for compositional analysis; and Titration and wet chemistry for acidity, moisture, fat, protein, and peroxide values. These techniques are highly accurate and reproducible, forming the backbone of regulatory compliance testing. However, instrumentation is expensive, requiring capital investment and maintenance; skilled technicians are needed to operate analytical equipment, and sample preparation can be time-consuming and labor-intensive (Kumar, 2025). Also, throughput is limited with most methods, only capable of analyzing a few dozen samples per day.

In addition, chemical contaminants in food (e.g., mycotoxins, heavy metals, plasticizers, or industrial adulterants) often occur at trace levels requiring sensitive detection. Mycotoxins have been highlighted by the Food and Agriculture Organization (FAO) and the WHO as critical food safety issues, urging nations to implement strict monitoring and control mechanisms (Adelere *et al.,* 2025). While modern spectroscopic systems can detect such contaminants, AI-driven multivariate pattern recognition is increasingly used because traditional chemometric tools (PCA and PLS-DA) struggle with nonlinear relationships present in complex food matrices (Behmadi *et al.,* 2025).

14.3.4 Packaging Inspection and Physical Contaminant Detection

Traditional packaging inspection typically relies on manual visual checks for sealing defects, label errors, and deformities; X-ray systems for detecting foreign materials (glass and metal), and metal detectors to ensure physical safety compliance. These technologies work well for certain contaminants but cannot reliably detect low-density materials (e.g., plastic fragments, wood pieces, insects). Additionally, X-ray systems require controlled environments and may generate false positives on complex packaging layers (Yahyaei *et al.*, 2025). The industry trend is moving toward AI-driven X-ray and hyperspectral systems capable of detecting subtle density differences or spectral fingerprints, but most conventional factories still use older, rule-based systems.

14.3.5 Limitations of Traditional Approaches

Across all categories – sensory, microbiological, chemical, and physical inspection – traditional methods share several systemic limitations (Chhetri, 2024) (Table 14.3).

TABLE 14.3 Limitations of traditional approaches to food quality control and safety assessment

LIMITATION	*DESCRIPTION*	*EXAMPLES OF IMPACT IN FOOD SYSTEMS*
Slow and Reactive Testing	Traditional microbiological and chemical methods require long processing times (hours to days), causing delays in decision-making.	Batch release delays; products remain in cold storage awaiting results; and increased energy and operational cost.
Labor-Intensive and High Operational Costs	Skilled technicians are required for sample preparation, instrument operation, and interpretation; equipment is expensive.	High testing costs for small producers; limited testing frequency; and challenges scaling testing at high throughput.
Limited Sampling Coverage	Only small subsamples can be tested from large batches, leaving room for undetected heterogeneity or localized contamination.	False negatives in pathogen testing; undetected spoilage pockets in bulk storage; and adulterants missed across large lots.
Lack of Continuous Monitoring	Conventional methods provide periodic results rather than real-time insights, making them unsuitable for dynamic supply chains.	Inability to detect rapid spoilage events; missed temperature excursions during transport; and delayed contamination alerts.
Difficulty with Complex or High-Dimensional Data	Traditional chemometric tools struggle with nonlinear relationships and large datasets (e.g., hyperspectral, multispectral, and genomic data).	Slow or inaccurate interpretation of spectral data and inability to detect subtle adulteration signatures without AI.
Human Subjectivity and Variability	Manual inspections depend heavily on inspector experience, physical conditions, and fatigue levels.	Inconsistent grading of fruits/vegetables; variation in defect detection; and inconsistent assessments during long shifts.

These constraints create a "detection gap" between sampling points that AI-enabled sensors and predictive models can help close. Major drivers for AI-based innovations include stricter regulation, consumer demand for traceability, reduction of food waste through better shelf-life prediction, and the falling cost of sensors and computation. Increased digitalization of supply chains also produces the data necessary for ML, making AI solutions more feasible than a decade ago (Liu *et al.*, 2023). Recent studies emphasize that as sensor technology advances, traditional methods alone cannot handle the data complexity generated by modern imaging, spectroscopy, and IoT systems. AI bridges this gap by enabling deep, fast, and scalable interpretation (Ahmed, 2024; Liu *et al.*, 2023; Sufi, 2025).

14.4 ROLE OF ARTIFICIAL INTELLIGENCE AND MACHINE LEARNING IN FOOD QUALITY ASSESSMENT

AI and ML have emerged as transformative tools in modern food quality assurance. Their capacity to analyze large datasets, detect subtle patterns, automate decision-making, and enhance predictive accuracy positions them as powerful alternatives to traditional food inspection approaches. Unlike manual or laboratory-based analysis, AI systems can rapidly process images, spectral signatures, sensor data, and real-time environmental metrics to provide reliable and scalable assessment of food quality (Liang *et al.*, 2025). Figure 14.1 gives a framework for AI-enabled food quality and safety.

AI-driven automation minimizes human intervention, reduces inspection time, and enhances throughput, especially in large-scale production environments. ML systems integrated with high-speed cameras and robotics can evaluate thousands of items per hour, which is well beyond human capabilities (Waqar *et al.*, 2024). Automated grading systems for fruits, vegetables, grains, and meat products allow producers to maintain uniform quality standards while lowering operational costs.

Traditional methods often struggle with subtle or nonlinear quality indicators. In contrast, AI models, particularly DL, excel at pattern identification, enabling high accuracy even under complex conditions. CNNs, for example, can classify meat freshness, detect mold on cereal grains, or identify bruises on fruits with >95% accuracy in many industrial deployments (Nazir *et al.*, 2025).

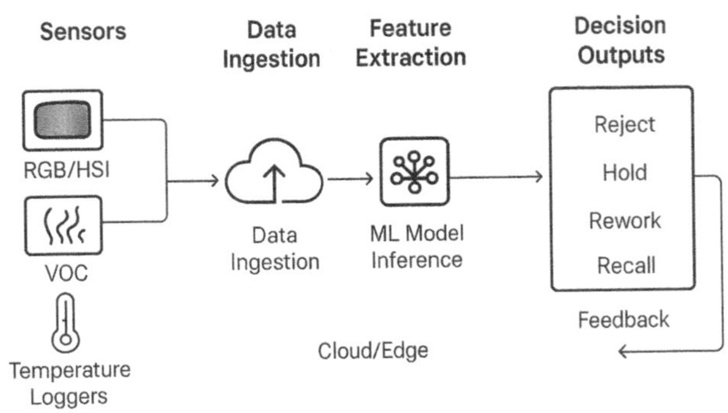

FIGURE 14.1 Conceptual framework for AI-enabled food quality and safety.

One of the most significant contributions of AI lies in predictive modeling for early detection. By analyzing environmental parameters (temperature, humidity, and pH), microbial growth kinetics, or time-series sensor data, ML models can forecast spoilage events before they are visible (Shehzad, 2025). Examples include time-series ML models (predicting bacterial growth in meat packages), gas-sensor-based AI systems (detecting volatile organic compounds (VOCs) associated with spoilage), and AI-enhanced biosensors, which identify trace levels of pathogens such as *E. coli, Salmonella,* or *Listeria* with high sensitivity (Zhao *et al.,* 2025). Predictive analytics support proactive decision-making, allowing producers to adjust storage conditions, modify processing parameters, or remove risky products before they reach consumers.

14.4.1 AI/ML Success in Food Quality Assessment

A CNN-based vision system was employed to detect defects such as rind diseases, pigmentation issues, and bruising in citrus. The AI system increased defect detection accuracy from 78% (manual inspection) to 96% (automated), reduced labor costs by 40%, and improved overall grading consistency (Yu *et al.,* 2025). A meat processor deployed an AI-driven E-nose system to detect spoilage VOCs. The ML model predicted spoilage 24–48 hours before traditional microbiological assays, enabling early removal of compromised batches and saving an estimated $1.2 million annually in waste reduction (Ramadan *et al.,* 2025). Researchers developed a hyperspectral imaging (HSI) + CNN model for detecting subtle adulterants (e.g., melamine and urea) in milk (Aqeel *et al.,* 2024). The model achieved >98% classification accuracy, outperforming FTIR chemical analysis while requiring no chemical reagents.

Dairy plants have applied ML for detecting milk adulteration, predicting spoilage based on fermentation kinetics, and monitoring cheese ripening stages using spectral sensors. Studies show improved early detection of contamination and better consistency in quality grading compared to manual inspection (Agiomavriti *et al.,* 2024; Liakos *et al.,* 2025). ML assists with microbial risk mapping in slaughterhouses, automated bruise and foreign body detection on carcasses, and monitoring cold chain integrity. Predictive models for microbial growth during storage can optimize holding times and freezing strategies, reducing recalls (Tarlak, 2023). Breweries and beverage manufacturers use ML to correlate chemical profiles with sensory outcomes, enabling flavor consistency across batches. Table 14.4 gives AI/ML contributions to food safety.

TABLE 14.4 Benefits of AI/ML in food quality assessment

AI/ML CONTRIBUTION	*BENEFITS*	*FOOD INDUSTRY APPLICATION EXAMPLES*	*REFERENCES*
Automation & Efficiency	High throughput, reduced cost	Sorting fruits/vegetables, automated grading	Tariq *et al.* (2025)
Higher Accuracy	Detects subtle defects/ contaminants	Meat freshness, cereal grain defects	Singh *et al.* (2025)
Predictive Modeling	Early spoilage/ contamination alerts	VOC-based spoilage prediction	Buwa *et al.* (2026)
Real-Time Monitoring	Continuous sensor-driven tracking	Smart packaging, IoT cold-chain monitoring	Nasrudin *et al.* (2025)
Non-Destructive Testing	Preserves product integrity	Hyperspectral inspection, imaging	Aqeel *et al.* (2024); Xu *et al.* (2024)

14.5 SPECIFIC APPLICATIONS OF AI AND ML IN FOOD SAFETY AND QUALITY CONTROL

AI and ML have become indispensable in addressing modern food safety challenges. They support early detection of hazards, ensure compliance with safety standards, and enhance traceability across supply chains. Unlike traditional inspection methods, which are often slow, reactive, and labor-intensive, AI-driven solutions enable proactive, real-time quality evaluation and risk prediction (Buwa *et al.*, 2026).

Food-borne pathogens such as *Salmonella, L. monocytogenes*, and *Escherichia coli* continue to cause significant global public health concerns. AI and ML enhance detection through rapid pattern recognition, laboratory automation, and interpretation of complex biological signals. A hybrid CNN-SVM model trained on hyperspectral data detected *Salmonella* contamination in poultry carcasses within seconds, outperforming optical inspection technologies and conventional plating (Soni *et al.*, 2022).

Food adulteration, whether economic (e.g., dilution and substitution) or malicious, poses significant safety risks. AI and ML support non-destructive adulteration detection using spectral, chemical, and visual data. Common targets include milk adulteration (e.g., melamine, urea, and water), olive oil dilution with cheaper vegetable oils, and grain and spice adulteration (e.g., Sudan dyes in chili powder). A deep-learning HSI model accurately identified melamine adulteration down to 0.1% concentration in milk, surpassing standard chemical methods (Aqeel *et al.*, 2024).

Furthermore, AI improves agricultural efficiency by optimizing crop management, pest control, and yield prediction. By enabling up to 25% more effective irrigation and lowering fertilizer consumption by 20%–30% through data-driven decision-making, AI-based systems enhance crop management, planting schedules, and resource utilization, all of which boost productivity (Kutyauripo *et al.*, 2023). Ferentinos (2018) noted that AI systems that use image recognition are able to detect diseases with up to 99.5% accuracy in pest control and have contributed to a 30–50% decrease in the use of pesticides. Similarly, You *et al.* (2017) discovered that AI models that use meteorological and remote sensing data to estimate yield have achieved R2 values of up to 0.89, outperforming conventional techniques in predicting accuracy. Table 14.5 presents the major application areas of AI/ML in food quality and safety management.

14.6 CHALLENGES, LIMITATIONS, AND ETHICAL CONSIDERATIONS

Despite the significant breakthroughs made possible by AI and ML, their application in food safety and quality management is not without obstacles. These challenges span technical, economic, infrastructural, regulatory, and ethical dimensions.

Reliable ML depends on representative, labeled datasets (Table 14.6). For many food safety events (e.g., contamination), labeled instances are rare, leading to class imbalance. Standardized, open datasets are scarce, hampering reproducibility and cross-site validation. Federated learning and curated public datasets could help but require governance frameworks (Liu *et al.*, 2023). A study by Ünal *et al.* (2024) showed that apple bruise detection models trained on laboratory images performed poorly when tested under real supply-chain lighting due to data mismatch.

Models trained on a narrow set of products or conditions can perform poorly when exposed to new varieties or geographies. DL methods such as CNNs often operate as "black boxes," making

TABLE 14.5 AI/ML applications in food safety and quality control

APPLICATION AREA	*AI/ML TECHNIQUES USED*	*DATA SOURCES/ INPUTS*	*DETAILED FUNCTIONS*	*EXAMPLES*	*KEY OUTCOMES & BENEFITS*	*REFERENCES*
Detection of Food-borne Pathogens	CNNs, SVM, Random Forests, Deep Neural Networks, ML-integrated biosensors.	Microbial images, biosensor signals, genomic data, hyperspectral signatures.	Identify microbes; classify pathogens; detect contamination; genomic fingerprinting.	CNN-SVM hybrid detected *Salmonella*; ML biosensors detected *E. coli*.	Accuracy >95%; faster than plating/PCR; on-site testing.	Soni *et al.* (2022); Chhetri (2024); Karim *et al.* (2025)
Prediction & Monitoring of Food Spoilage	LSTM, GRU, Time-Series Forecasting, ANN	VOC patterns, temp/humidity, smart packaging sensors.	Predict spoilage timeline; model microbial growth; detect VOCs.	LSTM predicted beef spoilage 48 h early; sensor-ML for fish freshness.	Early alerts; reduced waste; improved cold-chain safety.	Shehzad (2025); Zhao *et al.* (2025); Buwa *et al.* (2026)
Detection of Food Adulteration	CNNs+HSI, SVM, PCA+ML, Random Forests	NIR/FTIR spectra, HSI cubes, chemical fingerprints	Identify adulterants; quantify adulteration; classify purity	HSI-CNN detected melamine at 0.1%; SVM detected olive oil adulteration.	High sensitivity; non-destructive; authenticates foods	Liu *et al.* (2023); Aqeel *et al.* (2024)
Quality Grading of Fresh Produce	Computer Vision, CNNs, Transfer Learning, k-means	RGB, spectral images, 3D scans	Detect defects; color uniformity; size/shape; ripeness	Tomato grading ↑35%; CNN for mango ripeness.	Uniform grading; real-time sorting; reduced human error.	Soltani Firouz and Sardari (2022); Tariq *et al.* (2025)
Hazard Identification & Predictive Risk Modeling	Bayesian Networks, Decision Trees, RF, Gradient Boosting, ML-IoT	Sensor logs, contamination history, environmental data	Predict hazards; assess risks; model environment-biological interactions	ML reduced dairy contamination by 18%	Proactive mitigation; compliance; faster corrective action.	Liu *et al.* (2023); Taiwo *et al.* (2024)

APPLICATION AREA	*AI/ML TECHNIQUES USED*	*DATA SOURCES/ INPUTS*	*DETAILED FUNCTIONS*	*EXAMPLES*	*KEY OUTCOMES & BENEFITS*	*REFERENCES*
Optimization of Processing Parameters	Reinforcement Learning, ANN optimization, Genetic Algorithms	Temp, pressure, pH, irradiation dose	Optimize sterilization/ pasteurization; control fermentation	RL improved canned-vegetable safety, less nutrient loss	Safer processing; energy/cost savings; consistent quality	Jafari *et al.* (2025); Jha *et al.* (2026)
Traceability and Supply Chain Integrity	Blockchain+AI, LSTM, anomaly detection	Supply chain logs, RFID, IoT sensors	Detect anomalies; predict cold-chain failure; authenticate origin	Blockchain-AI improved seafood traceability	Better safety; reduced recall time; transparency	Almabrok (2023); Liu *et al.* (2023)
Allergen Detection & Risk Assessment	SVM, Deep Learning, biosensor-AI	Immunoassay signals, optical biosensors, NIR	Detect allergens; quantify levels; predict cross-contamination	AI biosensor detected peanut protein at trace levels	Improved allergen safety; supports sensitive consumers	Taiwo *et al.* (2024); Goumas *et al.* (2025)
Consumer Sensory Analysis	E-nose/E-tongue, PCA-ANN, Deep Learning	Gas/taste sensors, aroma patterns	Analyze aroma/flavor; predict acceptance; consistency control	E-nose+ML predicted coffee flavor >90%	Objective evaluation; faster product development	Ling and Heldman, (2025)
Packaging Integrity & Tamper Detection	CNNs, anomaly detection, sensor ML	X-ray, pressure sensors, vibration data	Detect leaks, punctures, seal failures	ML detected micro-leaks in vacuum-packed meat	Reduced failures; improved shelf stability	Chhetri (2024); Liu *et al.* (2023)

TABLE 14.6 Summary of challenges and limitations of AI/ML in food safety and quality control

CHALLENGE AREA	*DESCRIPTION OF THE LIMITATION*	*IMPLICATIONS*	*TYPICAL EXAMPLES*	*REFERENCES*
Data Quality & Availability Issues	Limited, inconsistent, or poorly annotated datasets hinder effective model training. Class imbalance and proprietary data restrictions further reduce model reliability.	Reduced model accuracy, overfitting, and poor generalization in real production environments.	Poor performance of bruise detection models trained only on lab images and inconsistent microbial datasets.	Deléglise *et al.* (2022); Qin *et al.* (2025)
Low Interpretability & Transparency	Many ML models (especially deep learning) operate as "black boxes," offering limited explainability.	Regulatory compliance difficulties, reduced trust, inability to justify or defend choices in audits.	Difficulty explaining why a batch of milk or meat is rejected by automated inspection systems.	Chhetri (2024); Liakos *et al.* (2025)
Infrastructure Limitations	Limited internet access, low compute power, lack of sensors, and unreliable electricity reduce AI feasibility, especially in low-resource regions.	Interrupted monitoring, unreliable predictions, difficulties scaling real-time inspection systems.	Incomplete cold-chain temperature monitoring due to connectivity issues.	Katende (2025)
High Implementation Costs	High costs for sensors, hyperspectral imaging systems, cloud deployment, and skilled personnel limit adoption among small and medium enterprises.	Slow adoption, financial reluctance, and uneven technological distribution in the food sector.	SMEs unable to afford high-resolution imaging platforms for defect detection.	Katende (2025)
Regulatory and Ethical Challenges	Unclear regulations for AI-driven systems, liability concerns, data governance issues, and risk of algorithmic bias.	Slow certification, safety concerns, legal ambiguity, and consumer mistrust.	In cases of misclassification, it is unclear who is legally responsible – the producer, AI system provider, or model developer.	Levina and Mattern (2023)
Environmental Sustainability Issues	AI models (especially deep learning) consume significant energy, require large computational resources, and increase e-waste.	Increased carbon footprint, sustainability concerns, pressure on producers to justify resource use.	Energy-intensive hyperspectral image processing during model training.	Chhetri (2024); Oise and Konyeha (2025)
Integration Challenges in Real Food Systems	Difficulties integrating AI tools with legacy equipment, seasonal variation, differing real-world conditions, and staff resistance.	Many AI pilot projects fail to scale; models perform well in trials but poorly on actual production lines.	Computer vision sorting systems failing under variable lighting and contamination.	Singh and Adhikari (2023)

it difficult to understand how they reach decisions. Black-box DL models complicate regulatory acceptance; XAI techniques must be adopted to provide traceable justifications for decisions used in enforcement or recalls (Şahin *et al.*, 2025). As highlighted by Chhetri (2024), low model interpretability poses barriers to certification and adoption in regulated sectors such as meat inspection and allergen detection.

Regulatory authorities are still adapting to AI outputs used for compliance. Questions remain around who is liable for false negatives/positives, how AI models must be validated for food safety decisions, and how to handle model updates. Clear standards for validation, traceability, and audit trails are needed. In addition, models trained on biased datasets may disproportionately misclassify specific product types or production regions (Van Giffen *et al.*, 2022). The European Food Safety Authority Emerging Risks, EFSA(2024) has begun setting global precedents, classifying food safety AI systems as "high-risk," requiring extensive documentation and explanations (Röhrs *et al.*, 2024).

14.6.1 Challenges and Barriers to AI/ML Adoption in Food Safety and Quality Control

AI and ML offer transformative capabilities for food safety, quality control, and supply chain monitoring. However, their adoption across the global and regional food industries, especially in low- and middle-income countries, faces practical, technical, economic, and regulatory constraints. These limitations must be understood to develop strategies that ensure equitable access and effective deployment of AI-driven food safety systems. Figure 14.2 presents the key barriers and challenges that limit the widespread implementation of AI/ML in the food sector.

FIGURE 14.2 Challenges and barriers to AI/ML adoption in food safety and quality control.

14.7 FUTURE TRENDS AND OPPORTUNITIES IN AI/ML FOR FOOD SAFETY AND QUALITY CONTROL

The next decade will witness accelerated integration of AI and ML within the global food system. While current applications focus mainly on defect detection, supply-chain monitoring, predictive microbiology, and sensor-based inspection, emerging research signals a shift toward more autonomous, intelligent, and interconnected food-safety ecosystems.

Advances in multimodal models, self-supervised learning, and small-footprint DL (for edge devices) will make robust models easier to deploy. Federated learning will enable cross-company model improvement while preserving privacy, while explainable and certified AI pipelines will be required for regulatory acceptance (Farahani and Monsefi, 2023). These advances will make spectral imaging accessible even to small and medium-scale food producers. Combining data streams from multiple sensors (e.g., NIR + VOC + thermal + acoustic) produces stronger and more robust predictions (Wang *et al.*, 2024). Future AI systems will integrate these multimodal data sources into unified quality scores and automated decision systems. AI and ML will transform food safety through advanced imaging, robotics, blockchain traceability, sensor fusion, and explainable systems. The future will emphasize autonomy, transparency, scalability, and inclusivity, enabling safer, more efficient, and more sustainable food systems worldwide.

There is a strong opportunity for AI to improve food safety in low- and middle-income countries by enabling low-cost sensor deployment and cloud analytics. Partnerships and open datasets are keys to ensuring inclusive benefits and preventing technology concentration in wealthy markets. Figure 14.3 shows the key technological, regulatory, and societal trends shaping the future of AI-driven food quality and safety.

FIGURE 14.3 Future trends and opportunities in AI/ML for food safety and quality control.

14.8 CONCLUSION

AI and ML are increasingly transforming global food safety and quality control. As food systems become more complex and consumers demand higher transparency and reliability, AI-driven tools offer opportunities that far exceed traditional inspection and monitoring methods. These technologies enable rapid data analysis, real-time surveillance, predictive modeling, and automated decision-making across the entire farm-to-fork chain. Current applications, such as computer-vision-based defect detection, hyperspectral imaging, microbial risk prediction, smart packaging, and supply-chain traceability, demonstrate substantial improvements in accuracy, efficiency, and regulatory compliance. Various food sectors confirm that AI not only enhances operational performance but also reduces human error and supports continuous improvement through IoT-enabled smart systems. Despite these benefits, several barriers hinder the widespread adoption of AI and ML. Challenges include poor data quality, interoperability issues, high deployment costs, limited technical expertise, and evolving regulatory frameworks. Overcoming these limitations will require coordinated collaboration among industry stakeholders, researchers, technology developers, and policymakers. Looking ahead, the future of AI in food safety is defined by deeper automation, stronger data ecosystems, and more integrated digital platforms. Advancements in robotics, advanced sensors, and privacy-preserving analytics will expand opportunities, especially in low- and middle-income countries where resource-efficient solutions are vital. Hence, AI and ML represent a central pillar in the modernization of global food safety management. With appropriate policy support, workforce development, and technological integration, these tools will play an increasingly critical role in ensuring food integrity, reducing waste, protecting consumers, and strengthening the resilience and sustainability of food systems worldwide.

REFERENCES

Adelere, I. A., Aboyeji, D. O., Auta, H. S., Barasarathi, J., & Ahmed, A. (2025). Mycotoxins and global food safety. In: Aransiola S. A., Kotun B. M. and Naga R. M. (Eds.). *Biotoxins in Food: Threats and Benefits* (pp. 96–134). Boca Raton, FL: CRC Press.

Agiomavriti, A. A., Nikolopoulou, M. P., Bartzanas, T., Chorianopoulos, N., Demestichas, K., & Gelasakis, A. I. (2024). Spectroscopy-based methods and supervised machine learning applications for milk chemical analysis in dairy ruminants. *Chemosensors*, *12*(12), 263.

Ahmed, R. M. (2024). Integration of wireless sensor networks, internet of things, artificial intelligence, and deep learning in smart agriculture: A comprehensive survey: Integration of wireless sensor networks, internet of things. *Journal of Innovative Intelligent Computing and Emerging Technologies (JIICET)*, *1*(01), 8–19.

Ain, Q. U., Nazir, R., Nawaz, A., Shahbaz, H., Dilshad, A., Mufti, I. U., & Iftikhar, M. (2024). Machine learning approach towards quality assurance, challenges and possible strategies in laboratory medicine. *Journal of Clinical and Translational Pathology*, *4*(2), 76–87.

Almaary, K. S. (2023). Food-borne diseases and their impact on health. *Biosciences Biotechnology Research Asia*, *20*(3), 745–755.

Almabrok, H. A. (2023). Blockchain for supply chain management: To enhance transparency, traceability, and efficiency. *African Journal of Advanced Pure and Applied Sciences (AJAPAS)*, 2(4), 239–253.

Aqeel, M., Sohaib, A., Iqbal, M., & Ullah, S. S. (2024). Hyperspectral identification of milk adulteration using advance deep learning. *IEEE Access*, *12*, 174965–174982.

Aransiola, S. A., Akinsola, R. O., Abioye, O. P., & Maddela, N. R., (Eds.). (2024). *Emerging Contaminants in Food and Food Products* (1st ed.). Boca Raton, FL, USA: CRC Press. Taylor and Francis Group. https://doi.org/10.1201/9781003438236

Aransiola, S. A., Kotun, B. C., & Maddela, N. R. (Eds.). (2025). *Biotoxins in Food: Threats and Detection*. CRC Press. https://doi.org/10.1201/9781003500605

Babaniyi, G. G., Babaniyi, B. R., Aransiola, S. A., & Sawicka, B. (Eds.). (2026). *Biotechnology of Climate-Smart Agriculture in Food Security and Sustainable Farming*. CRC Press. https://doi.org/10.1201/9781003619284

Behmadi, H., Azarpazhooh, E., Ramaswamy, H. S., & Ahmadauli, K. (2025). Artificial intelligence and machine learning in food analysis. In *Novel Food Analysis: Techniques and Applications* (pp. 391–418). New York, NY: Springer US.

Bollen, A. F., & Prussia, S. E. (2022). Sorting for defects. In *Postharvest Handling* (pp. 377–398). Cambridge, MA: Academic Press.

Buwa, V., Khan, N., Selvaraj, R. C. A., & Nlebedim, V. (2026). Quality control and inspection techniques with AI and ML. In *Artificial Intelligence in Food Science* (pp. 425–445). Cambridge, MA: Academic Press.

Chhetri, K. B. (2024). Applications of artificial intelligence and machine learning in food quality control and safety assessment. *Food Engineering Reviews*, *16*(1), 1–21.

Deléglise, H., Interdonato, R., Bégué, A., d'Hôtel, E. M., Teisseire, M., & Roche, M. (2022). Food security prediction from heterogeneous data combining machine and deep learning methods. *Expert Systems with Applications*, *190*, 116189.

Dhal, S. B., & & Kar, D. (2025). Leveraging artificial intelligence and advanced food processing techniques for enhanced food safety, quality, and security: A comprehensive review. *Discover Applied Sciences*, *7*(1), 75.

European Food Safety Authority Emerging Risks, EFSA (2024). [(accessed on 18 June 2024)]. Available online: https://www.efsa.europa.eu/en/topics/topic/emerging-risks?wtrl=01

Farahani, B., & Monsefi, A. K. (2023). Smart and collaborative industrial IoT: A federated learning and data space approach. *Digital Communications and Networks*, *9*(2), 436–447.

Ferentinos, K. P. (2018). Deep learning models for plant disease detection and diagnosis. *Computers and Electronics in Agriculture*, *145*, 311–318.

Ferone, M., Gowen, A., Fanning, S., & Scannell, A. G. (2020). Microbial detection and identification methods: Bench top assays to omics approaches. *Comprehensive Reviews in Food Science and Food Safety*, *19*(6), 3106–3129.

Ficili, I., Giacobbe, M., Tricomi, G., & Puliafito, A. (2025). From sensors to data intelligence: Leveraging IoT, cloud, and edge computing with AI. *Sensors*, *25*(6), 1763.

Garg, S., Kolli, V. S., & Shirkole, S. S. (2022). Sorting operations for the classification of agricultural crops. In *Postharvest and Postmortem Processing of Raw Food Materials* (pp. 53–76). Duxford, Cambridge, UK: Woodhead Publishing.

Gillani, S. N., Ahmad, T., Rehman, A., Saba, I., Irshad, S. B. E., Abbas, M., & Ali, K. (2024). The application of chemical analysis techniques in microbiology research: A review of methods, advancements, and implications for food and nutrition. *Applied Chemical Analysis Techniques in Microbiology Research*, *13*(01), 72–81.

Goumas, G., Vlachothanasi, E. N., Fradelos, E. C., & Mouliou, D. S. (2025). Biosensors, artificial intelligence biosensors, false results and novel future perspectives. *Diagnostics*, *15*(8), 1037.

Gupta, R., Tiwari, S., & Chaudhary, P. (2025). Generative AI techniques and models. In *Generative AI: Techniques, Models and Applications* (pp. 45–64). Cham: Springer Nature Switzerland.

Jafari, M., Nazari, R. R., Rezaei, M., Moazzen, M., & Shariatifar, N. (2025). Machine learning for detoxification of aflatoxin M1 by lactococcus lactis probiotic in kashk production. *Food Safety and Risk*, *12*(1), 4.

Jamali, V., Sachithanandhan, V. K., Nambi, E., Loganathan, M., Shanmugasundaram, S., & Chandrasekar, V. (2025). AI and machine learning applications. In *Emerging Non-Destructive Technologies for Food Quality Analysis* (pp. 501–555). Cham: Springer Nature Switzerland.

Jha, A., Kalaiselvi, K., Nandhini, R. S., Raj, A. S., & Karthik, P. (2026). Advancements in computational fluid dynamics-aided artificial intelligence and machine learning in food processing. In *Artificial Intelligence in Food Science* (pp. 359–376). Cambridge, MA: Academic Press.

Jha, S. B., & Babiceanu, R. F. (2023). Deep CNN-based visual defect detection: Survey of current literature. *Computers in Industry*, *148*, 103911.

Jiang, Q., Zhang, M., Mujumdar, A. S., & Wang, D. (2023). Non-destructive quality determination of frozen food using NIR spectroscopy-based machine learning and predictive modelling. *Journal of Food Engineering*, *343*, 111374.

Karim, S. N., Hew, P. S., Anwar, F., Sukor, R., Jambari, N. N., Sanny, M., & Khatib, A. (2025). Innovative technological approaches in the detection and mitigation of food toxicants. *Food Reviews International*, 1–45. https://doi.org/10.1080/87559129.2025.2544980

Katende, R. (2025). Rethinking data-efficient artificial intelligence for low-resource settings. *Machine Learning with Applications*, *25*,100796.

Kaushik, R., Mahajan, S., Dass, G., & Chatli, M. K. (2025). Introduction to molecular diagnostics and viral diseases. In *Molecular Diagnostics for Viral Diseases: Challenges and Emerging Concepts* (pp. 1–28). Singapore: Springer Nature Singapore.

Kazanskiy, N., Khabibullin, R., Nikonorov, A., & Khonina, S. (2025). A comprehensive review of remote sensing and artificial intelligence integration: Advances, applications, and challenges. *Sensors*, *25*(19), 5965.

Kumar, D. (2025). Spectroscopic techniques in chemical analysis: A review of emerging trends. *Indian Journal of Pharmaceutical Chemistry and Analytical Techniques (IJPCAT)*, *1*(3), 55–70.

Kutyauripo, I., Rushambwa, M., & Chiwazi, L. (2023). Artificial intelligence applications in the agrifood sectors. *Journal of Agriculture and Food Research*, *11*, 100502.

Levina, O., & Mattern, S. (2023). Ethical and legal analysis of machine learning based systems: A scenario analysis of a food recommender system. In *Recommender Systems: Legal and Ethical Issues* (pp. 165–185). Cham: Springer International Publishing.

Li, D., Bai, L., Wang, R., & Ying, S. (2024). Research progress of machine learning in extending and regulating the shelf life of fruits and vegetables. *Foods*, *13*(19), 3025.

Liakos, K. G., Athanasiadis, V., Bozinou, E., & Lalas, S. I. (2025). Machine learning for quality control in the food industry: A review. *Foods*, *14*(19), 3424.

Liang, Y., Li, Z., Shi, J., Zhang, N., Qin, Z., Du, L., & Huang, X. (2025). Advances in hyperspectral imaging technology for grain quality and safety detection: A review. *Foods*, *14*(17), 2977.

Ling, J., & Heldman, D. R. (2025). Integration of machine learning technologies in food flavor research: Current applications, challenges, and future perspectives. *International Journal of Agricultural and Biological Engineering*, *18*(3), 1–11.

Liu, M., Liu, M., Bai, L., Shang, W., Ren, R., Zhao, Z., & Sun, Y. (2023). Establishing a berry sensory evaluation model based on machine learning. *Foods*, *12*(18), 3502.

Mallapaty, S. (2022). Why are Pakistan's floods so extreme this year? *Nature*. https://doi.org/10.1038/d41586-022-02813-6

Mandal, S., Kiran, P. R., Mandal, S., & Aradwad, P. (2025). Enhancing food safety and quality assurance through artificial intelligence. *Smart Farming, Smarter Solutions: Revolutionizing Agriculture with Artificial Intelligence* (pp. 239–265). https://doi.org/10.1201/9781003536932-12

Mu, W., Kleter, G. A., Bouzembrak, Y., Dupouy, E., Frewer, L. J., Radwan Al Natour, F. N., & Marvin, H. J. P. (2024). Making food systems more resilient to food safety risks by including artificial intelligence, big data, and internet of things into food safety early warning and emerging risk identification tools. *Comprehensive Reviews in Food Science and Food Safety*, *23*(1), e13296.

Nasrudin, M. W., Ismail, I., Othman, H., Mustafa, W. A., Ali, S. A. S., & Arifin, N. A. M. (2025, August). IoT-driven real-time monitoring and reject analysis in beverage manufacturing: A systematic review. In *2025 9th International Conference on Man-Machine Systems (ICoMMS)* (pp. 453–458). IEEE.

Nazir, A., Hussain, A., & Assad, A. (2025). CNN in food industry: Current. In *Artificial Intelligence in the Food Industry: Enhancing Quality and Safety* (pp. 329–354). https://doi.org/10.1201/9781032633602-17

Ogidi, O. I., Wenapere, C. M., & Chukwunonso, O. A. (2025). Enhancing food safety and quality control with computer vision systems. In *Computer Vision Techniques for Agricultural Advancements* (pp. 51–88). Hershey, PA: IGI Global Scientific Publishing.

Oise, G., & Konyeha, S. (2025). Environmental impacts in e-waste management using deep learning. *Discover Artificial Intelligence*, *5*(1), 1–18.

Qin, K., Dai, X., & Wu, L. (2025). Hybrid sampling and ensemble learning for food safety sampling inspection classification. *Journal of Food Protection*, *88*(12), 100650.

Rajender, N., & Gopalachari, M. V. (2024). An efficient dimensionality reduction based on adaptive-GSM and transformer assisted classification for high dimensional data. *International Journal of Information Technology*, *16*(1), 403–416.

Ramadan, K. M., Bendary, E. S., Khalil, H. B., Ali, S. A., Ahmed, A. R., & Mahmoud, M. A. (2025). Smart detection of food spoilage using microbial volatile compounds: Technologies, challenges, and future outlook. *Journal of Agricultural and Food Chemistry*, *73*(31), 19222–19243.

Rashed, M., Fakhry, S., & Satour, R. (2025). Revolutionizing food quality with machine vision and machine learning techniques. In *Food in the Metaverse and Web 3.0 Era: Intersecting Food, Technology, and Culture* (pp. 71–124). Hershey, PA: IGI Global Scientific Publishing.

Röhrs, S., Rohn, S., & Pfeifer, Y. (2024). Risk classification of food incidents using a risk evaluation matrix for use in artificial intelligence-supported risk identification. *Foods*, *13*(22), 3675.

Şahin, E., Arslan, N. N., & Özdemir, D. (2025). Unlocking the black box: An in-depth review on interpretability, explainability, and reliability in deep learning. *Neural Computing and Applications*, *37*(2), 859–965.

Savary, S., Akter, S., Almekinders, C., Harris, J., Korsten, L., Rötter, R., & Watson, D. (2020). Mapping disruption and resilience mechanisms in food systems. *Food Security*, *12*(4), 695–717.

Shehzad, K. (2025). Predictive AI models for food spoilage and shelf-life estimation. *Global Trends in Science and Technology*, *1*(1), 75–94.

Singh, N., & Adhikari, D. (2023). Challenges and solutions in integrating AI with legacy inventory systems. *International Journal for Research in Applied Science and Engineering Technology*, *11*(12), 609–613.

Singh, T. B., Ghosh, Y., Kundu, R., & Vivek, K. (2025). Multispectral imaging technique: Principles and recent applications in food processing. In *Emerging Non-Destructive Technologies for Food Quality Analysis* (pp. 187–214). Cambridge, UK: Woodhead Publishing.

Soltani Firouz, M., & Sardari, H. (2022). Defect detection in fruit and vegetables by using machine vision systems and image processing. *Food Engineering Reviews*, *14*(3), 353–379.

Soni, A., Dixit, Y., Reis, M. M., & Brightwell, G. (2022). Hyperspectral imaging and machine learning in food microbiology: Developments and challenges in detection of bacterial, fungal, and viral contaminants. *Comprehensive Reviews in Food Science and Food Safety*, *21*(4), 3717–3745.

Sufi, F. (2025). Beyond the sensor: A systematic review of AI's role in next-generation machine health monitoring. *Applied Sciences*, *15*(19), 10494.

Taiwo, O. R., Onyeaka, H., Oladipo, E. K., Oloke, J. K., & Chukwugozie, D. C. (2024). Advancements in predictive microbiology: Integrating new technologies for efficient food safety models. *International Journal of Microbiology*, *2024*(1), 6612162.

Tariq, R., Qureshi, I., Farooq, S., Ayoub, W. S., Akhter, S., Javaid, R., & Sultana, A. (2025). Artificial intelligence in cutting and sorting of fruits and vegetables. In *Artificial Intelligence in the Food Industry* (pp. 134–152). Boca Raton, FL: CRC Press.

Tarlak, F. (2023). The use of predictive microbiology for the prediction of the shelf life of food products. *Foods*, *12*(24), 4461.

Trigka, M., & Dritsas, E. (2025). A comprehensive survey of deep learning approaches in image processing. *Sensors*, *25*(2), 531.

Ünal, Z., Kızıldeniz, T., Özden, M., Aktaş, H., & Karagöz, Ö (2024). Detection of bruises on red apples using deep learning models. *Scientia Horticulturae*, *329*, 113021.

Van Giffen, B., Herhausen, D., & Fahse, T. (2022). Overcoming the pitfalls and perils of algorithms: A classification of machine learning biases and mitigation methods. *Journal of Business Research*, *144*, 93–106.

Vidyalakshmi, G., Gopikrishnan, S., Boulila, W., Koubaa, A., & Srivastava, G. (2025). Digital twins and cyber-physical systems: A new frontier in computer modeling. *Computer Modeling in Engineering & Sciences*, *143*(1), 51.

Vimbi, V., Shaffi, N., & Mahmud, M. (2024). Interpreting artificial intelligence models: A systematic review on the application of LIME and SHAP in Alzheimer's disease detection. *Brain Informatics*, *11*(1), 10.

Walsh, K. B. (2018). Fruit and vegetable packhouse: Technologies for assessing fruit quantity and quality. In *Advances in Agricultural Machinery and Technologies* (pp. 367–395). Boca Raton, FL: CRC Press.

Wang, L., Cheng, Y., Parekh, G., & Naidu, R. (2024). Real-time monitoring and predictive analysis of VOC flux variations in soil vapor: Integrating PID sensing with machine learning for enhanced vapor intrusion forecasts. *Science of the Total Environment*, *924*, 171616.

Waqar, M., Bhatti, I., & Khan, A. H. (2024). Leveraging machine learning algorithms for autonomous robotics in real-time operations. *International Journal of Advanced Engineering Technologies and Innovations*, *4*(1), 1–24.

Wu, K., Ji, Z., Wang, H., Shao, X., Li, H., Zhang, W., & Bao, X. (2025). A comprehensive review of AI methods in Agri-food engineering: Applications, challenges, and future directions. *Electronics*, *14*(20), 3994.

Xu, S., Wang, H., Liang, X., & Lu, H. (2024). Research progress on methods for improving the stability of non-destructive testing of agricultural product quality. *Foods*, *13*(23), 3917.

Yahyaei, K., Khan, M. S. M., Ghosh, S., Roy, A., Craig, P., Varshney, N., & Asadizanjani, N. (2025, April). X-adapt: AI-driven design-based strategy to address x-ray compatibility challenges in advanced packaging metrology. In *Metrology, Inspection, and Process Control XXXIX* (Vol. 13426, pp. 865–875). SPIE.

Yang, R., Singh, S. K., Tavakkoli, M., Karami, M. A., & Rai, R. (2023). Deep learning architecture for computer vision-based structural defect detection. *Applied Intelligence*, *53*(19), 22850–22862.

You, J., Li, X., Low, M., Lobell, D., & Ermon, S. (2017, February). Deep Gaussian process for crop yield prediction based on remote sensing data. In *Proceedings of the AAAI conference on artificial intelligence* (Vol. 31, No. 1), pp. 4559–4565.

Yu, K., Zhong, M., Zhu, W., Rashid, A., Han, R., Virk, M. S., & Ren, X. (2025). Advances in computer vision and spectroscopy techniques for non-destructive quality assessment of citrus fruits: A comprehensive review. *Foods*, *14*(3), 386.

Yu, W., Ouyang, Z., Zhang, Y., Lu, Y., Wei, C., Tu, Y., & He, B. (2025). Research progress on the artificial intelligence applications in food safety and quality management. *Trends in Food Science & Technology*, *156*, 104855.

Zhao, Y., Sun, T., Zhang, H., Li, W., Lian, C., Jiang, Y., & Chen, S. (2025). AI-enhanced electrochemical sensing systems: A paradigm shift for intelligent food safety monitoring. *Biosensors*, *15*(9), 565.

15 Developments in Artificial Intelligence Analysis and Microbial Metagenomics for Forensic Identification

Nengak Yunana, Tsado Priscilla Yetu, Grace Galadima, Zion Charles, and Bitrus Rose Giwa

15.1 INTRODUCTION TO MICROBIAL FORENSICS

A scientific field that combines microbiology with forensic science with the goal of tracking and analyzing biocriminal activity is known as microbial forensics. Microbial forensics encompasses a wide range of forensic science combined with microbiology that includes, but is not restricted to, the examination of microorganisms or the toxins they produce, as well as the materials used to generate, store, and distribute an infectious agent or toxic substance. This new branch is crucial for the modern world, where terrorists and criminals could have disastrous access to technology. Although the attack's effects have been observed on several occasions throughout history, they weren't given much thought until the 2001 anthrax outbreak in the United States. Recognizing this field's prospective; the US government established "Microbial Forensics" as a field (McNeil, 2024).

An attack of this sort could be motivated by either ideological or personal factors, posing a local or international threat. It is expected that a large number of illicit private groups may have biological weapons in their possession. Highly transmissible and potentially fatal agents are used in biological assaults to target humans. The goal is to have an extensive effect on the target, which might cause significant harm locally due to its being self-sustaining. As can be seen, this strategy has the potential to be hazardous since it introduces novel organisms within the ecosystem that could have disastrous effects.

Some of the more well-reported biological crimes in recent times include a research laboratory employee purposefully infecting colleagues in Texas with a bacterium known as *Shigella dysenteriae*, the administration of "ricin" in Europe to kill a Bulgarian expatriate, and *Salmonella typhimurium* usage in

DOI: 10.1201/9781003727514-17

Oregon through politically driven contamination of salad bars in nearby eateries. The use of anthrax spores in New York in October 2001 is the most well-known bioterrorism incident of the twenty-first century (Parasidis, 2025). The Biological and Toxin Weapons Convention, or BTWC, was created and is currently ratified by over 65 states as a current effort to prevent the threat of bioweapons. Biological weapons usage and storage are prohibited by this treaty.

15.2 ROLE OF METAGENOMICS IN FORENSIC SCIENCE

Forensic analysis requires that the microbial makeup of the accused person's mouth cavity and obtained specimens be the same. There is a possibility of a match failure if salivary microbial communities fluctuate with time, in which case actual preparation is not included in the study. Salivary microbe characterization was initially investigated in the early 2000s using genetic fingerprinting techniques, such as denaturing gradient gel electrophoresis (DGGE). DGGE is frequently used for qualitative and somewhat quantitative data on microbial populations and changes in community structure for the identification and monitoring of bacterial populations. The enhanced band is then removed from the gel to extract taxonomic information (Sharma *et al.*, 2025). In situations where a researcher lacks enough human genetic material for identifying individuals, it was proposed that stable indigenous microbial profiles may be utilized for determining the distinctive characteristics of human beings. This has a significant impact on the field of forensic biology.

15.2.1 Tackling Bio-crime

Microbial forensics' primary objectives involve identifying and prioritizing biological hazards, identifying populations at risk, building a research database, developing detection protocols that include identifying distinct genetic and protein signatures, creating programs to ensure the accuracy of discoveries, and continuously updating based on previous research (Castro and De Ungria, 2022). It is harder than most people realize to identify and prioritize the target. It's possible that the organism that has to be given priority is completely unclear or could be a pathogen from plants that can cause significant economic harm rather than an infection to humans (Pandit *et al.*, 2022). It would be similarly difficult to identify the most vulnerable individuals.

Establishing a record of data would be useful in directing the outcomes, but on the other hand, it could also assist those who commit crimes in finding organisms that are not classified, allowing for the spread of more challenging-to-identify organisms. For all of these reasons, the legitimacy of using a record like this should be evident. Microbiology, genetics, forensic techniques, chemistry, and pure science are just a few of the traditional domains that will contribute to the database that is employed. Everyone involved should have easy access to a possible list of threats. This ought to be called a priority target.

15.2.2 Biological Sex Determination

When researchers are unable to obtain sufficient genetic information for personal authentication, microbiomic data can help determine their biological gender. For instance, airborne microbes were identified in indoor contexts (Elhaik *et al.*, 2021). Luongo and his associates examine the wide variety of airborne bacteria and fungi found in several campus dormitory spaces (Kramer et al., 2024). Using predictive machine learning (ML) approaches, the biological gender of the room occupants can be determined with 79% accuracy based on the relative number of microbes (Aliero *et al.*, 2022). Males were shown to have a higher proportion of microbiome; this could be because they shed higher biological material or because they apply less cosmetics barriers (Losol *et al.*, 2023).

15.2.3 Trace Evidence (Touch Evidence)

Every day, there is an increasing curiosity in forensic analysis of the microbiological profiles of various surfaces following contact. Multiple investigations have demonstrated that every individual's personal items have a significant amount of microbiome (Joos *et al.*, 2025). Thus, there is growing interest in researching these microbial profiles that have been left behind. Furthermore, it has been demonstrated that personal goods like shoes and cell phones support different microbiomes (Arundhathi *et al.*, 2021). In order to examine the potential utility of mobile phones as personal microbiome sensors, 17 people were chosen, and samples were collected from three different surfaces: the thumb, index finger, and contact surface of cell phones (Banadaki *et al.*, 2024).

But still there is a need to increase the sample size and the accuracy of results in the future. Skin microbial communities can be linked to personal items with an outstanding level of precision, according to postmortem analysis. While mobile devices, remote controls, and computers may be linked with 67% precision, bottles, medical equipment, spectacles, and automobile steering wheels can be linked with 100% precision. A trustworthy method of connecting items to the scene is the skin microbiome.

15.3 ARTIFICIAL INTELLIGENCE IN FORENSICS METAGENOMIC DATA ANALYSIS

In an exploration of the various uses of metagenomics in conjunction with artificial intelligence (AI), we go past the sequences. This journey reveals AI's promise in functional profiling, taxonomic composition analysis, finding new genes and microorganisms, understanding ecosystems, and using metagenomics to promote biotechnology (Negi *et al.*, 2024). AI combined with metagenomics enables us to forecast the functional potential of microbial populations. It enables us to interpret the biological processes, metabolic pathways, and enzyme functions that are encoded in the genes found in these types of communities. AI algorithms can offer important insights into the capacities of microbial habitats by examining the genetic composition.

Predicting functioning genes associated with biogeochemical processes, like carbon or nitrogen cycling, constitutes one of the important applications in this field. In order to answer ecological and environmental problems, researchers must comprehend how microorganisms affect the biogeochemical processes of their surroundings, which is made possible by AI-driven functioning profiling (Ghannam and Techtmann, 2021; Mcelhinney et al., 2022). The identification of new metabolic pathways in harsh conditions has been made possible in recent years by AI-driven functional profiling. AI analysis of metagenomic data from hydrothermal vents, for instance, revealed the existence of distinct genes that oxidize sulfur compounds, providing insight into the microbial mechanisms that underlie the ecosystem's energy flow. Additionally, metagenomics and AI have created new biotechnology opportunities.

Enzymes with possible industrial uses, including those involved in the breakdown of lignocellulose for the manufacture of biofuel, can be found by functional characterization. Investigators can choose the most effective candidates for biological prospecting efforts thanks to AI-driven performance prediction. AI is essential in this regard since it offers precise and effective ways to recognize and categorize germs. For example, ML algorithms can help characterize the microbial community by giving sequences taxonomic labels. AI-powered taxonomy untangling facilitates the investigation of unexplored habitats.

AI can be used to accurately analyze metagenomic information collected from a variety of contexts, including deep-sea ecosystems and harsh conditions, revealing the taxonomic variety present and illuminating the capability of microbes to evolve and adapt to different niches. Recent studies have revealed microbial richness in hostile conditions thanks to AI-powered phylogenetic analysis (Ogbuagu *et al.*, 2023). For example, metagenomic information gathered from hydrothermal vents beneath the seafloor showed the existence of hitherto unidentified archaeal lineages. Our knowledge of the diversity of life under harsh circumstances has increased thanks to AI's capacity to categorize these new creatures. Furthermore, taxonomic analysis powered by AI helps evaluate how environmental changes affect microbial populations. For instance, scientists have employed AI to investigate how climate change affects Arctic microbiomes.

15.4 FORENSIC APPLICATION OF HUMAN-ASSOCIATED MICROBIOME IN FORENSIC INVESTIGATIONS

One of the most crucial responsibilities in forensic science is individual identification. Numerous studies have demonstrated the clear polymorphisms and individual differences found in microecosystems such as the skin, mouth cavity, and gut. Microbial usage for the identification of individuals is based on variations in the quantity and composition of microbial communities in human microecosystems. According to the hypothesis, each person has a distinct set of microorganisms that sets them apart from other people. These bacteria can be detected using a microbiome study, and this specific microbial community can endure for extended periods of time. As a result, forensic identification of individuals may benefit from microbiome characterization. In a community of over 100 individuals, Franzosa et al. discovered that a person's microbiome is capable of accurately recognizing its original host. After a year, over 80% of people continue to be precisely identified thanks to the gut microbiome's extremely consistent performance (Shanahan *et al.*, 2021). The genotypic makeup of Cutibacterium acnes' 16S rRNA is specific to each person, according to a different study. The precision of each person's identification was over 90% when the 16S rRNA genotype of C. acnes was combined with skin-specific microbiome profile data using the random forest (RF) ML method (Dass *et al.*, 2025). The skin microbiome profile was less consistent over time than the C. acnes 16S rRNA genotype.

The Budowle group investigated the use of skin microbes for forensic human identification in a number of research projects. Clade-specific markers were found, and the skin's core microbiome was identified. The hidSkinPlex, a novel focused sequence panel, was created. It has 286 markers that span a variety of taxonomies of particular bacteria that are highly prevalent on human skin.

Polymorphic single-nucleotide genetic indicators tend to be more personalized than classification DNA markers, according to Sherier *et al.* (2022). They created an enhanced system called "hidSkinPlex+," which has fewer indicators compared to the original version of hidSkinPlex and contains 365

SNPs spread across 135 markers. ML performance can be increased and the technique of enrichment improved by removing the indicators that do not add to the precision of classification. Reanalyzing the same genome sequencing information, they discovered that the greatest Wright's fixation index (FST) in conjunction with support vector machines (SVMs) could increase the identification of individuals' accuracy ($p = 0.03$, chi-squared test).

15.5 ENVIRONMENTAL MICROBIAL SIGNATURES IN CRIME SCENE ANALYSIS

Individuals at a particular body site exhibit variance dictated by host DNA and lifestyle choices, despite the fact that human microbiome makeup differs predominantly between body locations, which reflect separate ecological niches (Joos *et al.*, 2025). By comparing the bacterial makeup of an individual with that of an object, for instance, such interpersonal variances may help identify donors. Bacterial community makeup features from air filters, keyboards, and dorm surfaces have been linked to persons using 16S rRNA amplicon data, even though microbial indicators are not predicted to equal forensic human STRs in terms of statistical accuracy (Wu *et al.*, 2021).

15.5.1 Soil Microbiome

Soil microbial community assessment has been seen as possibly helpful in investigations in the past decade, offering fine-scale resolution discriminating power. Fungal and bacterial communities have a very site-specific profile because the microbial community's structure is heavily influenced by a number of factors, including environmental conditions, vegetation cover, site management, seasonal variation, and soil type (Zhang *et al.*, 2020). A far more thorough picture of the community of microbes found in a particular soil sample can be obtained thanks to recent developments in MPS, which can provide millions of sequencing reads per sample.

One gram can understand up to 8.3 million distinct sequences, based on the type of soil being studied (Oliveira and Amorim, 2018). The outermost and underground soil microbes exhibited distinct characteristics during their breakdown process, according to recent investigations. Microbial communities in close proximity to buried cadavers exhibit the opposite traits (i.e., growing decreasing evenness, consistent diversity, and taxon richness), whereas microbial communities from surface soil show declining trends in taxon richness, variety, and evenness. Additionally, Proteobacteria has been identified as the most prevalent phylum in grave soil samples; in surface cadaver-soil communities, the proportion of species of Firmicutes increased, and Acidobacteria decreased, whereas the composition of the community of microbes in underground soil communities remained relatively constant (High, 2024).

15.5.2 Skin Microbiome

Numerous bacteria that are readily distributed and transmitted to surfaces upon contact are present on the surface of human skin (Baquero *et al.*, 2021). According to Franceschetti *et al.* (2024), the microbiome of human beings can be utilized to generate a unique signature, creating a connection between people and the items they interact with, including personal computers and cell phones (touch pads, keyboards, and other device surfaces). The extent of variability in the skin microbiome varies depending on the depth of the research; if identification is accomplished at the genus, species, and strain population levels, the diversity is highly specific for each individual, functioning as a unique signature with a discriminatory effect comparable to a unique fingerprint (Dass et al., 2025).

15.5.3 Hair Microbiome

One of the most frequently discovered pieces of evidence gathered at crime scenes is hair. But the great majority do not have the anagenic root (currently growing hair), which is necessary for posterior analysis using traditional typing techniques, such as DNA extraction and STR profiles (Rayimoglu *et al.*, 2025). The microbiota found in pubic hairs and human scalp was compared in a recent study, which also discussed the application of this evidence in both civil and criminal investigations (Gouello *et al.*, 2021). While pubic hairs are used in situations of sexual assault, most hairs obtained in forensic examinations are scalp hairs. The authors claim that information gathered from the microbiome analysis made it possible to distinguish between the two distinct sources of hair in addition to individualizing the contributor.

Couples who reside together exchange microbiomes during sexual activity, according to the previously cited study. Furthermore, it has been demonstrated that the microbiome on scalp hairs is more susceptible to compositional oscillations than pubic hairs (Dixon *et al.*, 2023).

15.5.4 Body Fluid Microbiome

Saliva, semen, and blood are the most frequently discovered bodily fluids in crime scenes. But other bodily fluids, such as perspiration, urine, and vaginal fluid, can also be found and serve important purposes in investigations, such as providing DNA evidence (Rajput, 2024). According to Issa (2024), it has been suggested that every form of bodily fluid exhibits a particular microbial composition that can be utilized as bioindicators, allowing one to deduce the type of bodily fluid present based on its microbial makeup. For example, certain bacteria like Veillonella atypica, Streptococcus mutans, and Streptococcus salivarius might be used to identify saliva, but certain bacteria like Lactobacillus gasseri and Lactobacillus crispatus could be used to identify vaginal fluid (Qi *et al.*, 2023). The human microbiome is completely unique to each individual when lower taxonomic levels (genus, species, and strains) are taken into account (Giliberti *et al.*, 2022). As a result, the fine resolution of the results obtained from MPS-based approaches will enable the identification of suspects by comparing the microbiomes or, at least in theory, establish a connection between victims and offenders.

15.6 AI-POWERED TOOLS FOR MICROBIAL SOURCE TRACKING

Microbial source tracking (MST) is a key development in forensic microbiology. It refers to the attribution of microbial communities to particular sources such as human individuals, objects, environmental reservoirs, or geographic locations with forensic relevance (Job *et al.*, 2023). Traditional methods of MST include marker gene amplification and culture-based assays, which often rely on host-specific genetic markers, library matching, or targeted assays (Jabri, 2025). These traditional methods often encounter limitations in sensitivity, specificity, and scalability when dealing with complex, mixed samples typical in forensic contexts. The surge in high-throughput sequencing data, alongside advances in AI and ML techniques, has transformed MST from a largely ecological or environmental monitoring tool into a prospective forensic instrument (Chaturvedi *et al.*, 2024). By integrating AI algorithms, forensic scientists can now analyze complex microbial community profiles, model probabilistic source assignments, and infer likely matches between evidence and suspects or crime scenes with greater accuracy and scalability. These AI-driven tools detect subtle patterns that traditional methods miss, thereby minimizing false positives and increasing the accuracy of forensic conclusions. Modern AI-enabled MST combines microbial community fingerprinting (using 16S rRNA gene sequencing or metagenomic shotgun sequencing) (Gupta *et al.*, 2024) with supervised or unsupervised learning approaches to predict the origin of the source.

One outstanding example is SourceApp, a bioinformatic pipeline designed for metagenomic fecal source tracking (FST) in environmental forensics, such as tracing pollution in water reservoirs that could relate to bio-crimes or tampering (Wani *et al.*, 2022). SourceApp uses read mapping algorithms like Burrows-Wheeler Aligner (BWA) to align shotgun sequencing reads to source-associated genome databases constructed from metagenome-assembled genomes (MAGs) and public references. ML components, an RF-based tool for assessing genome quality—ensure high-fidelity databases by evaluating completeness and redundancy, discarding low-quality MAGs (quality score <50%). Parameter tuning across 640 combinations optimizes sensitivity (0.902) and specificity (0.898), outperforming Bayesian-based alternatives like fast expectation-maximization (FEAST) for MST and metaSourceTracker, which exhibit lower specificity (0.30 and 0, respectively) due to false positives in mixed inputs. In forensic applications, SourceApp apportions microbial contributions in mesocosms simulating pollution events, correlating 0.86 with ground truth cell densities, thus aiding in linking suspects to contaminated sites via microbial transfer.

AI classifiers further enhance MST by differentiating microbial signatures in human-associated samples. RF models, for instance, have been applied to 16S rRNA sequencing data from skin microbiomes to predict individual identity with accuracies up to 94.77% using single nucleotide polymorphism (SNP) markers in panels like HidSkinPlex (Xu *et al.*, 2023). These models identify patterns in taxa such as Cutibacterium acnes, which exhibit host-specific polymorphisms stable over time. In environmental forensics, SVMs integrated with Wright's fixation index (F_ST) differentiate microbial profiles from touched objects, achieving p-values of 0.03 for donor attribution (Swayambhu, 2025). Deep learning (DL) approaches, such as convolutional neural networks (CNNs), extend MST to image-based microbial analysis, though primarily in complementary roles. For example, CNN derivatives classify bacterial species from microscopic images, achieving 96% accuracy for pathogens like *Staphylococcus aureus*, which could be adapted for source tracking in forensic trace evidence (Xu *et al.*, 2023). In metagenomics, taxonomy-independent DL models process abundance profiles to classify sources without reference databases, yielding area under the curve (AUC) values of 0.99–1.0 for body fluids like saliva and vaginal secretions (Budzinski, 2025). These models mitigate amplification biases in amplicon sequencing by directly analyzing shotgun data, improving source attribution in degraded forensic samples.

Challenges in AI-powered MST include cross-reactivity among sources, addressed through dereplication algorithms like dRep. Normalization techniques, such as Microbe Census for genome equivalents, ensure accurate cell fraction estimates, which are crucial for apportioning mixed signals in forensic mixtures. Case studies demonstrate practical utility: in simulated bio-crimes, RF models on soil microbiomes link suspects to scenes via unique taxa like Bacillus and Propionibacterium, with accuracies exceeding 90% (Robinson *et al.*, 2021). Similarly, SVM classifiers distinguish racial/ethnic microbial variations, aiding victim profiling in unidentified remains (Au-Yeung *et al.*, 2021).

15.7 INTEGRATION OF METAGENOMICS AND AI IN POSTMORTEM INTERVAL (PMI) ESTIMATION

Postmortem interval (PMI) estimation is a fundamental aspect of forensic pathology, which provides temporal context to death investigations. Methods such as algor mortis or rigor mortis assessment, which are conventional methods (Teixeira *et al.*, 2025), are imprecise or unreliable beyond early stages due to environmental variability. The integration of metagenomics (comprehensive sequencing of microbial communities) with AI offers a data-driven alternative, leveraging the predictable succession of the thanatomicrobiome as a "microbial clock." This approach analyzes shifts in bacterial taxa post-death, using ML models to predict PMI with enhanced accuracy.

Metagenomic sequencing, particularly 16S rRNA amplicon and whole-metagenome shotgun (WMS), reveals microbial dynamics in decomposing tissues (Chelliah *et al.*, 2022). For instance, early PMI features aerobic taxa like Proteobacteria, transitioning to anaerobes such as Firmicutes and Bacteroidetes as decomposition advances (Moitas et al., 2024). In infected cadavers, AI-assisted models account for bacterial perturbations; a deep pathomics framework using DenseNet121 and VGG16 on histological images from liver, kidney, and muscle achieves 93.75% internal accuracy and 87.5% external validation, with a mean absolute error (MAE) of 2.25 hours (Liang *et al.*, 2025). Stacking ensembles integrate multi-organ data, outperforming single-tissue models by capturing holistic degradation patterns.

AI algorithms process these metagenomic datasets to model succession. RF is widely employed; Wu *et al.* (2024) developed a microbial clock to accurately predict PMI based on the rat decomposition model. Sharma *et al.* (2022) reported MAE of 1.5 ± 0.8 hours within 24 hours in mice using cecum microbiota. In human studies, RF models on rib microbiomes predict PMI with ±34 days' accuracy over nine months (Al-Juhani *et al.*, 2025). For infected scenarios, AI estimates PMI across temperatures (4–37°C) with AUC ≥0.873 at the patch level, demonstrating robustness to pathogens like *S. aureus* and *Escherichia coli* (Liang *et al.*, 2025).

Multi-omics integration amplifies AI efficacy. Metabolomics complements metagenomics; a feed-forward neural network (FFNN) on femoral blood metabolomes (4,876 samples) predicts PMI with MAE 1.45 days and median 1.03 days, clustering features into decreasing (e.g., acylcarnitines) and increasing (e.g., amino acids) patterns (Magnusson *et al.*, 2025). Systematic reviews highlight proteomics dominance for short PMI (<7 days), with metabolomics for medium (7–120 days), using regression and ensemble learning to identify biomarkers like xanthine (Secco *et al.*, 2025). Bayesian classifiers have been used to solve the inverse problem of PMI by integrating prior knowledge and generating probabilistic PMI estimates, accounting for uncertainties and the cumulative impact of various factors affecting decomposition. Bayesian classifiers incorporate priors for probabilistic PMI, as in SourceTracker adaptations for microbial shifts in gravesoil (Giles *et al.*, 2023).

DL refines predictions; CNN models analyze succession images or profiles, achieving 97.33% validation for diatom detection in drowning-related PMI (Ma *et al.*, 2025). Ensemble methods, like stacking in multi-organ models, mitigate variability from host factors (e.g., age and BMI) and environment (e.g., temperature and humidity), improving accuracy over linear regression (Waddell *et al.*, 2023).

15.8 FORENSIC AUTHENTICATION OF BIOLOGICAL SAMPLES USING AI

In forensic microbiology and bio-evidence analysis, the authenticity and integrity of biological samples (human tissues, bodily fluids, or environmental microbial traces) have become increasingly critical. As the complexity of cases increases (degraded samples, mixed sources, and environmental contamination), there is a growing need for AI to strengthen the forensic authentication of biological samples, offering both enhanced detection of adulteration and improved verification of sample provenance. In the domain of microbiome or metagenomic data, recent reviews highlight the increasing role of AI in forensic identification, geolocation, and body-fluid source attribution (Zahrah *et al.*, 2025). One of the key authentication use cases is detecting whether a biological trace has been intentionally or unintentionally contaminated or whether it originates from the claimed source.

In microbial forensics, AI classifiers distinguish authentic microbial profiles from artifacts, safeguarding investigations against deliberate manipulation.

ML models like RF and SVM authenticate samples by analyzing microbial SNPs and abundance profiles. HidSkinPlex panels, processed via SVM, detect tampering in skin traces with 94.77% accuracy, flagging anomalies in taxa like Propionibacterium (Yao et al., 2024). In contaminated scenarios, AI reduces false positives by filtering noise; ensemble learning on multi-omics data identifies tampering markers like unexpected metabolite spikes (Magnusson *et al.*, 2025). Though not focused on microbial authentication, Yetgin (2025) conducted a systematic review of AI in forensic science and emphasized that AI is intended to assist the forensic expert in ensuring valid evidence.

AI offers several advantages in the authentication of biological samples, including sensitive detection of anomalies, where it can identify subtle deviations from expected microbial community structures that might escape human review (Alsulimani *et al.*, 2024); scalability, as large reference libraries of microbial profiles or genetic markers can be processed rapidly, enabling robust comparisons across many samples; probabilistic output, rather than strict binary answers, where AI models can provide probability distributions, supporting forensic expert judgments and the probabilistic nature of evidence; and adaptive learning, as new data are collected (e.g., novel contamination types and novel sample-handling artifacts), allowing AI models to be retrained and refined, continuously improving authentication capabilities.

15.9 CHALLENGES AND ETHICAL CONSIDERATIONS IN AI-DRIVEN FORENSIC MICROBIOLOGY

Despite exceptional advances, AI-driven forensic microbiology faces a number of ethical and legal challenges that must be addressed for responsible implementation (Shukla, 2024;Mondal *et al.*, 2025). Primary concerns include data quality and reproducibility. AI requires a lot of high-quality data to function well. In forensic science, such datasets are limited, especially for specific areas or uncommon crimes (Arthanari *et al.*, 2025). Data quality challenges, such as incomplete metagenomic assemblies and biases in training datasets, risk inaccurate predictions. For example, undersampled sources in SourceApp lead to false negatives for certain taxa (Kumar *et al.*, 2024). Reproducibility is compromised by variable sequencing protocols, necessitating standardization.

AI systems are constructed upon foundational datasets; consequently, any preexisting societal biases embedded within these data will typically be perpetuated and amplified by the resulting algorithms (Scatiggio, 2020). This is a big concern in forensic work, where biased algorithms can lead to wrongful convictions. The identification of bias in AI algorithms applied to forensic science is imperative, alongside the formulation of robust methodologies for bias mitigation.

Model transparency and explainability are major concerns in forensic contexts. Some AI algorithms are "black boxes," making it difficult to understand how they reach conclusions (Hassija *et al.*, 2024). A "black-box" neural network providing a probability without a clear explanation of contributing features may struggle to satisfy legal admissibility standards.

The incorporation of AI into established forensic workflows requires strategic coordination and interdisciplinary collaboration between AI engineers and forensic specialists (Akeiber, 2025). Comprehensive training and ongoing technical support for forensic personnel are indispensable to facilitate the adoption and effective utilization of AI-based tools. Moreover, interoperability remains significantly constrained due to the absence of standardized data schemas, analytical protocols, and mechanisms for disseminating comparable analytical outcomes across disparate forensic laboratories.

Ethical concerns include privacy breaches, as microbiomes reveal health/lifestyle data, potentially violating consent in investigations (Kassem *et al.*, 2025). A key concern is AI in mass surveillance or for harmful purposes. Strong laws and ethical rules are needed to guide AI use in forensic science (Wiesenthal, 2022).

15.10 FUTURE DIRECTIONS IN AI AND MICROBIAL METAGENOMICS FOR FORENSICS

AI and microbial metagenomics are contributing to forensic science by providing insights into the source, timing, and validity of biological evidence. The future of AI and microbial metagenomics in forensics lies largely at the intersection of technological innovation, ethical governance, and interdisciplinary collaboration. Advancements in explainable AI, real-time analytics, multi-omics integration, and global cooperation will drive forensic microbiology toward becoming an indispensable pillar of criminal justice, strengthening evidence while safeguarding civil liberties.

The first major direction involves advancing sequencing technologies to provide higher resolution, real-time, and field-deployable metagenomic analyses. For instance, portable nanopore sequencing already allows in situ microbial detection at crime scenes and disaster zones, dramatically shortening turnaround times compared to traditional laboratory workflows (Oliveira *et al.*, 2024). Future devices are expected to integrate directly with AI-based analytic modules, enabling "edge computing" where sequencing data are processed and interpreted on site. This could allow investigators to obtain rapid microbial fingerprints, authenticate biological traces, and even estimate PMIs in real time.

Metagenomic innovations such as single-cell genomics, long-read sequencing, and multi-omics integration will expand the data available for forensic interpretation. AI can play a central role in analyzing diverse datasets (Arora *et al.*, 2023), uncovering complex microbial-forensic relationships that remain hidden in conventional analyses.

Emerging technologies like explainable AI (XAI) and real-time sequencing should be integrated with metagenomics for enhanced forensics. XAI approaches should be prioritized to ensure model decisions can be traced and understood by forensic practitioners. While AI algorithms, particularly DL models, have demonstrated significant predictive power in analyzing complex microbial datasets, their "black-box" nature poses challenges for interpretability and courtroom acceptance. Future AI systems designed for forensic metagenomics must incorporate transparent decision-making processes that allow forensic practitioners and legal experts to understand and validate the reasoning behind microbial source attributions or PMI estimates. XAI frameworks that visualize feature importance, decision pathways, and uncertainty quantification will build trust and facilitate regulatory approval (Mahto, 2024).

REFERENCES

Akeiber, H. J. (2025). A comprehensive study of cybercrime and digital forensics through machine learning and AI. *Al-Rafidain Journal of Engineering Sciences*, 3(1), 369–395.

Aliero, M. S., Pasha, M. F., Smith, D. T., Ghani, I., Asif, M., Jeong, S. R. & Samuel, M. (2022). Non-intrusive room occupancy prediction performance analysis using different machine learning techniques. *Energies*, 15(23), 9231.

Al-Juhani, A. A., Gaber, A. M., Desoky, R. M., Binshalhoub, A. A., Alzahrani, M. J., Alraythi, M. S. & Aseeri, A. A. (2025). From microbial data to forensic insights: Systematic review of machine learning models for PMI estimation. *Forensic Science, Medicine and Pathology*, 21(3), 1447–1476.

Alsulimani, A., Akhter, N., Jameela, F., Ashgar, R. I., Jawed, A., Hassani, M. A. & Dar, S. A. (2024). The impact of artificial intelligence on microbial diagnosis. *Microorganisms*, 12(6), 1051.

Arora, A., Alderman, J. E., Palmer, J., Ganapathi, S., Laws, E., Mccradden, M. D. & Liu, X. (2023). The value of standards for health datasets in artificial intelligence-based applications. *Nature Medicine*, 29(11), 2929–2938.

Arthanari, A., Raj, S. S. & Ravindran, V. (2025). A narrative review in application of artificial intelligence in forensic science: Enhancing accuracy in crime scene analysis and evidence interpretation. *Journal of International Oral Health*, 17(1), 15–22.

Arundhathi, B. C., Prasanth, S., Sivaranjani, R., Abishek, P., Arumugam, A. & Subramani, S. (2021). Microbial Interactions with Accessories—A Complete Analysis. In International Conference on Advances and Innovations in Recycling Engineering (61–79). Singapore: Springer Nature Singapore.

Au-Yeung, W. T. M., Sevakula, R. K., Singh, J. P., Heist, E. K., Isselbacher, E. M. & Armoundas, A. A. (2021). State of the art in artificial intelligence and machine learning techniques for improving patient outcomes pertaining to the cardiovascular and respiratory systems. In *Cardiac Bioelectric Therapy: Mechanisms and Practical Implications* (pp. 335–352). https://doi.org/10.1007/978-3-030-63355-4_24

Banadaki, F. D., Rahimian, B., Moraveji, F. & Varmazyar, S. (2024). The impact of smartphone use duration and posture on the prevalence of hand pain among college students. *BMC Musculoskeletal Disorders*, 25(1), 574.

Baquero, F., Saralegui, C., Marcos-Mencía, D., Ballestero, L., Vañó-Galván, S., Moreno-Arrones, ÓM. & Del Campo, R. (2021). Epidermis as a platform for bacterial transmission. *Frontiers in Immunology*, 12, 774018.

Budzinski, L. (2025). Single-cell Based Microbiota Signatures in Chronic Inflammatory Diseases. Technische Universitaet Berlin (Germany).

Castro, A. E. & De Ungria, M. C. A. (2022). Methods used in microbial forensics and epidemiological investigations for stronger health systems. *Forensic Sciences Research*, 7(4), 650–661.

Chaturvedi, N., Yadav, M. K., & Sharma, M. (2024). Applications of artificial intelligence and machine learning in microbial diagnostics and identification. In *Methods in microbiology* (Vol. 55, pp. 213–230). Academic Press.

Chelliah, R., Banan-MwineDaliri, E., Khan, I., Wei, S., Elahi, F., Yeon, S. J. & Oh, D. H. (2022). A review on the application of bioinformatics tools in food microbiome studies. *Briefings in Bioinformatics*, 23(2), bbac007.

Dass, M., Abbai, N. S. & Ghai, M. (2025). The human skin microbiome: Factors affecting individuality and application in forensic investigations. *International Journal of Legal Medicine*, 140(2), 1067–1084.

Dixon, R., Egan, S., Hughes, S. & Chapman, B. (2023). The sexome – A proof of concept study into microbial transfer between heterosexual couples after sexual intercourse. *Forensic Science International*, 348, 111711.

Elhaik, E., Ahsanuddin, S., Robinson, J. M., Foster, E. M. & Mason, C. E. (2021). The impact of cross-kingdom molecular forensics on genetic privacy. *Microbiome*, 9(1), 114.

Franceschetti, L., Lodetti, G., Blandino, A., Amadasi, A., & Bugelli, V. (2024). Exploring the role of the human microbiome in forensic identification: Opportunities and challenges. *International Journal of Legal Medicine*, 138(5), 1891–1905. https://doi.org/10.1007/s00414-024-03217-z

Ghannam, R. B., & Techtmann, S. M. (2021). Machine learning applications in microbial ecology, human microbiome studies, and environmental monitoring. *Computational and Structural Biotechnology Journal*, 19, 1092–1107

Giliberti, R., Cavaliere, S., Mauriello, I. E., Ercolini, D., & Pasolli, E. (2022). Host phenotype classification from human microbiome data is mainly driven by the presence of microbial taxa. *PLoS Computational Biology*, 18(4), e1010066

Giles, S., Errickson, D., Harrison, K. & Márquez-Grant, N. (2023). Solving the inverse problem of post-mortem interval estimation using Bayesian belief networks. *Forensic Science International*, 342, 111536.

Gouello, A., Dunyach-Remy, C., Siatka, C., & Lavigne, J. P. (2021). Analysis of microbial communities: an emerging tool in forensic sciences. *Diagnostics*, 12(1), 1.

Gupta, S., Kumar, D., Negi, R., Singh, R., Ansari, M. J., Kashyap, S. & Mehrotra, S. (Eds.). (2024). *Genomic Intelligence: Metagenomics and Artificial Intelligence*. CRC Press. https://doi.org/10.1201/9781003570233.

Hassija, V., Chamola, V., Mahapatra, A., Singal, A., Goel, D., Huang, K. & Hussain, A. (2024). Interpreting black-box models: A review on explainable artificial intelligence. *Cognitive Computation*, 16(1), 45–74.

High, T. L. (2024). *A Study of the Postmortem Soil Microbial Communities Associated with Albino Rat Decomposition* (Master's thesis, Alabama State University).

Issa, S. Y. (2024). Microbial witness: Unraveling mysteries with forensic microbiomes. In *Unlocking the Mysteries of Death-New Perspectives for Post-Mortem Examination*. IntechOpen.

Jabri, H. (2025). Source Tracking of Bacterial Contamination Based on Ribosomal Sequence Analysis.

Job, O. S., Bala, J. D., Abdulraham, A. A., Friday, N. N., Ibekie, S. A., Tsebam, C. J. & Abudullahi, D. (2023). Microbial source tracking: An emerging technology for microbial water quality assessment: A review. *UMYU Journal of Microbiology Research*, 8(1), 109–121.

Joos, R., Boucher, K., Lavelle, A., Arumugam, M., Blaser, M. J., Claesson, M. J. & Ross, R. P. (2025). Examining the healthy human microbiome concept. *Nature Reviews Microbiology*, 23(3), 192–205.

Kassem, H., Beevi, A. A., Basheer, S., Lutfi, G., Cheikh Ismail, L. & Papandreou, D. (2025). Investigation and assessment of AI's role in nutrition—An updated narrative review of the evidence. *Nutrients*, 17(1), 190.

Kramer, A., Lexow, F., Bludau, A., Köster, A. M., Misailovski, M., Seifert, U. & Scheithauer, S. (2024). How long do bacteria, fungi, protozoa, and viruses retain their replication capacity on inanimate surfaces? A systematic review examining environmental resilience versus healthcare-associated infection risk by "fomite-borne risk assessment". *Clinical Microbiology Reviews*, 37(4), e00186-23.

Kumar, B., Lorusso, E., Fosso, B. & Pesole, G. (2024). A comprehensive overview of microbiome data in the light of machine learning applications: Categorization, accessibility, and future directions. *Frontiers in Microbiology*, 15, 1343572.

Liang, X., Wang, G., Wang, H., Zhu, Z., Zhang, W., Li, Y. & Wang, Z. (2025). Artificial intelligence-assisted estimation of postmortem intervals in bacterially infected cadavers using pathological imaging across variable temperature conditions. *Science & Justice*, 65(6), 101344.

Liang, X., Wang, G., Zhu, Z., Zhang, W., Li, Y., Luo, J. & Wang, Z. (2025). Using pathology images and artificial intelligence to identify bacterial infections and their types. *Journal of Microbiological Methods*, 232, 107131.

Losol, P., Sokolowska, M., Hwang, Y. K., Ogulur, I., Mitamura, Y., Yazici, D. & Akdis, C. A. (2023). Epithelial barrier Theory: The role of exposome, microbiome, and barrier function in allergic diseases. *Allergy, Asthma & Immunology Research*, 15(6), 705.

Ma, A., Chen, M., Yu, Q., Chen, L., Huang, P. & Zhang, J. (2025). A simple optical microscopy-based diatom classification approach for forensic drowning site inference: A pilot study. *International Journal of Legal Medicine*, 139(5), 2207–2218.

Magnusson, R., Söderberg, C., Ward, L. J., Arpe, J., Kugelberg, F. C., Elmsjö, A. & Nyman, E. (2025). The Human Metabolome and Machine Learning Improves Predictions of the Post-Mortem Interval. bioRxiv, 2025–03.

Mahto, M. K. (2024). Explainable artificial intelligence: Fundamentals, approaches, challenges, XAI evaluation, and validation. In *Explainable Artificial Intelligence for Autonomous Vehicles* (pp. 25–49). CRC Press.

McElhinney, J. M. W. R., Catacutan, M. K., Mawart, A., Hasan, A., & Dias, J. (2022). Interfacing machine learning and microbial omics: A promising means to address environmental challenges. *Frontiers in Microbiology*, 13, 851450. https://doi.org/10.3389/fmicb.2022.851450

McNeil, D. G. (2024). *The Wisdom of Plagues: Lessons from 25 Years of Covering Pandemics*. Simon and Schuster.

Moitas, B., Caldas, I. M. & Sampaio-Maia, B. (2024). Microbiology and postmortem interval: A systematic review. *Forensic Science, Medicine and Pathology*, 20(2), 696–715.

Mondal, R. S., Akter, L. & Bhuiyan, M. N. A. (2025). Integrating AI and ML techniques in modern microbiology. *Applied IT & Engineering*, 3(1), 1–10.

Negi, R., Singh, R., Kashyap, S., & Gupta, S. (2024). Exploring the intersection of metagenomics and artificial intelligence. In *Genomic Intelligence* (pp. 1–20). CRC Press.

Ogbuagu, O. O., Mbata, A. O., Balogun, O. D., Oladapo, O., Ojo, O. O., & Muonde, M. (2023). Artificial intelligence in clinical pharmacy: Enhancing drug safety, adherence, and patient-centered care. *International Journal of Multidisciplinary Research and Growth Evaluation*, 4(1), 814–822

Oliveira, M. & Amorim, A. (2018). Microbial forensics: New breakthroughs and future prospects. *Applied Microbiology and Biotechnology*, 102(24), 10377–10391.

Oliveira, M., Marszałek, K., Kowalski, M., Frolova, A., Łabaj, P. P., Branicki, W. & Dinis-Oliveira, R. J. (2024). Sequencing technologies in forensic microbiology: Current trends and advancements. *Forensic Sciences*, 4(4), 523–545.

Pandit, M. A., Kumar, J., Gulati, S., Bhandari, N., Mehta, P., Katyal, R. & Kaur, J. (2022). Major biological control strategies for plant pathogens. *Pathogens*, 11(2), 273.

Parasidis, E. (2025). *America's Military Biomedical Complex: Law, Ethics, and the Drive for Scientific Innovation*. Oxford University Press.

Qi, F., Fan, S., Fang, C., Ge, L., Lyu, J., Huang, Z. & Zhang, X. (2023). Orally administrated *Lactobacillus gasseri* TM13 and *Lactobacillus crispatus* LG55 can restore the vaginal health of patients recovering from bacterial vaginosis. *Frontiers in Immunology*, 14, 1125239.

Rajput, H. (2024). Biological evidence for crime investigation. *Journal of Digital Security and Forensics*, 1(1), 34–52.

Rayimoglu, G., Yonar, F. C. & Anılanmert, B. (2025). From one strand Dyed/Undyed hair with/without root to fast and successful STR profiling and evaluation with principle component analysis. *Electrophoresis*, 46(17), 1292–1307.

Robinson, J. M., Pasternak, Z., Mason, C. E. & Elhaik, E. (2021). Forensic applications of microbiomics: A review. *Frontiers in Microbiology*, 11, 608101.

Scatiggio, V. (2020). Tackling the issue of bias in artificial intelligence to design ai-driven fair and inclusive service systems. How human biases are breaching into AI algorithms, with severe impacts on individuals and societies, and what designers can do to face this phenomenon and change for the better.

Secco, L., Palumbi, S., Padalino, P., Grosso, E., Perilli, M., Casonato, M. & Viel, G. (2025). "Omics" and postmortem interval estimation: A systematic review. *International Journal of Molecular Sciences*, 26(3), 1034.

Shanahan, F., Ghosh, T. S. & O'Toole, P. W. (2021). The healthy microbiome—What is the definition of a healthy gut microbiome? *Gastroenterology*, 160(2), 483–494.

Sharma, D., Valmiki, H., Chayal, P., Kumar, S. & Chhotaray, S. (2025). Microbiome study of Murrah buffalo mastitis milk with emphasis on Acinetobacter species. *BMC Microbiology*, 25(1), 703.

Sharma, R., Bhute, A. R. & Bastia, B. K. (2022). Application of artificial intelligence and machine learning technology for the prediction of postmortem interval: A systematic review of preclinical and clinical studies. *Forensic Science International*, 340, 111473.

Sherier, A. J., Woerner, A. E. & Budowle, B. (2022). Determining informative microbial single nucleotide polymorphisms for human identification. *Applied and Environmental Microbiology*, 88(7), e00052-22.

Shukla, S. (2024). Principles governing ethical development and deployment of AI. *Journal of Artificial Intelligence Ethics*, 7(1), 45–60.

Swayambhu, M. (2025). *FORMics: Forensic Microbiome and Metabolome Analysis for Body Fluid Identification* (Doctoral dissertation, University of Zurich).

Teixeira, M. J., Barbosa, D. J., Dinis-Oliveira, R. J. & Freitas, A. R. (2025). Redefining postmortem interval estimation: The need for evidence-based research to bridge science and justice. *Frontiers in Microbiology*, 16, 1646907.

Waddell, T. H. (2023). *Applying Bayesian-networks to model multi-organ pathologies* (Doctoral dissertation, University of Oxford).

Wani, A. K., Akhtar, N., Sena, S. & Singh, J. (2022). Microbial forensics: A potential tool for investigation and response to bioterrorism. *Health Sciences Review*, 5, 100068.

Wiesenthal, M. (2022). *The Ethical Implications of AI-based Mass Surveillance Tools* (Master's thesis).

Wu, Z., Guo, Y., Hayakawa, M., Yang, W., Lu, Y., Ma, J. & Niu, J. (2024). Artificial intelligence-driven microbiome data analysis for estimation of postmortem interval and crime location. *Frontiers in Microbiology*, 15, 1334703.

Wu, Z., Lyu, H., Liang, W., Jing, X., Wang, Y. & Ma, X. (2021). Microbial community in indoor dusts from university dormitories: Characteristics, potential pathogens and influence factors. *Atmospheric Pollution Research*, 12(3), 321–333.

Xu, G., Teng, X., Gao, X. H., Zhang, L., Yan, H. & Qi, R. Q. (2023). Advances in machine learning-based bacteria analysis for forensic identification: Identity, ethnicity, and site of occurrence. *Frontiers in Microbiology*, 14, 1332857.

Yao, H., Wang, Y., Wang, S., Sun, C., Zhou, Y., Jiang, L. & Luo, H. (2024). A multiplex microbial profiling system for the identification of the source of body fluid and skin samples. *Forensic Science International: Genetics*, 73, 103124.

Yetgin, A. (2025). Revolutionizing multi-omics analysis with artificial intelligence and data processing. *Quantitative Biology*, 13(3), e70002.

Zahrah, A., Tri Meilana, A. N. S. & Auerkari, E. I. (2025). Microbiological analysis in forensic identification and machine learning: A review. *Bulletin of the International Association for Paleodontology*, 19(1), 55–63.

Zhang, K., Delgado-Baquerizo, M., Zhu, Y. G. & Chu, H. (2020). Space is more important than season when shaping soil microbial communities at a large spatial scale. *mSystems*, 5(3), e00783.

Effects of Data Quality on AI Decision-Making in Microbiology

16

Abdulrazaq Izuafa, Chiedu Epiphany Edoamodu, Daniel Abu Anyebe, Jamilu Sani, Shamsudeen Aliyu, and Oluwafemi Adebayo Oyewole

16.1 INTRODUCTION

In a busy clinical microbiology laboratory in London, an AI-powered diagnostic system flagged a blood culture sample as *Klebsiella pneumoniae* and recommended immediate carbapenem therapy (Habibur *et al.*, 2024). In reality, the pathogen was *Escherichia coli* with a resistance profile that rendered carbapenems ineffective. The failure was not rooted in the algorithm itself but in the data that trained it, bacterial images had been mislabeled, sequencing data carried undetected errors, and metadata was incomplete (Habibur *et al.*, 2024; Wu & Gadsden, 2023). This case illustrates a critical truth: artificial intelligence (AI) is only as reliable as the data it consumes. The adoption of AI in microbiology has accelerated rapidly, transforming diagnostics, drug discovery, and microbial ecology. Machine learning models now assist in pathogen identification with reported accuracies exceeding 90 percent under controlled conditions (Ekersular & Alkan, 2024). They predict antimicrobial resistance (AMR), classify bacterial colonies through convolutional neural networks with near-perfect performance on benchmark datasets, and analyze complex microbiome profiles for applications in precision medicine. Some AI tools can even predict antibiotic susceptibility within hours, cutting the time required by conventional culture methods from days to mere fractions. The potential to reshape microbiological practice is enormous (Ahmadi *et al.*, 2024; Ekersular & Alkan, 2024).

Beneath impressive AI performance metrics lies a fundamental vulnerability: the reliability of any AI system is wholly dependent on the quality of its data. When training datasets are noisy, incomplete, or biased, these flaws propagate throughout the system. In microbiology, such data shortcomings can cause misdiagnoses, inappropriate treatments, and adverse patient outcomes. An algorithm that

DOI: 10.1201/9781003727514-18

performs well on controlled, standardized datasets may struggle when applied to the diverse and variable nature of real-world clinical samples (Ferrara, 2024; Roy, 2024).

This chapter explores how data quality impacts the dependability of AI applications in microbiology. It evaluates factors such as accuracy, completeness, timeliness, and standardization, illustrating through both successful and failed examples how poor-quality data undermines AI decision-making. Additionally, it presents strategies to enhance data workflows, minimize errors, and design AI systems resilient to imperfect data. Overcoming these challenges is critical for AI to realize its full potential in microbiology and ensure patient safety.

16.2 THE SYMBIOSIS BETWEEN DATA AND AI

The relationship between data and AI in microbiology is best understood as a symbiotic partnership in which the strength of one determines the success of the other. Without large, accurate, and diverse datasets, even the most advanced algorithms are ineffective. Conversely, without powerful AI systems, the immense complexity of microbiological data remains underutilized. Big data has fundamentally reshaped how AI models learn, offering vast repositories of information that fuel the recognition of patterns, the prediction of outcomes, and the ability to generalize across new clinical scenarios (Huo & Wang, 2024; Tsitou *et al.*, 2024).

Among biomedical sciences, microbiology is especially data-intensive. Every experiment generates layers of high-dimensional information that must be carefully integrated for meaningful interpretation. Genomic sequencing produces millions of base pairs for each microbial isolate. Proteomics contributes thousands of dynamic protein expression profiles. Mass spectrometry platforms such as Matrix-Assisted Laser Desorption/Ionization Time-of-Flight Mass Spectrometry (MALDI-TOF MS) create protein fingerprints that must be matched against extensive databases, as reported by Huo and Wang (2024). High-resolution microscopic imaging demands advanced computer vision tools for classification and analysis (Hamadani *et al.*, 2024). Meanwhile, metagenomics generates terabytes of sequencing data that capture the complexity of entire microbial ecosystems. This complexity is magnified by variation in time and space. Microbial communities shift rapidly in response to environmental changes, antibiotic pressure, and host immune responses. Resistance profiles and pathogen prevalence vary between regions, so AI models trained in one context may not perform effectively in another. Clinical metadata adds further layers, patient demographics, treatment history, sampling techniques, and laboratory protocols, all of which influence microbial behavior and algorithmic accuracy (Ali *et al.*, 2023; Lu *et al.*, 2025).

In this context, the principle of "garbage in, garbage out" (GIGO) is especially important. Impressive results have been demonstrated by AI models trained on high-quality, carefully curated datasets. As an illustration, well-annotated images of bacterial colonies have been used to classify with high accuracy of over 95 percent. AI performance however is considerably compromised when the training data are not complete, mislabeled, or biased. The GIGO difficulties of microbiology are peculiar. Affected cultures may inaccurately train models when they are not removed before preprocessing (Barragan-Montero *et al.*, 2022; Maurer, 2021). The intersection between samples can cause spurious patterns, which algorithms will mistakenly believe to be significant. The absence of metadata (e.g., missing records of antibiotic exposure or vague times of when samples were collected) deprives AI systems of the valuable context they require to make accurate predictions. Factors are complicated by time. Obsolete resistance monitoring information can lead AI to prescribe treatments that are now out of step with the current epidemiology. Without datasets that span a long period of time, seasonal changes in the prevalence of the pathogen may not be detected (Alsulimani *et al.*, 2024; Palla & Iwunwa, 2025).

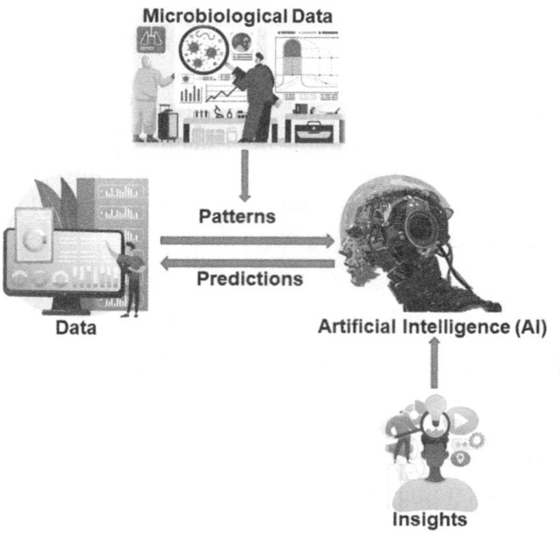

FIGURE 16.1 Interaction between data and AI in microbiological contexts.

The future of AI in microbiology depends on establishing strong data foundations. High-quality datasets are essential for generating clinically relevant predictions, improving patient outcomes, and advancing scientific discovery.

On the other hand, the quality of data may destroy even the most sophisticated algorithms, turning even the potential advantages of AI into threats. This critical relationship is not only challenging to maintain technically but also a clinical and ethical responsibility (Tsitou *et al.*, 2024). Figure 16.1 illustrates the cyclical interaction between microbiological data and AI.

16.3 DIMENSIONS OF DATA QUALITY IN MICROBIOLOGY

Microbiological data quality has many dimensions that are interconnected, each one of which can disrupt AI system functionality on its own. These dimensions are important in creating robust AI applications that can resist real-life microbiological practices (Huo & Wang, 2024). The most basic dimension of data quality is accuracy, which is used to define whether the information that has been recorded reflects the biological reality in the right way. In microbiology, there are often problems with accuracy during manual annotation procedures in which human professionals assign labels to bacterial species, resistance phenotypes, or microscopic observables. Training datasets with mislabeled microbial strains are an exceptionally pernicious type of inaccuracy, as they train AI systems to perceive wrong trends. As an example, training data is *Klebsiella pneumoniae* misunderstood as *E. coli*; AI models are taught to associate morphological or molecular features with those of *Klebsiella* and identify it as *E. coli* in clinical practice (Chaturvedi *et al.*, 2024; Kotwal *et al.*, 2022).

Labeling inaccuracies affect the issue of misclassification, which is not the only effect. Misclassification of resistance in the course of training in antimicrobial susceptibility testing results in the AI models giving inappropriate treatment recommendations. It has been demonstrated that even a few percentages of incorrectly labeled training samples can substantially worsen the performance of the model, especially in cases when the error is systematic as opposed to random (Kim *et al.*, 2022). Completeness considers the presence of all the information that is required in datasets. Microbiological datasets are often characterized by a lack of genomic sequences, incomplete metadata, or lack of clinical correlations, which constrain AI model usefulness. The unfinished assemblies of genomes tend to occur with difficult organisms with repetitive DNA patterns or poor-quality sequencing libraries and deny AI systems access to important genetic determinants of pathogenicity or resistance. The lack of patient metadata, including previous antibiotic exposure, immunocompromised condition, or sample collection location, denies AI models the contextual data needed to make the correct clinical prediction (Kumar *et al.*, 2024; Wang *et al.*, 2024).

Longitudinal studies are especially susceptible to the issue of completeness, as patients' follow-up data might not be complete. To train meaningful patterns, AI models that make predictions on the outcome of treatment or disease progression need full datasets with sufficient time intervals. When significant portions of patient outcomes are missing, models may develop biased predictions that overestimate or underestimate treatment efficacy (Terranova & Venkatakrishnan, 2024). Consistency refers to uniformity of data across different sources, time points, and measurement systems.

Across the globe, microbiology labs operate with a wide range of protocols, instruments, and interpretation standards. This diversity poses a major challenge when trying to integrate data for AI training. Even when working with identical specimens, labs may use different culture media, incubation settings, or identification methods, leading to systematic variations in results. These inconsistencies make it harder for AI models to learn reliably and generalize across datasets, ultimately affecting the accuracy of their predictions.

Mass spectrometry databases may contain different reference spectra for the same organism, leading to conflicting identification results when AI systems attempt to integrate information from multiple sources (Ain *et al.*, 2024).

Temporal consistency is an additional problem as the methods in the laboratory continue to evolve. New sequencing technologies, in particular, modern long-read technologies, generate data that do not have the same qualitative characteristics as older data. AI models that are trained on the combination of these datasets can become confused. Similarly, antimicrobial susceptibility clinical breakpoints are revised with time. The data may seem paradoxical in old datasets when history-based and stricter standards conflict with older and more permissive criteria (Chandrashekar *et al.*, 2024).

Timeliness is the up-to-dateness and timeliness of the data in the AI training and inference. Microbiology is a rapidly advancing subject. There is a rapid change in the prevalence of pathogens, resistance, and practices. When an AI model is trained using the old surveillance data, it may prescribe outdated antibiotics, which are no longer effective, or overlook new forms of resistance. As an example, carbapenem-resistant *Enterobacteriaceae* have swept the globe in the last ten years. Pre-2015 data-based models could also underestimate resistance and provide incorrect treatment recommendations (Gupta & Bhandary, 2024; Pennisi et al., 2025).

Timeliness is complicated by the seasonal and geographic variations in the epidemiology of pathogens. The patterns that AI needs to consider include a seasonal increase of respiratory viruses or epidemic strain spread. Models with poor predictive performance in unfamiliar time periods or regions are generated by datasets that do not contain these temporal patterns (Kraemer *et al.*, 2025).

Standardization refers to the use of common formats, terminologies, as well as protocols in order to ensure data can interact and be read the same way. A wide variety of standards, such as taxonomic nomenclature, antimicrobial susceptibility standards, and genomic annotation standards, have been developed by the microbiology community. However, the rate of adherence to it is not consistent between research groups and labs (Riesco & Trujillo, 2024). The absence of harmonized laboratory protocols is a significant problem. Various bacterial identification techniques, biochemical tests, and modern molecular techniques can sort organisms differently or with different accuracy.

The coverage depth, library preparation, and quality control of genomic sequencing protocols differ as well and result in datasets with varying error profiles and detection limits. Variations in AI systems can also learn spurious relationships because the data unites non-standardized protocols, which encompass methodology-related variables instead of biological variables (Arroyo Mühr *et al.*, 2021; To, 2024).

16.4 CASE STUDIES: WHEN DATA QUALITY SHAPES AI DECISIONS

As real-life experience demonstrated in microbiology, AI decisions depend on the quality of data; good data can provide amazing outcomes, whereas bad data can be dangerous. One of the pioneering studies of colorectal cancer detection proves the effectiveness of high-quality microbiome data in AI diagnostics. To identify the difference between healthy controls and patients with CRC, researchers constructed an explainable AI framework based on curated data of gut microbiota and demographic data. Quality control was done on the dataset by filtering out low-abundance taxa and standardizing it using compositional data analysis tools. Then it was confirmed on a variety of populations (Al-Kabani *et al.*, 2025; Greathouse & Choudhury, 2025). This high-quality work of the AI, due to scrupulous data processing, eliminated contaminants and remedied technical variability among sequencing runs. Taxa, the presence of which occurred in at least 10 percent of the samples, were retained to guarantee good statistical power. Microbiome data issues were solved with the help of centered log-ratio transformations; at the same time, biological relationships were maintained (Levy *et al.*, 2022; Nandan Prasad, 2024).

Due to this careful curation, the AI recognized biologically significant biomarkers, including *Fusobacterium, Porphyromonas, Peptostreptococcus*, and *Parvimonas*, bacteria already implicated in CRC in the knowledge base. The accuracy of the model was 93.2 percent, and the precision and recall were high, which makes the model applicable in clinical settings. Noteworthy, the predictions of the AI were interpretable so that the clinicians could observe what microbial features were used to make the diagnosis (Conde-Pérez *et al.*, 2024). The quality of the data was a priority throughout the whole process: the samples were gathered in accordance with a standardized protocol, DNA extraction and sequencing were standardized, metadata were completely recorded and assessed to eliminate bias, and low-quality sequences were filtered out to minimize noise (Jansirani *et al.*, 2025).

A multi-center study on AI-based AMR prediction showed that, in the case of inaccurate resistance databases, which caused incorrectly recommended drugs, the consequences were catastrophic, demonstrating the devastating role of bad data quality. In their study, researchers constructed machine-learning models to forecast the susceptibility of bacteria to antibiotics based on the susceptibility data of bacteria in different laboratories. The training data however had many quality flaws that undermined clinical utility (Nafchi *et al.*, 2025). The factual problem was mainly systematic errors in labeling: various tests of susceptibility, benchmarks, and quality controls were applied in laboratories. Several labs had not revised their clinical breakpoints, and the same isolates were found resistant in one lab and sensitive in the other. These inconsistencies also gave the AI wrong assumptions about the association of genomic characteristics with false phenotypes, making it susceptible to making life-threatening mistakes (Gajic *et al.*, 2022). Another challenge lies in the time gap between older samples processed with outdated methods and newer ones handled by automated systems. These differences introduce systematic bias, making it harder for AI models to learn consistently. In addition, fragmented genomic assemblies often lack key resistance genes or regulatory regions, which limits the model's ability to make accurate predictions and compromises its clinical relevance. The AI was not able to learn clinically relevant patterns because of the lack of metadata in access to antibiotic exposure, treatment response, and clinical outcomes (Alsulimani *et al.*, 2024).

In the clinical validation of the AI system, the system did not have the ability to generalize to hospitals and patients. The models that were trained using data of one institution only achieved 40 percent accuracy when used with isolates in other institutions, because of differences in laboratories and demographics. The false predictions of resistance led the clinicians to prescribe these broad-spectrum antibiotics when they were not needed, contributing to resistance and increased expenses. The ineffective narrow-spectrum antibiotics and clinical failures were caused by false-negative predictions (Rabbani *et al.*, 2022). The failure of the AI was due to endemic issues in the data quality at each phase of model development and implementation. The researchers came to the conclusion that harmonized data is insufficient to predict AMR since lab protocols must be standardized and the quality of such predictions should be thoroughly controlled (McGuire *et al.*, 2021).

There are unique data quality issues with environmental microbiology applications, leading to uneven AI performance. A comprehensive discussion of marine microbial communities shows that data on incomplete metagenomics leads to uncertainty in ecological models but, in certain cases, can provide useful data (Cerk *et al.*, 2024). The aim of scientists was to find AI models, which would forecast the reaction of microbial communities to environmental upheavals like temperature changes, the addition of nutrients, and pollution. Millions of reads of complete microbial communities were included in the metagenomic datasets, but AI execution was made difficult by many quality limitations (Wijaya *et al.*, 2023). One of the biggest issues was taxonomic incompleteness because the majority of environmental microbes do not have reference genomes in public databases. Metagenomic sequences that could not be placed into a particular taxon were approximately one in five, and gaps in the taxonomic composition of the community were very large. It was this lack of complete information that made AI models unable to form a connection between specific microorganisms and environmental responses, lowering their ability to predict a novel state of the ecosystem (Xu *et al.*, 2021). This geographic sampling bias has rendered it hard to come up with strong models. Most metagenomic data is based on well-studied marine environments in developed areas, with the tropical, polar, or deep-sea systems underrepresented. The models of AI, trained on such biased datasets, do not work in such environments and collapse disastrously when exposed to areas that are underrepresented (Walsh *et al.*, 2024). Irrespective of these limitations, AI systems have been able to provide valuable outputs in certain applications. When trained on taxa with complete reference genomes and well-characterized microbial groups, the models can predict the shifts of communities due to changes in temperature. The metabolic pathway prediction, even in the case of the absence of taxonomic completeness of the functional gene analysis, can predict the change in the conditions of various nutrients (Leggieri *et al.*, 2021). These anomalies highlight the fact that AI results are situational and depend on the quality of the data. Complete, high-quality data can be used to make reliable forecasts when operating in well-investigated settings, but incomplete or biased data can only generate minor forecasts when operating in uncharted environments (Xu *et al.*, 2021). This case brings to the fore the need to be aware of the data constraints in real-life application of AI systems.

All these studies are united by the fact that data quality is the decisive factor of the success of AI in the field of microbiology. High quality, if well-distributed datasets, allows transformative diagnostics, and poor-quality data may result in clinical failure. Even partially degraded datasets may be helpful if the weaknesses are recognized, included in the model development and deployment plans (Naveed *et al.*, 2024).

16.5 DATA BIAS AND ITS RIPPLE EFFECT

The significant threat to AI reliability in microbiology is data bias. It causes systematic errors, which propagate in machine learning models, causing biased predictions with severe effects. Contrary to random errors that are likely to be neutralized when there are many data, bias creates trends. These patterns are recognized as real biological relations by AI systems, which can lead to discriminatory or unsuitable results in both clinical and research environments (Alsulimani *et al.*, 2024).

In microbiology, sampling bias is primarily manifested by the overrepresentation of pathogenic strains over commensal or environmental microorganisms in training sets. Clinical laboratories specialize in pathogenic microbes, and thereby, the database is greatly biased in favor of those pathogens that are not necessarily the most representative of the entire community of microbes. It is the pathogen-confirming bias that trains AI to pay excessive attention to pathogenic signals and ignore the contributions of nonpathogenic or neutral microorganisms to health and illness (Lloréns-Rico *et al.*, 2021). Excessive representation of widespread species, the common ones that are not restricted to, are *E. coli*, *Staphylococcus aureus*, and *Streptococcus pneumoniae*, which have cascading effects. These common pathogens allow AI models to become highly sensitive to these pathogens yet are less sensitive to rare organisms or uncharacteristic manifestations of common diseases (Kool *et al.*, 2023).

Additional sampling bias is hospital-acquired infections. Clinical isolates are enriched with organisms that have a particular resistance profile or virulence characteristics to survive in the healthcare environment. Models that have been trained primarily on hospital isolates might not be useful in community-acquired infections where the pathogen populations and the antimicrobial susceptibility patterns differ. The situation is complicated by the presence of temporal sampling bias. Microbial populations that are measured in research datasets are usually sampled within a certain time or outbreak mode. Models that are trained on data during epidemic periods can overestimate the prevalence of pandemic strains or resistance mechanisms that temporarily increased during the crises but return to baseline after the outbreak ends (Abu-El-Ruz *et al.*, 2025; Harris, 2023). Figure 16.2 shows how bias starts in microbiological data and flows into AI predictions, affecting their accuracy and fairness.

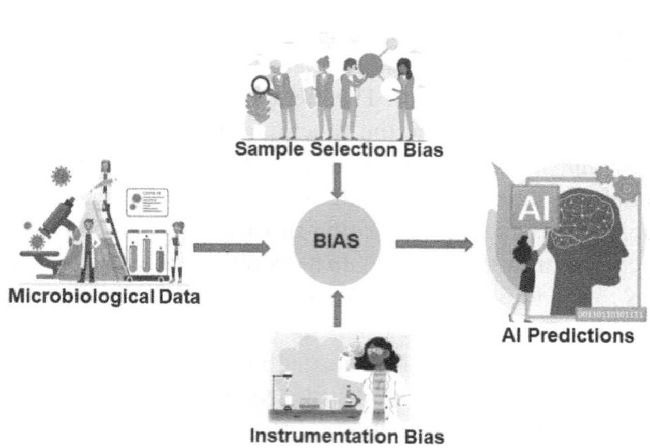

FIGURE 16.2 Sources and propagation of bias in microbiological data and AI predictions.

The most significant yet underestimated issue of microbiological AI is geographic bias. It increases the AI training disparity in health across the globe. The vast majority of high-quality genomic and clinical data is available in the well-endowed health systems of North America, Europe, and some developed Asian countries. The fact that AI models are great at identifying microbes prevalent in developed countries and terrible at identifying ones in low- and middle-income countries (LMICs) where the patterns of diseases, the variety of pathogens, and resistance to antimicrobials vary significantly is due to that geographic concentration (Tsitou *et al.*, 2024). As a case in point, tropical illnesses brought by parasites, bacteria, and viruses that are endogenous to particular areas are not sufficiently represented in training sets on the planet. AI diagnostic tools that have been predominantly trained on temperate-climate data might fail to identify or mislabel tropical pathogens as a more common organism associated with the training dataset (Kumar *et al.*, 2024).

AMR also has significant geographic variation, which is largely caused by local prescribing practices, policies, and access to healthcare. Models that have been trained on the information of countries with stringent stewardship programs can underestimate resistance in those areas where there is extensive abuse. On the other hand, high-resistance-based models can provide excessive generalized therapy in regions that have less resistance burden.

Another geographic bias is host genetics. Population genetics has an influence on the susceptibility to diseases, immune responses, and drug metabolism. AI models developed using genomic data of the majority of European descent might not be extendable to individuals of African, Asian, and Native American descent, whose genetic makeup alters host-pathogen interaction (Aijaz *et al.*, 2023; Ravi & Singh, 2024).

The division of clinical and environmental microbial data generates biases in a systematic manner, curbing the applicability of AI in different contexts. Clinical data are associated with isolated pathogens cultured in laboratory environments, whereas environmental data are complex microbial communities in nature. This disparity implies that AI models can perform well in certain fields but fail to do so in others (Huo & Wang, 2024). Clinical isolation selects the best organisms that thrive well in the lab. Diversity is usually skewed by excluding fastidious, slow-growing, or unculturable microbes. Cultures of such data could be unrecognized and not culturable by the models trained on such data, or they could be classified with the error of new species (Marcos-Zambrano *et al.*, 2021).

There are other biases that are brought by environmental data. Efficiency of DNA extraction varies by taxa; tough-cell-walled organisms end up over-represented, whereas weak-cell-walled organisms are underrepresented. PCR amplification biases also cause additional distortions in taxonomic representation, which form artificial patterns of abundance that AI is misled to perceive as reality (Pagano *et al.*, 2023). Temporal dynamics are also different: Samples in clinical studies are usually one snapshot in time, and environmental communities vary seasonally and nutritionally. This variability in environmental monitoring or longitudinal research studies may not be considered in models that are trained based on static clinical data (Hughes *et al.*, 2021).

Such biases cause distorted AI prediction and poor representation of new microbes, and these effects have far-reaching implications that are more severe than statistical performance figures. They have adverse impacts on the practice of patient care, scientific discovery, and health choices of the population. As the training data are not representative, AI predictions are biased, causing healthcare disparities: some groups of people are given inappropriate diagnoses and treatment (Haider *et al.*, 2024). Emerging microbes without detection form blind spots. New pathogens, newly described species, or rare organisms can be incorrectly classified or overlooked by models that have been trained using biased data. This risk was revealed during the COVID-19 pandemic because AI systems that were trained on previous respiratory trends did not detect SARS-CoV-2's new patterns (Naqvi *et al.*, 2025).

Prejudice exacerbates itself with AI. Prejudiced training results in prejudiced forecasts, making clinical judgments and subsequent information gathering, further supporting the initial prejudice. Misdiagnosis of some of the conditions among some population groups decreases the rate of detection among new patients, which further distorts datasets. The research priorities are being shifted to well-represented organisms, and poor representations are abandoned. This loop broadens the lack of knowledge, and it forms some technological blind spots instead of what is actually biological (Chinta *et al.*, 2024; Siddique *et al.*, 2023).

Dealing with bias cannot be fixed with technical solutions alone. It requires very strong data-collection policies, global partnerships, and ethical systems, which are based on equity and representativeness. Systematic bias detection and mitigation are the only ways the microbiological community will be able to create AI systems that benefit all the populations equally and can develop science without reinforcing the current disparities (Chinta et al., 2024).

16.6 NOISE AND ERRORS IN MICROBIOLOGICAL DATA AND THEIR IMPACT ON AI SYSTEMS

Microbiological data noise and errors are a threat to AI system reliability. Noise can have the appearance of natural biological variability, unlike such blatant data quality issues as missing values, mislabeled samples, etc. This renders it difficult to detect and resolve, but it silently corrupts predictive model performance (Mairi *et al.*, 2025). Genomic sequencing has transformed microbiology; however, it also presents technical noise, which can be damaging to AI training.

Errors introduced by DNA polymerase during sequencing can lead to random base substitutions, which may be mistaken for genuine mutations linked to AMR or virulence factors. This misinterpretation can compromise the accuracy of downstream analyses and hinder reliable AI-driven insights.

Although most errors are random, systematic bias in error pattern can confuse AI in the identification of artifacts as actual biological signals. Each sequencing platform has its own error profile. Illumina systems tend to introduce substitution errors into short sequences, and Oxford Nanopore introduces insertion and deletion errors in homopolymer sequences. Thus, models that are trained on one platform might not generalize well to data on a different learning platform and end up learning platform-specific errors, which are mistakenly interpreted as biological variation (Huo & Wang, 2024).

Preparation of the library adds additional noise. Genomic representation may be distorted by polymerase chain reaction amplification to provide AI systems with the illusion of a significant gene-expression signal due to uneven coverage. The bias due to GC content can miss AT-rich areas that can harbor vital virulence genes or regulatory sequences that remain undetected by AI models. Systematic variation is also introduced by the traditional culture-based methods (Wei *et al.*, 2024). Compositions of the media, incubation conditions, and laboratory practices can influence the growth of microbes, expression of phenotypic traits, and antimicrobial-susceptibility outcomes. These variations bring inconsistencies, which an AI model can misinterpret as either a diagnostic marker or a taxonomic marker. The risk of cross-contamination will only contribute to the risk, providing false-positive indicators that train AI systems to identify contaminants as clinical conditions. Due to the intermittent nature of contamination and its automatic invisibility, it may go unnoticed and undermine the credibility of the models (Al-Shuhaib & Hashim, 2023).

Errors of human annotation of microbial images are significant. Supervised AI involves high amounts of manually labeled data that is usually prepared by trained microbiologists. Even the best of annotations may include errors because of fatigue, time constraints, subjective opinions, and interobserver differences. The same image might be classified differently by different experts depending on the way they are trained, or what is required by the institutions, or their experience. Prisons of cell shape and colony morphology can also be on a continuum and not in distinct categories, making annotation choices subjective. Conflicts between annotations give AI conflicting signals that reduce the accuracy of the classification. Bias may also occur where annotators use varied standards for the various scenarios. The samples of high-risk patients can be subjected to tight scrutiny, whereas normal surveillance samples can be marked not as carefully. These human biases are internalized by AI models as being biological (Burns *et al.*, 2023; Goodswen *et al.*, 2021). Big projects in annotation are a burden. Error rates are also higher in case of fatigue when the annotators are editing thousands of pictures with strict deadlines. Such mistakes can exhibit the characteristics of time, making AI learn even more confused. A close relationship between such closely related species as those of *Klebsiella pneumoniae* and *Klebsiella oxytoca* results in morphological similarity and makes distinguishing between them a challenging undertaking (Burns *et al.*, 2023).

The automated systems have sensor and instrument limitations, which present special technical noise. Mass spectrometry-based identification systems, such as the MALDI-TOF MS, are very sensitive to the sample preparation, the use of the matrix, and also the calibration, which impacts on the accuracy and reliability. Weak or distorted spectra that result in inaccurate classifications can be generated by poor preparation, and spectral data can be displaced systematically by calibration drift. These changes will be interpreted as biological variation by AI models (Raju *et al.*, 2024). The automation of antimicrobial-susceptibility testing is based on the measurements of bacterial growth. Noise may be due to temperature, optical density variation, or plate position. These mistakes cause systematic biases that AI will incorrectly consider as resistance traits. There are also additional complications presented by digital imaging platforms: noisy pixels on camera sensors, spotty lighting, and inconsistent focus damage microbial image datasets. The trained AI with such images can mistake artifacts with colony features. Likewise, incubator or culture-monitoring environmental sensors are also sources of uncertainty. Temperature variations, changes in humidity, or pH influence the growth and metabolism of microbes, which affect the training data (Salam *et al.*, 2023).

The part of the noise that is most frightening in terms of microbiological AI systems is that the noise can propagate down the whole pipeline of the analytics. Although errors may be minor during the sequencing, culture, or annotation process, they can be transmitted to preprocessing, training, and prediction, causing significant systematic biases. Noise can often be enhanced by preprocessing algorithms that selectively discard low-quality data, a process that forms hidden bias. The noise of various AI model architectures is responsive to noise. Deep neural networks are sensitive to small perturbations, despite the fact that they are powerful. Although the noise can be too insignificant to detect, dramatic misclassification can occur, and in clinical practice, such can be the source of dangerous consequences (Gajic *et al.*, 2022). This can be exacerbated by feedback loops in the development of AI in an iterative fashion. When a model is trained to take noisy data as real information, active-learning plans can be used to amplify such trends, gradually poisoning training samples with false correlations. To handle these challenges, there is the need to have systematic approaches that cut across the whole pipeline of data collection to prediction. In the absence of strong methods to identify and deal with noise, the outputs of AI systems will be prone to causing outputs that are more technical than biological truth (Majumder *et al.*, 2025).

16.7 THE CONSEQUENCES OF POOR DATA QUALITY IN MICROBIOLOGICAL AI SYSTEMS

Understandably, poor data quality of microbiological AI systems has implications beyond the academic metrics. It is detrimental to patient safety, population health, and scientific advancement. These implications are obvious: loss of life, increased healthcare expenditures, and lost opportunities. Being aware of these high-stakes consequences is essential in order to warrant the investments in the improved data quality (Alsulimani *et al.*, 2024).

The risks of poor-quality data are the most immediate ones, among which misdiagnosis can be identified. In case the AI systems do not detect the pathogens accurately, patients will be provided with incorrect treatments, which can slow down the recovery process, increase the number of complications, or even cause a failure of the treatment. To illustrate, wrong identification of bacteria may lead to the wrong antibiotics being given. An AI being trained on mislabeled data may mix up *Enterococcus faecium* with *Enterococcus faecalis* and prescribe ampicillin to vancomycin-resistant *Enterococcus faecium*, which is not effective. These mistakes may result in septic shock and even death unless they are dealt with quickly (Gerush *et al.*, 2024).

Fungal infections have the same threats as well. The drug-resistant yeast Candida auris has been falsely labeled as a less dangerous organism, causing patients to be given ineffective antifungals and develop a rapid clinical decline. In the times of the COVID-19 pandemic, small-sized datasets frequently did not allow even AI models to distinguish between SARS-CoV-2 and other respiratory viruses. This is what contributed to the spread of the virus by facilitating its behavior.

Errors during diagnosis are particularly damaging when time is a factor, like with sepsis or meningitis. Even several hours of delay can determine the survival of a patient and his/her long-term outcomes (Černáková *et al.*, 2021). Figure 16.3 displays that errors in AI-based microbiology systems can lead to misdiagnosis, undermine confidence, and reduce general reliability.

Wrong decisions in antimicrobial treatment due to low data quality compromise AI systems that inform antibiotic or antifungal treatments. Poor-quality data on susceptibility may result in systematic treatment mistakes, augment AMR, and lead to higher expenses. In case AI models do not consider the resistance, they can prescribe narrow-spectrum medications that cannot work and leave infections to develop further. On the other hand, resistance can be overestimated to cause unnecessary broad-spectrum treatment, which disrupts the microbiome and accelerates the proliferation of multidrug-resistant organisms. Such errors prolong the duration of stay, healthcare spending, and put patients at

FIGURE 16.3 Consequences of errors in AI-based microbiological decision-making.

risk of new complications like device-related infections and extended treatment complications (Pennisi et al., 2025).

The poor quality of data is the primary cause of false results in microbiome research and personalized medicine. In the case of errors in the datasets, researchers and clinicians can detect artificial associations between microbes and disease. By way of example, false associations caused by contamination or biased sampling may be interpreted by AI as biological truths, resulting in ineffective or damaging treatments. Even worse, population bias: in case the training data contains representatives of one demographic group in disproportion, AI models may fail with underrepresented individuals. Another complexity that increases with time is that microbiomes change over time. AIs that are trained on snapshot data can make predictions based on short-term conditions and ignore long-term trends that are of real significance. These issues are a part of a reproducibility crisis due to which false results propagate uncontrollably in research (Liu *et al.*, 2024).

The undergoing of biotechnology and biofuel production is also poor because of poor data quality. AI is finding application in microbiology within the industrial sector to optimize microbial strains, control fermentation, and speed up biofuel advancement. Silence or noise in training sets may predispose AI to suggest genetic or processing modifications that may, in fact, lower yields. These errors may cause large-scale biofuel plants to be nonprofit and slow the transition to sustainable energy. Equally, AI, which has been trained using biased information, may redirect drug-discovery initiatives and deprive years of investment in research (Sun *et al.*, 2024).

16.8 ACHIEVING DATA INTEGRITY IN MICROBIOLOGICAL AI APPLICATIONS

Microbiological AI will need to be designed to keep data intact, and it will combine comprehensive and methodological approaches to quality-related concerns across the data lifecycle. This starts with the collection of data, followed by preprocessing, model deployment, and constant monitoring. Due to the complexity and heterogeneity of microbiological data, it should maintain a high quality via complex technical solutions, effective organizational policies, and international partnerships that lower data integrity as the main focus of any efficient AI systems (Huo & Wang, 2024).

Strong data cleaning and preprocessing pipelines are the initial defense mechanism against bad data quality. These pipelines have to be optimized for microbiological data and maintain biological signatures. As an example, sequence data needs special pipelines that remove technical artifacts and capture real biological variation. This method can also be seen with the CLEAN pipeline: it removes platform-specific control reads of Illumina and Nanopore data, decontaminates host DNA in metagenomic data, and removes ribosomal RNA in transcriptomic data (Harfouche *et al.*, 2023). These prevent learning of AI models as artifacts of biology. Balanced filtering is very essential—the filters should not eliminate microbial diversity to eliminate low-quality data. Excessively strict thresholds result in bias in taxonomy, and excessively relaxed thresholds introduce technical noise. Dynamically changing the thresholds according to the specifics of the dataset and the purpose of its keepers (adaptive quality filtering) is a good alternative (Aizpurua *et al.*, 2024).

Identical in importance is contamination detection. The sampling cross-contaminations have the capacity to form false associations, which AI models will misinterpret as actual biology. Contemporary statistical data analysis methods and algorithms determine unusual taxonomic trends and relationships between abundance and sequence similarity to indicate possible contamination.

This approach allows each sample to be carefully verified, ensuring that any contaminated ones are identified and removed. Additionally, normalization techniques help preserve biological integrity by correcting for variations across sequencing runs, different labs, and time periods, making the data more consistent and reliable for AI analysis.

Statistical design allows one to take into consideration and compensate for batch effects without biasing real biological differences. Normalization should be properly validated, thereby making sure that the meaningful signals are not erased (Goumas *et al.*, 2025).

Adopting FAIR data principles, which include Findable, Accessible, Interoperable, and Reusable, offer a system to keep data in microbiological AI of high quality and accessible. Findability is based on rich metadata and persistent identifiers enabling datasets to be found and to provide long-term access. The infrastructure needed to be accessible should offer support for large-scale datasets and the safety of sensitive information, and federated learning could allow the sharing of insights without storing the information at a central location. Interoperability is based on common standards and formats that allow data integration across institutions, but there is still a challenge in the consistency of implementation. The process of reusability requires well-documented, clearly licensed, and transparent quality measurements such that data may be trusted and utilized in a wide range of research settings (Inau *et al.*, 2021).

Imperfect data AI methods are used to deal with imperfect data, where perfect data quality is not always achievable. The systems have to work with inputs that are noisy or incomplete. Strong machine-learning algorithms include uncertainty estimates and tolerance to noise, and ensemble techniques minimize the effects of flaws in a particular piece of data. The Bayesian methods explicitly deal with the uncertainties in prediction, which is particularly important in clinical decision-making. Transfer learning also allows models trained on large, high-quality datasets to be adapted to smaller or lower-quality microbiological datasets. Noise injection or adversarial training are examples of data augmentation that make models more resilient by exposing them to real-world distortions. Active learning gives more importance to the most informative and credible samples to train on and hence can reduce the impact of bad-quality data (Roy, 2024).

The use of open science and global data warehouse is also important to scale AI in microbiology with the collaboration of different countries. There are large archives such as GenBank, EMBL, and DDBJ that provide comprehensive collections of genomes under automated quality checks and curations by specialists to provide accuracy. Collaborative surveillance networks are networks that distribute data on AMR, pathogen genomes, and clinical outcomes in real-time. These systems allow researchers to identify potential threats, test AI models in large populations, and maintain a high-quality standard. Standardization efforts enhance such initiatives and bring about global consistency in methods of laboratories, data forms, and quality measures. This makes it less variable and may confound AI training and provide comparability between regions (Leigh *et al.*, 2024). Open science is sustainable over the long term through financial backing, data sharing incentives, and international data stewardship agreements. Finally, data integrity in microbiological AI cannot solely be a technical problem, and researchers, institutions, and policymakers should make high-quality data their priority in advancing science and successfully transferring its knowledge to clinical practice (Taddese *et al.*, 2025).

16.9 THE FUTURE: INTELLIGENT AI NEEDS INTELLIGENT DATA

The future of AI in microbiology is based on a single concept. The smart data ecosystems will flourish with AI that are learning, adaptive, accurate, ethical, and reliable. It is not merely a vision of tools but of self-improving systems that will ensure that data quality is automatic, supporting science, and clinical care with AI. The key to this future is unified, quality microbial databases (Vashishth *et al.*, 2025). New systems will not be so fragmented as they are nowadays by connecting data in hospitals, research centers, and public health agencies across the globe. They will give real-time images of the diversity of microbes, evolution of pathogens, and resistance mechanisms. The advanced AI will constantly seek out errors, biases, and inconsistencies, and indicate or fix them, and refine through human feedback. Data of various types will be combined, and insights will be drawn out of clinical reports and scientific papers using semantic technologies and natural language processing (Cabrera *et al.*, 2021).

International surveillance will alter the way we monitor the menace of microbes. Genomic, clinical, environmental, and epidemiological data combined with AI will also forecast outbreaks and propose interventions before issues occur. Privacy-preserving and federated learning allow countries to work together and preserve patient data and provide equitable access to AI resources (Ali, 2024). AI and laboratory automation are also used to support this vision. New laboratories will include robotic systems, automated imaging, and AI-based analysis pipelines that will become better, as more samples are processed. Machine learning will optimize incubation conditions and testing protocols, and predictive maintenance will minimize downtime and forecast equipment failures. It will focus on quality in all levels, and once incorporated with electronic health records, such systems will produce personalized recommendations to each patient (Ashraf *et al.*, 2025).

Technology is not enough. The moral aspect of microbiological AI is essential. Open, responsible, and fair credible AI should be open, responsible, and fair. Explainable AI enables clinicians to trust and understand recommendations, and bias auditing ensures that there is a consistent performance between populations and health systems. Continuity of real-world validation assures reliability, patient autonomy, and confidentiality by strong privacy models. To achieve this future, there must be collaboration on the international level (Zhang & Zhang, 2023). The stewardship of data needs to be a profession that is systematically trained, equipped, and supported in terms of social resilience to support microbiologists. Clinicians and data scientists are supposed to acquire the basics of microbiology and be actively involved in developing and testing AI. A policymaker needs to come up with more creative, adaptive, and protective mechanisms that promote international cooperation (Dumm *et al.*, 2024).

The path forward is clear. An intelligent data ecosystem requires cross-functional collaboration and special investment. Its advantages are significant: better patient care and quicker pathogenic results, as well as greater preparedness to microbial threats. Although AI is potentially the brain, data is still the lifeblood of microbiology. It is a scientific and ethical responsibility to safeguard and improve this lifeblood, which is done fairly, reliably, and compassionately by AI (Azam, 2024).

16.10 CONCLUSION

AI is transforming microbiology, and its ability to succeed depends on the quality of data. Through quality information, AI provides almost perfect cancer diagnosis, precise identification of bacteria, and prediction of antimicrobial resistance, which previously required days to complete but now requires hours. These results indicate that AI can save lives, reduce healthcare expenses, and make advances in science. Nonetheless, bad data may cause grave mistakes. Improper lab protocols and biased naming of samples make AI fail to detect pathogens, give inaccurate predictions, and overlook at-risk populations. These issues demonstrate that AI is not a tool; it is an obligation that requires effective data management. Accuracy, completeness, timeliness, and standardization are all aspects of data quality that rely on both technical support (e.g., automated quality control tools, machine-learning frameworks, etc.)

and organization-wide commitment to professionalism, cooperation, and ethical standards. Equity and credibility are the two issues that should be addressed to avoid harming patients and avoiding bias. Finally, AI can become the lifeblood of contemporary microbiology, although information is its blood. Enhancing the integrity of the data is a scientific and ethical goal that will enhance healthcare and inspire discovery.

REFERENCES

Abu-El-Ruz, R., AbuHaweeleh, M. N., Hamdan, A., Rajha, H. E., Sarah, J. M., Barakat, K., & Zughaier, S. M. (2025). Artificial intelligence in bacterial infections control: A scoping review. *Antibiotics*, 14(3), 256.

Ahmadi, A., Courtney, M., Ren, C., & Ingalls, B. (2024). A benchmarked comparison of software packages for time-lapse image processing of monolayer bacterial population dynamics. *Microbiology Spectrum*, 12(8), e00032-24.

Aijaz, M., Ahmad, M., Ansari, M. A., & Ahmad, S. (2023). Antimicrobial resistance in a globalized world: Current challenges and future perspectives. *International Journal of Pharmaceutical Drug Design*, 1(1), 7–22.

Ain, Q. U., Nazir, R., Nawaz, A., Shahbaz, H., Dilshad, A., Mufti, I. U., & Iftikhar, M. (2024). Machine learning approach towards quality assurance, challenges and possible strategies in laboratory medicine. *Journal of Clinical and Translational Pathology*, 4(2), 76–87.

Aizpurua, O., Dunn, R. R., Hansen, L. H., Gilbert, M. T. P., & Alberdi, A. (2024). Field and laboratory guidelines for reliable bioinformatic and statistical analysis of bacterial shotgun metagenomic data. *Critical Reviews in Biotechnology*, 44(6), 1164–1182.

Ali, H. (2024). AI for pandemic preparedness and infectious disease surveillance: Predicting outbreaks, modeling transmission, and optimizing public health interventions. *International Journal of Research Publication and Reviews*, 5(8), 4605–4619.

Ali, T., Ahmed, S., & Aslam, M. (2023). Artificial intelligence for antimicrobial resistance prediction: Challenges and opportunities towards practical implementation. *Antibiotics*, 12(3), 523.

Al-Kabani, A., Huda, B., Haddad, J., Yousuf, M., Bhurka, F., Ajaz, F., & Banerjee, Y. (2025). Exploring experimental models of colorectal cancer: A critical appraisal from 2D cell systems to organoids, humanized mouse avatars, organ-on-chip, CRISPR engineering, and AI-driven platforms—Challenges and opportunities for translational precision oncology. *Cancers*, 17(13), 2163.

Al-Shuhaib, M. B. S., & Hashim, H. O. (2023). Mastering DNA chromatogram analysis in Sanger sequencing for reliable clinical analysis. *Journal of Genetic Engineering and Biotechnology*, 21(1), 115.

Alsulimani, A., Akhter, N., Jameela, F., Ashgar, R. I., Jawed, A., Hassani, M. A., & Dar, S. A. (2024). The impact of artificial intelligence on microbial diagnosis. *Microorganisms*, 12(6), 1051.

Arroyo Mühr, L. S., Guerendiain, D., Cuschieri, K., & Sundström, K. (2021). Human papillomavirus detection by whole-genome next-generation sequencing: Importance of validation and quality assurance procedures. *Viruses*, 13(7), 1323.

Ashraf, A. A., Rai, S., Alva, S., Alva, P. D., & Naresh, S. (2025). Revolutionizing clinical laboratories: The impact of artificial intelligence in diagnostics and patient care. *Diagnostic Microbiology and Infectious Disease*, 111(4), 116728.

Azam, U. (2024). A new era in cosmetology. *European Journal of Biomedical*, 11(5), 91–102.

Barragan-Montero, A., Bibal, A., Dastarac, M. H., Draguet, C., Valdes, G., Nguyen, D., & Lee, J. A. (2022). Towards a safe and efficient clinical implementation of machine learning in radiation oncology by exploring model interpretability, explainability and data-model dependency. *Physics in Medicine & Biology*, 67(11), 11TR01.

Burns, B. L., Rhoads, D. D., & Misra, A. (2023). The use of machine learning for image analysis artificial intelligence in clinical microbiology. *Journal of Clinical Microbiology*, 61(9), e02336-21.

Cabrera, ÁA., Druck, A. J., Hong, J. I., & Perer, A. (2021). Discovering and validating AI errors with crowdsourced failure reports. *Proceedings of the ACM on Human-Computer Interaction*, 5(CSCW2), 1–22.

Cerk, K., Ugalde-Salas, P., Nedjad, C. G., Lecomte, M., Muller, C., Sherman, D. J., & Frioux, C. (2024). Community-scale models of microbiomes: Articulating metabolic modelling and metagenome sequencing. *Microbial Biotechnology*, 17(1), e14396.

Černáková, L., Roudbary, M., Brás, S., Tafaj, S., & Rodrigues, C. F. (2021). Candida auris: A quick review on identification, current treatments, and challenges. *International Journal of Molecular Sciences*, 22(9), 4470.

Chandrashekar, K., Niranjan, V., Vishal, A., & Setlur, A. S. (2024). Integration of artificial intelligence, machine learning and deep learning techniques in genomics: Review on computational perspectives for NGS analysis of DNA and RNA Seq data. *Current Bioinformatics*, 19(9), 825–844.

Chaturvedi, N., Yadav, M. K., & Sharma, M. (2024). Applications of artificial intelligence and machine learning in microbial diagnostics and identification. *In* Srivastava, A., & Mishra, V. (Eds.), *Methods in microbiology* (Vol. 55, pp. 213–230). Academic Press, Cambridge, MA, USA. https://doi.org/10.1016/bs.mim.2024.05.013.

Chinta, S. V., Wang, Z., Yin, Z., Hoang, N., Gonzalez, M., Quy, T. L., & Zhang, W. (2024). FairAIED: Navigating fairness, bias, and ethics in educational AI applications. arXiv preprint arXiv:2407.18745.

Conde-Pérez, K., Aja-Macaya, P., Buetas, E., Trigo-Tasende, N., Nasser-Ali, M., Rumbo-Feal, S., & Poza, M. (2024). The multispecies microbial cluster of fusobacterium, parvimonas, bacteroides and faecalibacterium as a precision biomarker for colorectal cancer diagnosis. *Molecular Oncology*, 18(5), 1093–1122.

Dumm, R. E., Marlowe, E. M., Patterson, L., Larkin, P. M., She, R. C., & Filkins, L. M. (2024). The foundation for the microbiology laboratory's essential role in diagnostic stewardship: An ASM laboratory practices subcommittee report. *Journal of Clinical Microbiology*, 62(10), e00960-24.

Ekersular, M. N., & Alkan, A. (2024). Detection of COVID-19 disease with machine learning algorithms from CT images. *Gazi University Journal of Science*, 37(1), 169–181.

Ferrara, E. (2024). The butterfly effect in artificial intelligence systems: Implications for AI bias and fairness. *Machine Learning with Applications*, 15, 100525.

Gajic, I., Kabic, J., Kekic, D., Jovicevic, M., Milenkovic, M., Mitic Culafic, D., & Opavski, N. (2022). Antimicrobial susceptibility testing: A comprehensive review of currently used methods. *Antibiotics*, 11(4), 427.

Gerush, K. I., Pashkovska, N. V., & Ukrainets, O. Z. (2024). Errors in the diagnosis of types of diabetes mellitus: Causes and prevention strategies (literature review and own research results). *International Journal of Endocrinology (Ukraine)*, 20(4), 307–315.

Goodswen, S. J., Barratt, J. L., Kennedy, P. J., Kaufer, A., Calarco, L., & Ellis, J. T. (2021). Machine learning and applications in microbiology. *FEMS Microbiology Reviews*, 45(5), fuab015.

Goumas, G., Vlachothanasi, E. N., Fradelos, E. C., & Mouliou, D. S. (2025). Biosensors, artificial intelligence biosensors, false results and novel future perspectives. *Diagnostics*, 15(8), 1037.

Greathouse, K. L., & Choudhury, A. (2025). Precision nutrition and the gut microbiome: Harnessing AI to revolutionize cancer prevention and therapy. *Cell Host & Microbe*, 33(6), 766–776.

Gupta, Y. D., & Bhandary, S. (2024). Artificial Intelligence for Understanding Mechanisms of Antimicrobial Resistance and Antimicrobial Discovery: A new age model for translational research. *In* A. Khanna, M. El Barachi, S. Jain, M. Kumar, & A. Nayyar (Eds.), *Artificial intelligence and machine learning in drug design and development* (pp. 117–156). John Wiley & Sons, Ltd., Hoboken, New Jersey, USA. doi: 10.1002/9781394234196.ch5

Habibur, M., Ara, S. A., Moshiur, M., Hasan, M. M., Chandra, D., Shahana, M., & Al, A. (2024). Artificial intelligence for improved diagnosis and treatment of bacterial infections. *Microbial Bioactives*, 7(1), 1–18.

Haider, S. A., Borna, S., Gomez-Cabello, C. A., Pressman, S. M., Haider, C. R., & Forte, A. J. (2024). The algorithmic divide: A systematic review on AI-driven racial disparities in healthcare. *Journal of Racial and Ethnic Health Disparities*, 13(1), 188–217.

Hamadani, A., Ganai, N. A., Henna, H., & Bashir, J. (Eds.). (2024). *A biologist's guide to artificial intelligence: Building the foundations of artificial intelligence and machine learning for achieving advancements in life sciences. In* Hamadani, A., & Ganai, N. A. (Eds.). Academic Press (Elsevier), London, United Kingdom. doi: 10.1016/C2023-0-01341-3

Harfouche, A. L., Nakhle, F., Harfouche, A. H., Sardella, O. G., Dart, E., & Jacobson, D. (2023). A primer on artificial intelligence in plant digital phenomics: Embarking on the data to insights journey. *Trends in Plant Science*, 28(2), 154–184.

Harris, M. (2023). Machine learning and artificial intelligence for pathogen identification and antibiotic resistance detection: Advancing diagnostics for urinary tract infections. *BioMed*, 3(2), 246–255.

Hughes, A. C., Orr, M. C., Ma, K., Costello, M. J., Waller, J., Provoost, P., & Qiao, H. (2021). Sampling biases shape our view of the natural world. *Ecography*, 44(9), 1259–1269.

Huo, D., & Wang, X. (2024). A new era in healthcare: The integration of artificial intelligence and microbial. *Medicine in Novel Technology and Devices*, 23, 100319.

Inau, E. T., Sack, J., Waltemath, D., & Zeleke, A. A. (2021). Initiatives, concepts, and implementation practices of FAIR (findable, accessible, interoperable, and reusable) data principles in health data stewardship practice: Protocol for a scoping review. *JMIR Research Protocols*, 10(2), e22505.

Jansirani, D., Begum, S. S., Velmurugan, G. G., Ramalakshmi, K., & Indiraleka, M. (2025). Standardizing sample preparation: Ensuring integrity in urinary omics. *In* Giri, A., Sharma, I., & Ojha, R. (Eds.), *Exploring noninvasive disease biomarkers with urinary omics analysis* (pp. 95–115). Academic Press (Elsevier). doi: 10.1016/B978-0-443-33574-7.00011-0

Kim, J. I., Maguire, F., Tsang, K. K., Gouliouris, T., Peacock, S. J., McAllister, T. A., & Beiko, R. G. (2022). Machine learning for antimicrobial resistance prediction: Current practice, limitations, and clinical perspective. *Clinical Microbiology Reviews*, 35(3), e00179-21.

Kool, J., Tymchenko, L., Shetty, S. A., & Fuentes, S. (2023). Reducing bias in microbiome research: Comparing methods from sample collection to sequencing. *Frontiers in Microbiology*, 14, 1094800.

Kotwal, S., Rani, P., Arif, T., Manhas, J., & Sharma, S. (2022). Automated bacterial classifications using machine learning based computational techniques: Architectures, challenges and open research issues. *Archives of Computational Methods in Engineering*, 29(4), 2469–2490.

Kraemer, M. U., Tsui, J. L. H., Chang, S. Y., Lytras, S., Khurana, M. P., Vanderslott, S., & Bhatt, S. (2025). Artificial intelligence for modelling infectious disease epidemics. *Nature*, 638(8051), 623–635.

Kumar, B., Lorusso, E., Fosso, B., & Pesole, G. (2024). A comprehensive overview of microbiome data in the light of machine learning applications: Categorization, accessibility, and future directions. *Frontiers in Microbiology*, 15, 1343572.

Leggieri, P. A., Liu, Y., Hayes, M., Connors, B., Seppälä, S., O'Malley, M. A., & Venturelli, O. S. (2021). Integrating systems and synthetic biology to understand and engineer microbiomes. *Annual Review of Biomedical Engineering*, 23(1), 169–201.

Leigh, D. M., Vandergast, A. G., Hunter, M. E., Crandall, E. D., Funk, W. C., Garroway, C. J., & Paz-Vinas, I. (2024). Best practices for genetic and genomic data archiving. *Nature Ecology & Evolution*, 8(7), 1224–1232.

Levy, J., Lu, Y., Montivero, M., Ramwala, O., McFadden, J., Miles, C., & Vaickus, L. (2022). Artificial intelligence, bioinformatics, and pathology: Emerging trends part II––Current applications in anatomic and molecular pathology. *Advances in Molecular Pathology*, 5(1), e25–e52.

Liu, B., Jiang, H., & Li, L. (2024). The innovative applications and prospects of artificial intelligence in agricultural research: Gene editing, crop improvement, and intelligent data analysis. *Advances in Resources Research*, 4(4), 624–653.

Lloréns-Rico, V., Vieira-Silva, S., Gonçalves, P. J., Falony, G., & Raes, J. (2021). Benchmarking microbiome transformations favors experimental quantitative approaches to address compositionality and sampling depth biases. *Nature Communications*, 12(1), 3562.

Lu, J., Zhang, W., He, Y., Jiang, M., Liu, Z., Zhang, J., & Wan, C. (2025). Multi-omics decodes host-specific and environmental microbiome interactions in sepsis. *Frontiers in Microbiology*, 16, 1618177.

Mairi, A., Hamza, L., & Touati, A. (2025). Artificial intelligence and its application in clinical microbiology. *Expert Review of Anti-Infective Therapy*, 23(7), 469–490.

Majumder, A., Mandal, B., Dalal, S., Sahoo, R., & Samanta, S. (2025). In vitro susceptibility assays for bacterial response to antimicrobial agents. *Laboratory Techniques for Fish Disease Diagnosis* (pp. 55–77).

Marcos-Zambrano, L. J., Karaduzovic-Hadziabdic, K., Loncar Turukalo, T., Przymus, P., Trajkovik, V., Aasmets, O., & Truu, J. (2021). Applications of machine learning in human microbiome studies: A review on feature selection, biomarker identification, disease prediction and treatment. *Frontiers in Microbiology*, 12, 634511.

Maurer, L. G. (2021). Managing the medical matrix: A "DAIS" for artificial intelligence in health care (and beyond). *Case Western Reserve Journal of Law, Technology & the Internet*, 13, 107.

McGuire, R. J., Yu, S. C., Payne, P. R., Lai, A. M., Vazquez-Guillamet, M. C., Kollef, M. H., & Michelson, A. P. (2021). A pragmatic machine learning model to predict carbapenem resistance. *Antimicrobial Agents and Chemotherapy*, 65(7), e0006321.

Nafchi, H. M., Solatzadeh, H., Hajimaghsoudi, E., & Babakhanzadeh, E. (2025). Personalizing cancer therapy: The role of pharmacogenetics in overcoming drug resistance and toxicity. *Molecular Biology Reports*, 52(1), 785.

Nandan Prasad, A. (2024). Data quality and preprocessing. *In Introduction to data governance for machine learning systems: Fundamental principles, critical practices, and future trends* (pp. 109–223). Apress, Berkeley, CA.

Naqvi, S. A. H., Abbas, A., Hasnain, A., Bilal, Z., Hakim, F., Shabbir, M., & Iqbal, M. U. (2025). Advancing fungal phylogenetics: Integrating modern sequencing, dark taxa discovery, and machine learning. *Archives of Microbiology*, 207(9), 1–27.

Naveed, M., Memon, Z. N., Abdullah, M., Makhdoom, S. I., Azeem, A., Mehmood, S., & Majeed, M. (2024). Use cases and future aspects of intelligent techniques in microbial data analysis. *In Microbial data intelligence and computational techniques for sustainable computing* (pp. 259–280). Singapore: Springer Nature.

Pagano, T. P., Loureiro, R. B., Lisboa, F. V., Peixoto, R. M., Guimarães, G. A., Cruz, G. O., & Nascimento, E. G. (2023). Bias and unfairness in machine learning models: A systematic review on datasets, tools, fairness metrics, and identification and mitigation methods. *Big Data and Cognitive Computing*, 7(1), 15.

Palla, U., & Iwunwa, M. G. C. (2025). Leveraging predictive models to enhance infection control and reduce cross-contamination in public health settings. *International Journal of Research Publication and Reviews*, 6(1), 407.

Pennisi, F., Pinto, A., Ricciardi, G. E., Signorelli, C., & Gianfredi, V. (2025). The role of artificial intelligence and machine learning models in antimicrobial stewardship in public health: A narrative review. *Antibiotics*, 14(2), 134.

Rabbani, N., Kim, G. Y., Suarez, C. J., & Chen, J. H. (2022). Applications of machine learning in routine laboratory medicine: Current state and future directions. *Clinical Biochemistry*, 103, 1–7.

Raju, C. M., Elpa, D. P., & Urban, P. L. (2024). Automation and computerization of (bio) sensing systems. *ACS sensors*, 9(3), 1033–1048.

Ravi, K., & Singh, B. (2024). ESKAPE: Navigating the global battlefield for antimicrobial resistance and defense in hospitals. *Bacteria*, 3(2), 76–98.

Riesco, R., & Trujillo, M. E. (2024). Update on the proposed minimal standards for the use of genome data for the taxonomy of prokaryotes. *International Journal of Systematic and Evolutionary Microbiology*, 74(3), 006300.

Roy, P. (2024). Enhancing real-world robustness in AI: Challenges and solutions. *Journal of Recent Trends in Computer Science and Engineering*, 12(1), 34–49.

Salam, M. A., Al-Amin, M. Y., Pawar, J. S., Akhter, N., & Lucy, I. B. (2023). Conventional methods and future trends in antimicrobial susceptibility testing. *Saudi Journal of Biological Sciences*, 30(3), 103582.

Siddique, S., Haque, M. A., George, R., Gupta, K. D., Gupta, D., & Faruk, M. J. H. (2023). Survey on machine learning biases and mitigation techniques. *Digital*, 4(1), 1–68.

Sun, L., Lai, M., Ghouri, F., Nawaz, M. A., Ali, F., Baloch, F. S., & Shahid, M. Q. (2024). Modern plant breeding techniques in crop improvement and genetic diversity: From molecular markers and gene editing to artificial intelligence—A critical review. *Plants*, 13(19), 2676.

Taddese, A. A., Addis, A. C., & Tam, B. T. (2025). Data stewardship and curation practices in AI-based genomics and automated microscopy image analysis for high-throughput screening studies: Promoting robust and ethical AI applications. *Human Genomics*, 19(1), 16.

Terranova, N., & Venkatakrishnan, K. (2024). Machine learning in modeling disease trajectory and treatment outcomes: An emerging enabler for model-informed precision medicine. *Clinical Pharmacology & Therapeutics*, 115(4), 720–726.

To, K. L. (2024). History, Present and Prospect of Homeopathy. In History, Present and Prospect of World Traditional Medicine (pp. 301–393).

Tsitou, V. M., Rallis, D., Tsekova, M., & Yanev, N. (2024). Microbiology in the era of artificial intelligence: Transforming medical and pharmaceutical microbiology. *Biotechnology & Biotechnological Equipment*, 38(1), 2349587.

Vashishth, T. K., Sharma, V., Sharma, K. K., Vidyant, S., & Bhardwaj, A. (2025). Future trends and challenges in pharmaceutical microbiology with the integration of artificial intelligence. *In* A. Shaik (Ed.), *AI-powered advances in pharmacology* (pp. 315–344). IGI Global Scientific Publishing. https://doi.org/10.4018/979-8-3693-3212-2.ch012

Walsh, C., Stallard-Olivera, E., & Fierer, N. (2024). Nine (not so simple) steps: A practical guide to using machine learning in microbial ecology. *mBio*, 15(2), e02050-23.

Wang, X. W., Wang, T., & Liu, Y. Y. (2024). Artificial intelligence for microbiology and microbiome research. arXiv preprint arXiv:2411.01098.

Wei, Y., Liu, X., Xu, C., Zhang, G., Yuan, W., Ho, H. P., & Xu, M. (2024). Artificial intelligence enhanced digital nucleic acid amplification testing for precision medicine and molecular diagnostics. arXiv preprint arXiv:2407.21080.

Wijaya, W., Suhaimi, Z., Chua, C. X. E., Sunil, R. S., Kolundžija, S., Bin Rohaizat, A. M., & Lauro, F. M. (2023). Frequent pulse disturbances shape resistance and resilience in tropical marine microbial communities. *ISME Communications*, 3(1), 55.

Wu, Y., & Gadsden, S. A. (2023). Machine learning algorithms in microbial classification: A comparative analysis. *Frontiers in Artificial Intelligence*, 6, 1200994.

Xu, Y., Liu, X., Cao, X., Huang, C., Liu, E., Qian, S., & Zhang, J. (2021). Artificial intelligence: A powerful paradigm for scientific research. *The Innovation*, 2(4), 100179.

Zhang, J., & Zhang, Z. M. (2023). Ethics and governance of trustworthy medical artificial intelligence. *BMC Medical Informatics and Decision Making*, 23(1), 7.

Microbial AI and Sustainable Food Systems

17

Rodrigo Duarte-Casar, Marlene Rojas-Le-Fort, Fabricio Andocilla-Oleas, and Darío Mondavi-Sobby

17.1 INTRODUCTION

The merging of artificial intelligence (AI) and microbiology marks a major shift in tackling systemic challenges in the agri-food sector. This new area, which we call Microbial AI, redefines solutions from large-scale field management to the detailed understanding and control of microbial populations and processes (Visuña et al., 2025). By using computational tools to model, analyze, and engineer the microbial communities that support our food system, we can redesign vital agricultural production and waste-recovery methods (Rugji et al., 2025). This chapter explores the potential of Microbial AI to build a more resilient, efficient, and sustainable global food system.

17.1.1 The Global Food System Challenge

The global food system faces a tripartite challenge: it must secure adequate nutrition for a projected population approaching 10 billion by mid-century, while simultaneously adapting to the profound disruptions of climate change and mitigating its own significant contributions to environmental degradation (Rosegrant et al., 2024). This creates a node of competing demands. On one hand, increasing agricultural output is a demographic imperative. On the other hand, this objective is fundamentally complicated by the increasing volatility of the global climate, which threatens production models through drought, flooding, and shifting phenological patterns.

Compounding this challenge, conventional agricultural practices are a principal driver of the very environmental constraints that threaten their viability. Agrarian land-use changes, primarily through deforestation, coupled with the intensive application of synthetic fertilizers and the extraction of freshwater resources, constitute a major source of greenhouse gas emissions and broader ecological degradation (Li et al., 2024). Pesticides are another source of pollution that threatens the integrity of the agrifood system (Duarte-Casar et al., 2025). Consequently, these practices can lead to soil depletion, water scarcity, and biodiversity loss, creating a paradoxical feedback loop:

DOI: 10.1201/9781003727514-19

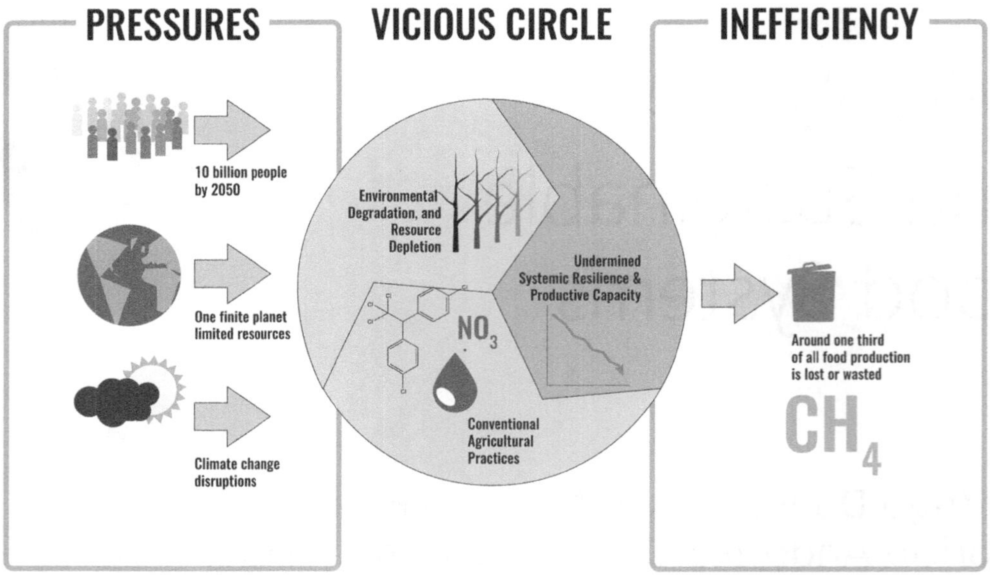

FIGURE 17.1 Pressures facing the global food system in 2050. Climate change, intensive cultivation, soil degradation, plus food loss and waste.

the methods deployed to expand the near-term food supply actively undermine the long-term resilience and productive capacity of the agroecosystems upon which they depend. Figure 17.1 summarizes the food crisis we are facing.

This systemic tension is further exacerbated by inefficiencies in the food chain, most notably the scale of post-harvest loss and consumer waste. Current estimates indicate that approximately one-third of all food produced for human consumption is lost or wasted (Casonato et al., 2023). This represents a sizable misallocation of finite inputs—land, water, energy, and capital—and their associated environmental costs (Rojas-Le-Fort et al., 2025). Therefore, the central challenge extends beyond increasing gross production. It necessitates a re-engineering of the food system toward greater resource efficiency, circularity, and resilience (Kumareswaran & Jayasinghe, 2022). The ultimate objective is to create a system capable of delivering nutrition security within planetary boundaries, optimizing resource use, and drastically reducing its aggregate environmental footprint.

17.1.2 The Unseen Power of Microbes

Microorganisms form the unseen, often-ignored foundation of the global food system, primarily functioning in soil. The soil microbiome, a complex community of bacteria, fungi, and archaea, is critical to plant health and agricultural success. These microbial communities facilitate nutrient cycling, fix atmospheric nitrogen, solubilize phosphorus, and break down organic matter, making vital elements accessible to crops. Additionally, they help build soil structure and improve water retention, while some species create symbiotic relationships with plant roots, expanding their ability to absorb nutrients (Bashir et al., 2025; Sharma & Salwan, 2023). The future of terrestrial farming is therefore closely connected to the health and activity of these microscopic organisms' ecosystems.

Beyond the soil, microbial activity drives food production and processing. For thousands of years, humans have relied on controlled fermentation, a microbial-driven process, to preserve and transform raw agricultural materials. This method is applied to a wide variety of staples, including bread, cheese, yogurt, beer, and wine, improving their shelf life, nutritional content, and sensory qualities. Recently, advanced fermentation techniques using precisely cultivated microorganisms and conditions are being developed to create alternative proteins, fats, and other food ingredients, marking a significant innovation in food technology with potential sustainability advantages (Hilgendorf et al., 2024; Vinestock et al., 2024).

The influence of microbes reaches the final stages of the food chain and affects both human health and the environment. Before consumption, the goal is to reduce microbial activity in foods to maximize shelf life and preserve food safety. After this stage, microbes are desirable once again. The human gut microbiota is essential for digesting complex carbohydrates, producing vitamins and other beneficial compounds such as short-chain fatty acids, and training the immune system (Bagheri et al., 2025). On the other hand, at the end of the food lifecycle, microbial communities are the main agents of decomposition, breaking down organic waste and recycling nutrients back into the environment. This process is vital for reducing food waste, closing nutrient cycles, and lowering the environmental impact of landfills (Hilger & Humer, 2003; Sagar et al., 2022). Thus, from soil to gut to decomposition, microbial processes are key to the entire food system.

17.1.3 Microbial AI and Chapter Roadmap

Examples of Microbial AI are: A field proactively healed by a bespoke probiotic blend (Xi et al., 2025), a shipping container that dynamically reroutes its cargo based on real-time microbiome data (Shunmugathammal et al., 2024), or a biorefinery that transforms food waste into bioplastics or biofuel using a designed microbial consortium (Daimary et al., 2025). By using computational tools to model, analyze, and engineer the microbial communities that support our food system, we can redesign vital agricultural production and waste-recovery methods (Rugji et al., 2025).

This chapter critically explores the potential of Microbial AI to build a more resilient, efficient, and sustainable global food system. This chapter will first establish the fundamentals of this interdisciplinary approach (Section 17.2). It will then explore its applications in sustainable food production (Section 17.3) and in reducing loss and waste (Section 17.4). A critical discussion of the ethical and societal implications follows (Section 17.5) before concluding with future directions and a summary of the field's transformative potential (Section 17.6).

17.2 THE MICROBIAL AI TOOLKIT: DATA, MODELS, AND WORKFLOWS

The power of Microbial AI emerges from the synergistic integration of three elements: comprehensive data, powerful analytical models, and a closed-loop workflow. Before exploring its applications, it is crucial to understand this toolkit. This section introduces the key data types derived from microbial communities, explains the machine learning (ML) techniques adept at deciphering their complex patterns, and outlines the fundamental feedback loop that transforms static data into dynamic, intelligent intervention.

17.2.1 The Multi-Omics Landscape in Food Environments

The efficacy of any Microbial AI system is contingent on the quality, resolution, and breadth of the data it ingests. In this context, data is derived from a suite of advanced molecular techniques collectively known as "multi-omics," which provide a layered, systems-level view of microbial communities. These technologies move beyond culturing-dependent methods to capture the genetic potential, functional activity, and metabolic output of entire microbiomes directly from their environmental context—be it soil, a fermentation tank, a food surface, or the gastrointestinal tract (Su et al., 2024). This multi-omics foundation enables AI models to move from mere correlation to mechanistic understanding and predictive power. An overview of the omics layers considered in this chapter is presented in Table 17.1.

Microbial ecology often begins by profiling community composition, asking the fundamental question: "Who is there?" This is typically achieved through marker gene sequencing, which targets conserved genomic regions that act as unique taxonomic barcodes—such as 16S rRNA for bacteria and archaea, 18S for eukaryotes, or internal transcribed spacer (ITS) for fungi. By analyzing these barcodes, researchers can assess the diversity of a microbiome. For instance, in soil health, a shift in the fungal-to-bacterial biomarker ratio can indicate the potential for carbon sequestration. While powerful, this method reveals identity but offers only inferred, rather than direct, knowledge of functional capabilities.

To directly access the functional genetic potential of a community, metagenomics is employed. This technique sequences all the DNA from an environmental sample, providing a glimpse into the community's collective genome. Unlike marker gene sequencing, metagenomics can pinpoint genes for processes like nitrogen fixation, antibiotic resistance, or pathogen virulence. In food production, for example, a metagenomic analysis of a cheese rind can reveal the genetic machinery for flavor

TABLE 17.1 The multi-omics data landscape in microbial AI

OMICS LAYER	*CORE QUESTION*	*TECHNOLOGY EXAMPLE*	*DATA OUTPUT*	*APPLICATION EXAMPLE IN FOOD SYSTEMS*
Metagenomics	Who is there and what can they do?	Shotgun Sequencing	Catalog of genes & taxonomic profiles	Identifying pathogens in a fermentation batch or nitrogen-fixing genes in soil.
Metatranscriptomics	What are they actively doing?	RNA-Seq	Profile of expressed genes	Monitoring a yeast consortium's stress response during beer fermentation.
Metabolomics	What are they producing?	Mass Spectrometry (MS)	Inventory of small molecules	Detecting spoilage volatiles in packaged meat or flavor compounds in cheese.
Marker Gene (16S/ ITS)	Who is there? (Census)	16S rRNA Amplicon Seq.	Relative abundance of taxa	Tracking soil health through shifts in fungal-to-bacterial ratio.

synthesis or spoilage prevention, mapping what the community could potentially do. AI is increasingly used for metagenomics analysis, contributing to the emerging field of genomic intelligence (Gupta et al., 2024).

However, the presence of a gene does not guarantee its expression. This gap is bridged by metatranscriptomics, which involves sequencing the total RNA (specifically messenger RNA) from a sample. This provides a snapshot of the genes that are actively being transcribed at the moment of sampling, answering the question, "what are they doing?" By analyzing metatranscriptomic data, researchers can discern the real-time metabolic activity of a microbiome through gene expression patterns (Hu et al., 2019). In a precision fermentation vat, for example, this could reveal how a microbial consortium responds to a change in temperature or nutrient feed, showing which pathways are upregulated or downregulated to produce the desired metabolite, thereby enabling dynamic process control.

The final and most direct measure of microbial activity comes from metabolomics, the large-scale study of small molecule metabolites. These metabolites—including organic acids, vitamins, antibiotics, and flavor compounds—represent the ultimate functional output of the microbiome. Metabolomic profiling, often via mass spectrometry or NMR, quantifies these biochemical footprints, effectively capturing the chemical conversation between microbes and their environment. In food safety, the detection of specific microbial toxins (e.g., mycotoxins) is a direct metabolomic readout of spoilage. When integrated with genomic and transcriptomic data, metabolomics closes the loop, connecting genetic potential and transcriptional activity to the tangible biochemical products that determine food quality, safety, and nutritional value.

The true power for Microbial AI lies not in these omics layers individually but in their integration. An AI model trained on correlated metagenomic, metatranscriptomic, and metabolomic data can learn the complex, nonlinear relationships between a community's genetic makeup, its environmental stimuli, and its functional output, as already used in plant science (Cembrowska-Lech et al., 2023). This integrated multi-omics approach enables accurate prediction of crop yield from soil microbiome data, the real-time optimization of a fermentation process, and the dynamic forecasting of a food product's shelf life, forming the essential data bedrock upon which intelligent food systems are built.

17.2.2 Essential AI/ML Techniques

The vast, high-dimensional data generated by multi-omics technologies presents a fundamental analytical challenge that traditional statistical methods are poorly equipped to handle. This is where AI, and specifically ML, becomes indispensable. ML algorithms are computational tools designed to identify complex, nonlinear patterns and relationships within large datasets, transforming raw omics data into actionable insights, predictions, and decisions (Chaturvedi et al., 2024). The selection of an appropriate ML technique depends on the nature of the available data and the specific biological question at hand, thereby forming a diverse toolkit for deciphering microbial complexity. An overview of the techniques is presented in Table 17.2.

A primary category of ML techniques is supervised learning, which is used when the goal is to predict a known, predefined outcome. In these models, the algorithm is trained on a labeled dataset where the input data (e.g., a soil microbiome profile) is paired with the corresponding output or "label" (e.g., crop yield, disease presence, or soil nitrogen level). Through this training process, the model learns the underlying function that maps the inputs to the outputs. Once trained, it can predict the label for new, unseen data. In Microbial AI, supervised learning is widely used for tasks such as diagnosing plant diseases from leaf microbiome data, predicting the shelf life of a food product based on its initial microbial load, or classifying the health status of an animal from its gut microbiome composition (Zhang et al., 2025).

TABLE 17.2 A toolkit of machine learning techniques in microbial AI

ML PARADIGM	*PRIMARY GOAL*	*WHEN TO USE*	*KEY STRENGTHS*	*EXAMPLE APPLICATIONS IN FOOD & AGRICULTURE*
Supervised Learning	Predict a known, pre-defined outcome or label.	You have a labeled dataset (inputs & known outputs).	High predictive accuracy for well-defined tasks.	Disease diagnosis from microbiome, food shelf-life prediction, and health status classification.
Unsupervised Learning	Discover hidden patterns, groupings, or structures in data.	No labels exist; exploratory analysis is needed.	Hypothesis generation, biomarker discovery, and data stratification.	Identifying soil subtypes, defining fermentation stages, and patient enterotyping.
Deep Learning	Model complex, hierarchical patterns in high-dimensional data.	Dealing with very complex data (e.g., images, sequences, multi-omics).	Superior pattern recognition, ability to integrate diverse data types.	Gene function prediction, metabolomic spectral analysis, and multi-sensor fusion for spoilage detection.
Temporal Modeling	Forecast future states in time-series data.	Data is collected over time, and temporal dynamics are critical.	Captures sequential dependencies and trends.	Forecasting fermentation endpoints, pathogen outbreak prediction, and early warning of system dysbiosis.

In many exploratory analysis scenarios, however, such pre-defined labels do not exist. Here, unsupervised learning techniques are employed to uncover the inherent structure within the data itself. These algorithms analyze input data without any reference to known outcomes, seeking to identify natural groupings or clusters. When applied to microbial community data, such as 16S rRNA sequencing results, unsupervised learning can reveal previously unknown subtypes of soil, distinct stages of fermentation, or patient enterotypes based solely on the compositional similarities of their microbiomes. This ability to discover hidden patterns is crucial for hypothesis generation, biomarker discovery, and stratifying complex samples into meaningful categories that can later be studied.

For the most intricate and high-dimensional datasets, such as those integrating multiple omics layers, neural networks, and particularly deep learning architectures, offer superior pattern recognition capabilities. Inspired by the structure of the human brain, these models consist of interconnected layers of nodes that can learn hierarchical representations of data. A simple model might learn to identify individual microbial taxa. In contrast, a deeper model could discern interactions between taxa, their combined metabolic pathways, and finally, the emergent property that influences a crop trait. Deep learning excels at tasks such as predicting gene function directly from DNA sequence data, interpreting complex spectral data from metabolomics, and integrating image data from sensors with genomic information to enable robust spoilage detection.

Finally, many microbial processes in food systems are inherently dynamic, unfolding over time. Monitoring fermentation or tracking soil health across a growing season requires time-series analysis. Specialized ML models, such as recurrent neural networks (RNNs) or long short-term memory (LSTM) networks, are designed to handle this temporal dimension. They can retain memories of previous states, allowing them to model sequences and forecast future events. This makes them ideal for predicting the optimal endpoint of a beer fermentation based on real-time metabolite and microbiome data, forecasting the outbreak of a plant pathogen by modeling its population dynamics, or providing early warnings of dysbiosis in aquaculture systems before clinical signs appear. By capturing these temporal patterns, ML enables not just retrospective analysis but truly predictive and proactive management of food systems.

The goal in applied Microbial AI is often to move beyond passive analysis and create a closed-loop intelligent system. This involves integrating these ML techniques into a cohesive framework where a model's prediction directly informs an intervention—for instance, an unsupervised model identifies an anomalous microbial state, a supervised model predicts the optimal probiotic amendment, and a time-series model then monitors the system's response, generating new data to refine the entire loop. It is this synergistic application of diverse ML paradigms that unlocks the potential to steer microbial communities toward desired, sustainable outcomes intelligently.

A paramount challenge in Microbial AI is moving from correlative predictions to causal understanding. While an ML model can reliably predict crop yield based on a specific microbiome signature, it does not necessarily reveal the mechanistic, causal interactions between the microbes and the plant. The "black box" nature of some complex models exacerbates this. Future research is focused on developing interpretable AI and employing causal inference techniques to move from knowing that a microbiome is beneficial to understanding why and how it functions, which is essential for reliable engineering.

17.2.3 The Microbial AI Feedback Loop: From Insight to Intervention

The individual components of multi-omics data and ML models achieve their full potential only when integrated into a dynamic, self-improving process known as the Microbial AI feedback loop. This concept moves beyond one-off analysis to establish a continuous cycle of measurement, learning, and intervention, transforming static observations into a responsive system capable of managing microbial communities toward predefined goals. It is this iterative, closed-loop architecture that distinguishes Microbial AI as a paradigm for active engineering rather than passive observation, enabling a level of precision and adaptability previously unattainable in food system management.

The loop begins with the data acquisition and integration phase. Here, heterogeneous data is continuously gathered from the target environment. This includes multi-omics data detailed in Section 17.2.1: genomic snapshots of community structure, transcriptomic profiles of active genes, and metabolomic readings of biochemical outputs. This biological data is fused with contextual metadata from other sensors, such as pH, temperature, humidity, and nutrient levels. This rich, multimodal dataset provides a comprehensive digital representation of the microbial ecosystem at a specific point in time, serving as the foundational input for all subsequent analyses.

This aggregated data stream then feeds into the AI analysis and predictive modeling phase. Here, the ML techniques from Section 17.2.2 are deployed to process complex data and generate actionable insights. A supervised learning model might predict an imminent spoilage event based on a specific metabolite signature. In contrast, an unsupervised model could detect an anomalous microbial shift, indicating the early stages of a pathogen outbreak. The output of this phase is not a description of the current state but a forecast of a future state or a specific recommendation for action. This transforms raw data into a decisive, data-driven insight.

The third phase, Intervention and Implementation, is where the system acts upon the AI's insight. The predictive output is translated into a precise, targeted intervention in the physical world. This could be the automated adjustment of a temperature setting in a fermentation bioreactor, the targeted release of a custom probiotic blend into a soil irrigation system, or an alert to a supply chain manager to reroute a specific batch of produce with a high predicted spoilage risk. The key characteristic of this stage is that the intervention is no longer based on a fixed schedule or generic best practices but is dynamically calibrated by the AI's real-time analysis of the microbial state.

Following the intervention, the loop is closed through monitoring and validation. The system returns to the data acquisition phase, meticulously monitoring the environment to capture the effects of the intervention. This generates a new, post-intervention dataset that answers the critical question: "Did the action have the intended effect?" The resulting outcome—whether the fermentation yield improved, the pathogen was suppressed, or the shelf life was extended—is then fed back into the AI models.

This final step is what creates the closed-loop system for continuous improvement. The new data, which now includes the results of the previous intervention, is used to retrain and refine the ML models. Over successive cycles, the system learns from its successes and failures, progressively improving its predictive accuracy and the efficacy of its interventions. This creates a cybernetic system that adapts to changing conditions, compensates for unforeseen disturbances, and relentlessly optimizes the process toward the ultimate goal of a more productive, efficient, and sustainable food system. The process is schematized in Figure 17.2.

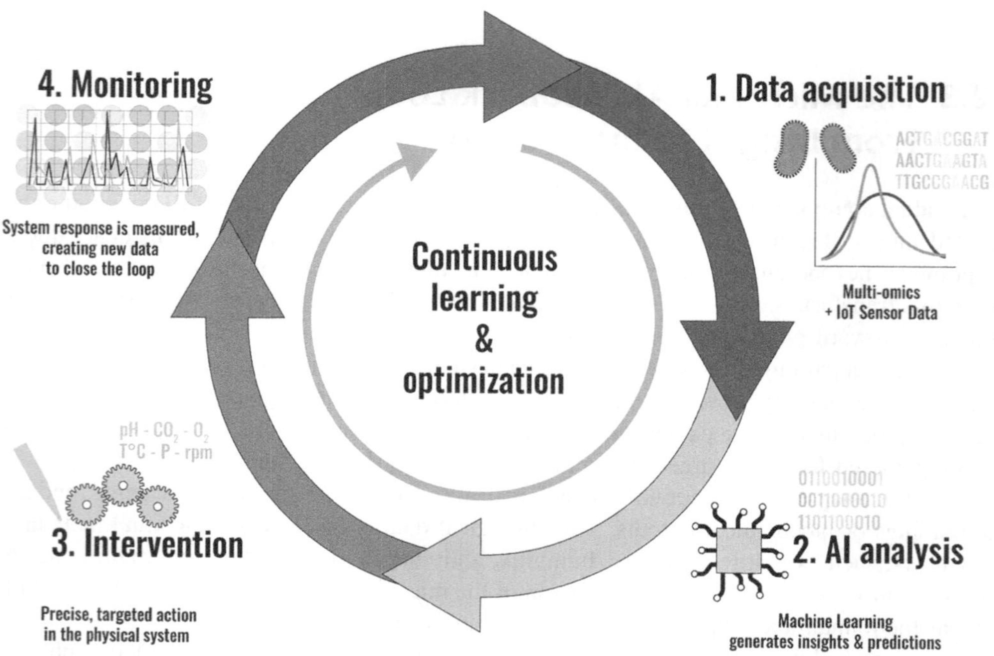

FIGURE 17.2 The microbial AI feedback loop. A continuous cycle illustrating how (1) multi-omics and sensor data are acquired, (2) analyzed by AI models to generate insights, (3) used to execute a precise intervention, and (4) monitored to validate outcomes and generate new data, which is fed back into the system to refine the AI models.

17.3 APPLICATIONS IN SUSTAINABLE FOOD PRODUCTION

The theoretical framework of Microbial AI finds its most immediate and impactful application in revolutionizing food production. By moving from broad-stroke agricultural practices to precise, microbiome-level management, this approach offers a pathway to intensify production sustainably, reduce environmental footprints, and enhance resilience across farming, processing, and husbandry. Table 17.3 summarizes notable applications of Microbial AI in sustainable food production.

TABLE 17.3 Applications of microbial AI in sustainable food production

APPLICATION DOMAIN	*PRIMARY OBJECTIVE*	*MICROBIAL AI APPROACH & ENABLING DATA*	*SPECIFIC USE-CASE EXAMPLE*
Smart Agriculture & Soil Health	Enhance crop resilience and yield while reducing synthetic inputs.	Diagnosis: ML models analyze soil metagenomic/metabolomic data to create a "soil health fingerprint." Prediction: AI detects early microbial signatures of pathogen outbreaks from phyllosphere/soil microbiome time-series data. Prescription: AI-driven systems recommend spatially variable application of biofertilizers/biopesticides.	Predicting a fungal disease outbreak in a wheat field weeks before symptoms appear and prescribing a targeted probiotic spray to suppress the pathogen.
Precision Fermentation & Food Design	Optimize traditional fermentations and engineer novel microbial foods.	Process Control: Real-time integration of metabolomic and sensor data to dynamically guide fermentation parameters. Strain Design: Generative AI models design novel metabolic pathways; supervised learning predicts high-yield microbial strains from genotype-phenotype data.	Dynamically adjusting temperature and nutrient feed in a beer fermentation based on real-time yeast activity to ensure perfect flavor consistency. Designing a yeast strain optimized to produce a dairy-identical protein in a bioreactor.
Animal Health & Aquaculture	Improve animal welfare, productivity, and reduce antibiotic use.	Health Monitoring: ML identifies gut microbiome "enterotypes" linked to optimal feed conversion and animal health. Disease Prevention: AI models recognize microbial dysbiosis preceding clinical disease, enabling early, targeted intervention.	Detecting the pre-clinical microbial shift associated with shrimp early mortality syndrome and administering a specific prebiotic to restore a healthy gut balance.

17.3.1 Smart Agriculture

The health of the soil microbiome, particularly the rhizosphere—the narrow zone of soil directly influenced by plant roots—is the bedrock of sustainable agriculture. Microbial AI enables a transition from treating soil as a passive substrate to managing it as a complex, living ecosystem. The first step involves high-resolution diagnosis. By applying ML models to metagenomic and metabolomic data from soil cores, it is now possible to move beyond basic chemical assays to generate a holistic "soil health fingerprint." These models can integrate microbial diversity, the abundance of key functional genes (e.g., for nitrogen fixation), and metabolic activity to accurately diagnose nutrient deficiencies, predict crop yield potential, and even assess a field's resilience to drought or disease, providing farmers with a deeply insightful starting point.

This diagnostic power is powerfully coupled with predictive defense. A key application lies in forecasting plant-pathogen outbreaks. By continuously monitoring shifts in the soil and phyllosphere (leaf surface) microbiome, AI models can identify the early-warning signatures of a burgeoning pathogen population long before visual symptoms manifest. This is not merely a detection system but a prescriptive one. The model can analyze which beneficial microbes are suppressed during the pathogen's rise and recommend a tailored, microbiome-based intervention. This could involve applying a specific consortium of probiotic bacteria or fungi that are known to antagonize the predicted pathogen, effectively pre-empting disease through ecological competition and stimulation of the plant's own immune responses, thereby drastically reducing the need for broad-spectrum chemical pesticides.

Finally, Microbial AI brings precision to organic inputs, optimizing the use of biofertilizers and biopesticides. Instead of uniform application, AI-driven decision support systems can analyze spatial soil microbiome variability across a field and prescribe variable-rate applications of microbial inoculants. For instance, a zone low in phosphorus-solubilizing bacteria would receive a targeted application of a corresponding biofertilizer, while an area with a history of fungal disease would receive a higher dose of biocontrol fungi. This ensures that these often-costly biological inputs are deployed only where and when they are needed, maximizing their efficacy, improving economic returns, and enhancing the overall stability and productivity of the agricultural ecosystem.

17.3.2 Precision Fermentation and Food Design

Fermentation, one of humanity's oldest food processing techniques, is being reborn through Microbial AI. In traditional food and beverage production, such as for yogurt, cheese, and beer, slight variations in the starting microbial culture or environmental conditions can lead to significant differences in the final product's yield, flavor, and texture. Precision fermentation uses AI to master this complexity. By integrating real-time metabolomic and metatranscriptomic data from fermentation tanks with sensor data on temperature and pH, ML models can dynamically predict the optimal pathway to a desired outcome. They can guide adjustments to the process in real-time, ensuring consistent, high-quality production, minimizing failed batches, and reducing resource waste.

Beyond optimizing existing processes, Microbial AI is a powerful engine for food innovation, particularly in the realm of alternative proteins. The discovery and engineering of novel microbial strains for producing proteins, fats, and nutritional compounds can be dramatically accelerated. Generative AI models can be trained on vast databases of microbial genomes and metabolomes to in silico design novel enzymes or metabolic pathways for a specific function, such as producing a milk protein or a health-promoting fatty acid. Subsequently, supervised learning models can analyze the genotype and phenotype of thousands of engineered microbial strains to predict which will have the highest yield and stability in a bioreactor. This AI-driven pipeline shortens the development cycle from years to months, paving the way for a new generation of sustainable, biomanufactured food ingredients.

17.3.3 Enhancing Animal Health and Aquaculture

The pressure to reduce antibiotic use in livestock and aquaculture has made gut health a paramount concern. Microbial AI offers a sophisticated approach to managing the gut microbiome for improved animal welfare and productivity. By regularly sequencing the gut microbiota of herds or schools of fish, farmers can move from reactive treatment to proactive health management. ML models can identify microbial signatures, or "enterotypes," that are correlated with optimal feed conversion efficiency, allowing for the selection of animals with naturally superior gut health or the formulation of prebiotic feeds that encourage a beneficial microbiome, thus reducing feed costs and environmental impact.

Furthermore, this technology is pivotal for disease prevention. AI models can be trained to recognize the subtle microbial dysbiosis that precedes clinical signs of common illnesses like poultry necrotic enteritis or shrimp early mortality syndrome. An early alert from such a system allows for swift, targeted intervention, such as the administration of a specific probiotic regimen designed to restore a healthy microbial balance. This shift from blanket antibiotic prophylaxis to microbiome-based, precision health management not only addresses the global crisis of antimicrobial resistance but also promotes stronger, more resilient animals, creating a more sustainable and ethical foundation for animal production.

17.4 APPLICATIONS IN REDUCING FOOD LOSS AND WASTE

A sustainable food system must not only produce efficiently but also conserve what is produced. An estimated third of all food is lost or wasted, representing a catastrophic misallocation of resources. Microbial AI is poised to drastically reduce this figure by introducing predictive intelligence into the post-harvest supply chain and transforming waste from a disposal problem into a valuable resource, thereby advancing a circular bio-economy. Applications of Microbial AI in reducing food loss and waste are summarized in Table 17.4.

17.4.1 Predictive Shelf Life and Quality Monitoring

The static "best before" date is a crude and often inaccurate tool, leading to both premature discarding of safe food and unexpected spoilage. Microbial AI enables a shift to dynamic, condition-aware shelf-life prediction. This approach involves monitoring the product's specific microbiome and its metabolic activity throughout the supply chain. By combining metagenomic data to identify specific spoilage organisms with metabolomic sensors that detect volatile organic compounds (VOCs) associated with decay, ML models can learn the unique spoilage trajectory of a product (Palma et al., 2018). For a packaged salad, for instance, the AI can integrate initial microbial load, temperature history, and real-time VOC emissions to predict its remaining fresh life with high accuracy, empowering retailers and consumers with precise, data-driven freshness indicators.

TABLE 17.4 Applications of Microbial AI in reducing food loss and waste

STRATEGIC FOCUS	*CORE CHALLENGE*	*MICROBIAL AI SOLUTION & KEY DATA INPUTS*	*PRACTICAL APPLICATION EXAMPLE*
Predictive Shelf Life & Quality Monitoring	Static expiration dates are inaccurate, leading to premature waste or unexpected spoilage.	Dynamic Prediction: ML models integrate real-time microbiome data (metagenomics) and volatile organic compound (VOC) profiles (metabolomics) with storage condition history to predict remaining freshness. Automated Inspection: Computer vision and hyperspectral imaging, trained on microbial contamination datasets, enable real-time, high-accuracy defect detection.	A smart sensor package in a shipping container predicts the exact remaining shelf life of strawberries based on their specific microbial load and current temperature, enabling dynamic pricing and prioritized sale.
Optimizing Supply Chains	Logistics are "blind" to the biological state of perishable goods, leading to inefficient routing and handling.	Intelligent Logistics: A digital twin of the supply chain integrates predicted shelf life, real-time IoT sensor data, and external factors to dynamically reroute shipments. Condition-Aware Stewardship: AI triggers automated, granular adjustments to the storage environment (e.g., temperature) based on a product's specific microbial risk profile.	A logistics AI reroutes a pallet of ripe avocados to a nearer distribution center and simultaneously lowers its storage temperature to extend its salable life by 48 hours.
Valorization of Waste Streams	Organic waste is a disposal cost rather than a resource, failing to close nutrient loops.	Microbial Discovery: ML models screen genomic databases to identify or design optimal microbial consortia for converting specific waste feedstocks into target products (e.g., biofuels and bioplastics). Process Optimization: Time-series AI models monitor and control bioreactors in real-time, adjusting parameters to maximize conversion efficiency and product yield.	An AI system selects a consortium of bacteria to convert brewery spent grains into bio-based plastic (PHA) and then optimizes the reactor's aeration and pH to maximize output.

This predictive capability is complemented by AI-powered inspection at critical control points. Early detection of microbial contamination, often invisible to the human eye, is crucial for preventing localized spoilage from escalating into a full-blown recall. Advanced sorting systems now leverage hyperspectral imaging and machine vision trained on vast datasets of contaminated and clean food products. These systems can identify subtle spectral signatures of microbial colonies, such as early-stage mold on a nut or a bacterial biofilm on a meat surface, with a speed and accuracy far surpassing human capabilities (Medina–García et al., 2025). By automatically ejecting contaminated items at processing facilities, this technology acts as a powerful filter, ensuring higher-quality produce enters the supply chain and reducing downstream loss.

17.4.2 Optimizing Supply Chains

The modern food supply chain is a complex logistical network, but it has largely been blind to the biological state of the products it transports. Microbial AI introduces a new layer of "microbial intelligence" into logistics planning. By creating a digital twin of the supply chain that integrates real-time data—including the predicted shelf life of individual batches, real-time temperature and humidity readings from IoT sensors in shipping containers, and external traffic and weather data—logistics AI can make supremely informed decisions (Sanfiya & Sabitha Banu, 2025). It can dynamically reroute a shipment of ripe fruit to a closer distribution center to prevent over-ripening or prioritize a batch of fish with a shorter predicted shelf life for immediate delivery.

This goes beyond simple prioritization to active condition management. The system can proactively adjust the storage environment of a product based on its microbial risk profile. If the AI predicts a spike in spoilage organisms for a particular pallet of cheese based on its current temperature and microbiome data, it can automatically instruct the climate control system in the warehouse to lower the temperature for that specific unit, effectively putting it into a state of suspended animation to extend its life. This granular, product-aware logistics system minimizes loss not through faster movement alone, but through intelligent, condition-specific stewardship of food as a living, perishable commodity.

17.4.3 Valorization of Waste Streams

For the organic waste that is inevitably generated, Microbial AI offers a pathway to valorization, closing the loop in the food system. The first challenge is identifying the most efficient microbial agents for breaking down complex waste streams into valuable products (Singh & Shyu, 2024). AI dramatically accelerates this discovery process (Jiang et al., 2024). ML models can be trained on genomic databases to predict which microbes, or synthetic consortia of microbes, possess the enzymatic machinery to efficiently convert specific food waste substrates (e.g., lignocellulosic matter from crop residues or lipids from slaughterhouse waste) into target products like biofuels (e.g., ethanol and butanol), biodegradable plastics (e.g., PHAs), or industrial enzymes (Xu et al., 2025).

Once promising microbial candidates are identified, AI further optimizes the bioconversion process itself. Fermentation or anaerobic digestion for waste valorization is a complex biochemical process influenced by numerous variables. ML models, particularly those designed for time-series analysis, can continuously monitor and control bioreactors. They analyze inputs such as feedstock composition, microbial density, pH, and gas output to dynamically adjust parameters like aeration, mixing, and nutrient feed. This ensures the microbial community operates at peak efficiency, maximizing the yield and quality of the end product while minimizing energy input and processing time. By transforming food waste from a cost-intensive disposal problem into a reliable feedstock for a circular economy, Microbial AI unlocks a future where the food system's waste streams become a wellspring of sustainable value.

17.5 ETHICAL CONSIDERATIONS AND SOCIETAL IMPLICATIONS

The transformative potential of Microbial AI is matched in magnitude by the profound ethical and societal questions it raises. The power to deliberately engineer the microbial world demands a parallel commitment to responsible innovation. Navigating this complex landscape is not a secondary concern but a prerequisite for ensuring that this technology benefits all of humanity without exacerbating existing inequalities or creating new, unforeseen risks. This section critically examines the paramount challenges that must be addressed through proactive governance and inclusive dialogue.

17.5.1 Biosecurity and Dual Use

The same tools used to design beneficial microbes for agriculture or waste processing could, in principle, be co-opted to create biological agents for harmful purposes. This "dual-use" dilemma is a primary biosecurity concern. A microbial strain engineered to enhance crop resilience could be maliciously redesigned into a hyper-virulent pathogen capable of triggering widespread crop failure and threatening food security through agroterrorism. The relative accessibility of gene-editing technologies and AI design tools lowers the barrier for such activities, necessitating robust oversight. Furthermore, the risk is not solely one of malicious intent (Gupta et al., 2024). There is a persistent danger of accidental release, where a microbe engineered for a confined industrial fermentation process escapes into the environment, potentially outcompeting native species or transferring its engineered genes to wild populations, with unpredictable ecological consequences.

17.5.2 Environmental Impact and Unintended Consequences

The deliberate release of AI-designed microbes into open environments, such as agricultural fields, represents a significant ecological experiment. The precautionary principle—which advocates for caution in the face of uncertain risks—must be central to deployment strategies. A microbe engineered to fix nitrogen in a cereal crop efficiently might spread to adjacent ecosystems, disrupting natural nutrient cycles and altering plant community dynamics. Another, designed to break down a specific waste polymer, might inadvertently degrade similar, essential natural polymers in the soil, compromising its structure. The complex, interconnected nature of microbial ecosystems means that even a well-intentioned intervention could have cascading, unintended effects. Comprehensive, long-term environmental risk assessments, informed by ecological modeling, are essential to anticipate and mitigate potential disruptions before open-air applications are permitted.

17.5.3 Data Ownership, Privacy, and Governance

Microbiome data is a valuable digital resource, but its ownership is a legal and ethical gray area. Does the microbiome of a farmer's field belong to the farmer, the company that sequenced it, or is it part of a shared genetic commons? This question of "bio-sovereignty" is critical. A large agribusiness could potentially patent a beneficial microbial consortium discovered through the analysis of thousands of farm microbiomes, effectively restricting farmers' access to the very biological resources that originated from their land (Ribeiro et al., 2018). Similarly, the unique microbial signature of a geographic region known for a specific terroir (e.g., for wine or cheese) could be digitally appropriated, leading to biopiracy and economic exploitation (Bokulich et al., 2016). Clear international frameworks are needed to define data rights, ensure equitable benefit-sharing, and protect the digital sovereignty of communities and nations over their biological data.

17.5.4 Equity and Access

There is a significant risk that Microbial AI could exacerbate the "digital divide" in agriculture, widening the chasm between high-tech and traditional farming (Javaid et al., 2023). The high costs of omics sequencing, AI infrastructure, and specialized consultants are likely to be initially accessible only to large-scale agribusinesses and wealthy nations. This could grant them an unprecedented advantage in productivity and cost-efficiency, potentially pushing smallholder farmers who cannot compete out of the market and consolidating corporate control over the global food supply. To avoid this, proactive policies are required. This includes public investment in open-source AI tools and low-cost diagnostic platforms, the development of farmer cooperatives to share data and resources, and international aid programs aimed at ensuring these transformative technologies contribute to, rather than undermine, global food sovereignty and the livelihoods of the world's most vulnerable food producers.

17.5.5 Regulatory Frameworks and Public Acceptance

The pace of innovation in Microbial AI far outstrips the development of corresponding regulatory frameworks. Current regulations for genetically modified organisms (GMOs) are often ill-suited to assess the nuanced risks of microbes designed by AI, which may involve complex community interactions rather than single, modified traits. Regulatory agencies worldwide face the challenge of creating agile, science-based pathways that can ensure safety without stifling innovation. Parallel to this is the critical challenge of public acceptance. Historical controversies surrounding "GMOs" have demonstrated that technological superiority alone is insufficient for adoption (Kaluđerović, 2019). Building public trust in "AI-designed" food and microbes requires radical transparency, inclusive public engagement, and clear communication about the benefits, risks, and ethical boundaries of the technology. Without a social license to operate, even the most promising Microbial AI applications may fail to reach their potential.

The rise of Microbial AI will inevitably shake up traditional agribusiness and food production economics. Value will increasingly come from data, algorithms, and specialized microbial communities, potentially shifting power away from traditional input suppliers. This could lead to new business models, such as "Microbiome-as-a-Service," where companies continuously monitor and manage a farm's soil biology, or the licensing of AI-designed starter cultures for precision fermentation. Navigating this economic shift will be as important as managing the technological one.

17.6 FUTURE DIRECTIONS

The field of Microbial AI stands at a pivotal juncture, having demonstrated significant potential while simultaneously revealing the scale of the challenges ahead. Its path depends on both technological breakthroughs and our collective ability to steer these advancements toward equitable and sustainable outcomes. Here, we outline the emerging trends poised to redefine the field and the critical research hurdles that must be overcome to realize the full promise of this powerful technological convergence.

17.6.1 Emerging Trends

The next wave of innovation in Microbial AI will be driven by more sophisticated AI architectures and deeper integration with other digital technologies. Generative AI represents a paradigm shift, moving from analytical models to creative ones. Instead of merely predicting the behavior of existing microbial communities, generative models can be used to design novel, synthetic microbial consortia from first principles. By learning the "rules" of microbial metabolism, symbiosis, and competition from vast omics datasets, these systems could propose optimal combinations of strains engineered to perform specific, complex functions, such as a consortium that sequentially breaks down agricultural waste into a valuable chemical, all within a single bioreactor.

Furthermore, the context for Microbial AI will be enriched by its integration with other Industry 4.0 technologies. The proliferation of low-cost Internet of Things (IoT) sensors will create a continuous, real-time data stream on environmental conditions (e.g., soil moisture, bioreactor temperature, and atmospheric gases), which, when fused with periodic omics data, will provide an unprecedented holistic view of bio-physical systems. This will enable truly dynamic and adaptive control loops. Concurrently, blockchain technology offers a mechanism to establish trust and transparency. By creating an immutable ledger that tracks a food product's journey—from a field with a verified beneficial microbiome, through an optimized supply chain, to the consumer—blockchain can provide verifiable proof of sustainability, safety, and ethical provenance, building consumer confidence and adding value to responsibly produced goods.

17.6.2 Key Research Challenges

Bridging the gap between current potential and widespread application requires addressing several foundational research challenges. A primary obstacle is the lack of standardized data collection and sharing protocols. The field is currently populated by heterogeneous methods for sampling, sequencing, and metadata reporting, which hinder the creation of large, high-quality datasets necessary for training robust, generalizable AI models. The development of universal standards is a critical, community-wide endeavor (Parks et al., 2022).

Secondly, the "black box" nature of many complex ML models, particularly deep learning, poses a significant challenge to scientific adoption and regulatory approval. Improved model interpretability is essential. Researchers need tools to understand not just what a model predicts but why: which microbial features and interactions were most important in reaching a decision (Asnicar et al., 2024). This interpretability is crucial for validating biological mechanisms, building trust with end-users like farmers and food processors, and ensuring that model recommendations are biologically sound and safe.

Finally, the prospective environmental and health risks associated with AI-designed microbes demand the development of robust risk-assessment protocols. Current frameworks are inadequate for evaluating the complex, multi-trophic interactions of engineered microbial communities (Bohua et al., 2023). New, predictive ecotoxicological models are needed to forecast long-term impacts on soil food webs, water quality, and biodiversity. This requires interdisciplinary research that combines microbial ecology, systems biology, and AI to develop safety-by-design principles embedded in the technology development process from the outset.

17.7 CONCLUSION

In conclusion, Microbial AI offers a paradigm-shifting toolkit to address the tripartite challenge outlined at the outset. It enables us to nourish a growing population by enhancing crop resilience and creating new foods through precision fermentation; to adapt to climate change by developing drought-tolerant crops and optimizing resource use; and to mitigate environmental impact by reducing pesticide/fertilizer runoff, creating circular waste streams, and building healthier soils. By learning the language of the microbial world, we can finally design food systems that work in concert with, rather than against, planetary biology.

However, this technological promise is inextricably linked to a complex web of ethical, social, and regulatory considerations. The questions of biosecurity, equity, data governance, and public trust are not ancillary; they are central to the responsible development of the field. The future of Microbial AI will therefore be written not only by computer scientists and microbiologists but also by ethicists, policymakers, farmers, and consumers. Its ultimate success will be measured not by its algorithmic sophistication alone, but by its tangible contribution to a more resilient, equitable, and sustainable global food system for generations to come. The journey has begun, and our collective commitment to responsible and inclusive innovation will determine its course.

REFERENCES

Asnicar, F., Thomas, A. M., Passerini, A., Waldron, L., & Segata, N. (2024). Machine learning for microbiologists. *Nature Reviews Microbiology, 22*(4), 191–205. https://doi.org/10.1038/s41579-023-00984-1

Bagheri, M., Zahmatkesh, A., Hussain, A., Taj, T., Ashique, S., & Mojgani, N. (2025). Gut Microbiota: Role in Health and Disease. In *Cancer Drug Discovery and Development: Vol. Part F812* (pp. 1–14). Humana Press Inc. https://doi.org/10.1007/978-3-031-86065-2_1

Bashir, Z., Majeed, S., Jan, A., Hamid, B., Wani, P. A., Rasol, R., Nisa, M., & Dar, B. A. (2025). Soil Health: Role of Microorganisms in Soil Health. In Z. Bashir, R. Kumar, & M. Nisa (Eds.), *Organic Farming: A Comprehensive Guide to Sustainable Agriculture* (pp. 119–140). CRC Press. https://doi.org/10.1201/9781003491460-6

Bohua, L., Yuexin, W., Yakun, O., Kunlan, Z., Huan, L., & Ruipeng, L. (2023). Ethical framework on risk governance of synthetic biology. *Journal of Biosafety and Biosecurity, 5*(2), 45–56. https://doi.org/10.1016/j.jobb.2023.03.002

Bokulich, N. A., Collins, T. S., Masarweh, C., Allen, G., Heymann, H., Ebeler, S. E., & Millsa, D. A. (2016). Associations among wine grape microbiome, metabolome, and fermentation behavior suggest microbial contribution to regional wine characteristics. *mBio, 7*(3). https://doi.org/10.1128/mBio.00631-16

Casonato, C., García-Herrero, L., Caldeira, C., & Sala, S. (2023). What a waste! Evidence of consumer food waste prevention and its effectiveness. *Sustainable Production and Consumption, 41*, 305–319. https://doi.org/10.1016/j.spc.2023.08.002

Cembrowska-Lech, D., Krzemińska, A., Miller, T., Nowakowska, A., Adamski, C., Radaczyńska, M., Mikiciuk, G., & Mikiciuk, M. (2023). An integrated multi-omics and artificial intelligence framework for advance plant phenotyping in horticulture. *Biology, 12*(10). https://doi.org/10.3390/biology12101298

Chaturvedi, N., Yadav, M. K., & Sharma, M. (2024). Applications of Artificial Intelligence and Machine Learning in Microbial Diagnostics and Identification. In *Methods in Microbiology* (Vol. 55, pp. 213–230). Academic Press Inc. https://doi.org/10.1016/bs.mim.2024.05.013

Daimary, N., Deb, B., Roy, B., Ranjan, R. K., & Mukherjee, A. (2025). Sustainable biorefinery approach for the transformation of biowaste into biofuels and chemicals for the circular economy: A review. *Sustainable Energy Technologies and Assessments, 82*. https://doi.org/10.1016/j.seta.2025.104457

Duarte-Casar, R., Romero-Benavides, J. C., Bailon-Moscoso, N., Navarro-Rojas, M., & Rojas-Le-Fort, M. (2025). Synergies between Green Chemistry Principles and the Sustainable Development Goals (SDGs) in Developing Eco-Friendly Pesticides. In B. R. Babaniyi, & E. E. Babaniyi (Eds.), *The Interplay of Pesticides and Climate Change: Environmental Dynamics and Challenges* (pp. 291–312). Springer Nature Switzerland. https://doi.org/10.1007/978-3-031-81669-7_11

Gupta, S., Kashyap, S., & Kumar, D. (2024). AI-Driven Analysis of Environmental Metagenomic Data. In S. Gupta, D. Kumar, R. Negi, R. Singh, M. J. Ansari, S. Kashyap, & S. Mehrotra (Eds.), *Genomic Intelligence: Metagenomics and Artificial Intelligence* (pp. 188–206). CRC Press. https://dx.doi.org/10.1201/9781003570233

Gupta, S., Kumar, C. B., & Kumar, D. (2024). Ethical Issues in AI-Enhanced Metagenomic Research. In S. Gupta, D. Kumar, R. Negi, R. Singh, M. J. Ansari, S. Kashyap, & S. Mehrotra (Eds.), *Genomic Intelligence: Metagenomics and Artificial Intelligence* (pp. 276–290). CRC Press. https://dx.doi.org/10.1201/9781003570233

Hilgendorf, K., Wang, Y., Miller, M. J., & Jin, Y.-S. (2024). Precision fermentation for improving the quality, flavor, safety, and sustainability of foods. *Current Opinion in Biotechnology, 86*. https://doi.org/10.1016/j.copbio.2024.103084

Hilger, H., & Humer, M. (2003). Biotic landfill cover treatments for mitigating methane emissions. *Environmental Monitoring and Assessment*, *84*(1–2), 71–84. https://doi.org/10.1023/A:1022878830252

Hu, A., Lu, Y., Hernández García, M., & Dumont, M. G. (2019). Targeted Metatranscriptomics of Soil Microbial Communities with Stable Isotope Probing. In *Methods in Molecular Biology* (Vol. 2046, pp. 163–174). Humana Press Inc. https://doi.org/10.1007/978-1-4939-9721-3_13

Javaid, M., Haleem, A., Khan, I. H., & Suman, R. (2023). Understanding the potential applications of artificial intelligence in agriculture sector. *Advanced Agrochem*, *2*(1), 15–30. https://doi.org/10.1016/j.aac.2022.10.001

Jiang, R., Yue, Z., Shang, L., Wang, D., & Wei, N. (2024). PEZy-miner: An artificial intelligence driven approach for the discovery of plastic-degrading enzyme candidates. *Metabolic Engineering Communications*, *19*. https://doi.org/10.1016/j.mec.2024.e00248

Kaluđerović, Ž (2019). Genetically modified crops – Bioethical approach. *Jahr*, *10*(2), 397–412. https://doi.org/10.21860/j.10.2.7

Kumareswaran, K., & Jayasinghe, G. Y. (2022). Systematic review on ensuring the global food security and covid-19 pandemic resilient food systems: Towards accomplishing sustainable development goals targets. *Discover Sustainability*, *3*(1). https://doi.org/10.1007/s43621-022-00096-5

Li, L., Awada, T., Zhang, Y., & Paustian, K. (2024). Global land use change and its impact on greenhouse gas emissions. *Global Change Biology*, *30*(12). https://doi.org/10.1111/gcb.17604

Medina–García, M., Amigo, J. M., Martínez-Domingo, M. A., Valero, E. M., & Jiménez–Carvelo, A. M. (2025). Strategies for analysing hyperspectral imaging data for food quality and safety issues – A critical review of the last 5 years. *Microchemical Journal*, *214*. https://doi.org/10.1016/j.microc.2025.113994

Palma, S. I. C. J., Traguedo, A. P., Porteira, A. R., Frias, M. J., Gamboa, H., & Roque, A. C. A. (2018). Machine learning for the meta-analyses of microbial pathogens' volatile signatures. *Scientific Reports*, *8*(1). https://doi.org/10.1038/s41598-018-21544-1

Parks, D. F., Voitiuk, K., Geng, J., Elliott, M. A. T., Keefe, M. G., Jung, E. A., Robbins, A., Baudin, P. V., Ly, V. T., Hawthorne, N., Yong, D., Sanso, S. E., Rezaee, N., Sevetson, J. L., Seiler, S. T., Currie, R., Pollen, A. A., Hengen, K. B., Nowakowski, T. J., & Haussler, D. (2022). IoT cloud laboratory: Internet of Things architecture for cellular biology. *Internet of Things (Netherlands)*, *20*. https://doi.org/10.1016/j.iot.2022.100618

Ribeiro, C. S., van Roode, M. Y., Haringhuizen, G. B., Koopmans, M. P., Claassen, E., & van de Burgwal, L. H. M. (2018). How ownership rights over microorganisms affect infectious disease control and innovation: A root-cause analysis of barriers to data sharing as experienced by key stakeholders. *PLoS ONE*, *13*(5). https://doi.org/10.1371/journal.pone.0195885

Rojas-Le-Fort, M., Colpari-Pozzo, M., Navarro-Rojas, M., Romero-Benavides, J. C., Ordóñez-Zambrano, T., Mondavi-Sobby, D., & Duarte-Casar, R. (2025). Environmental Impact of Traditional Energy Sources. In *Fuel Cells in Environmental Sustainability* (pp. 18–33). Boca Raton, FL: CRC Press. https://doi.org/10.1201/9781003645405-2

Rosegrant, M. W., Sulser, T. B., Dunston, S., Mishra, A., Cenacchi, N., Gebretsadik, Y., Robertson, R., Thomas, T., & Wiebe, K. (2024). Food and nutrition security under changing climate and socioeconomic conditions. *Global Food Security*, *41*. https://doi.org/10.1016/j.gfs.2024.100755

Rugji, J., Erol, Z., Taşçı, F., Musa, L., Hamadani, A., Gündemir, M. G., Karalliu, E., & Siddiqui, S. A. (2025). Utilization of AI – reshaping the future of food safety, agriculture and food security – a critical review. *Critical Reviews in Food Science and Nutrition*, *65*(26), 5136–5180. https://doi.org/10.1080/10408398.2024.2430749

Sagar, I., Nimonkar, Y., Dhotre, D., Shouche, Y., Ranade, D., Dewala, S., & Prakash, O. (2022). A microcosm model for the study of microbial community shift and carbon emission from landfills. *Indian Journal of Microbiology*, *62*(2), 195–203. https://doi.org/10.1007/s12088-021-00995-7

Sanfiya, S., & Sabitha Banu, A. (2025). Digital Twins & Industry 5.0 on Food Management. In *Sustainable Development Goals Series: Vol. Part F102* (pp. 87–105). Springer. https://doi.org/10.1007/978-3-031-76758-6_7

Sharma, V., & Salwan, R. (2023). Soil Microbiology. In R. Salwan, & V. Sharma (Eds.), *Laboratory Methods in Microbiology and Molecular Biology: Methods in Molecular Microbiology* (pp. 59–73). Elsevier. https://doi.org/10.1016/B978-0-323-95078-7.00001-2

Shunmugathammal, M., Subramoniam, S. S., Bharathy, N., & Gowtham, K. (2024). The Smart Shipping Container with Environmental Monitoring and Location Tracking System. *3rd International Conference on Advances in Computing, Communication and Materials, ICACCM 2024*. https://doi.org/10.1109/ICACCM61117.2024.11058988

Singh, S., & Shyu, D. J. H. (2024). Profiling of Microbial Community and Their Role in Solid Waste Treatment. In V. Kumar, S. A. Bhat, S. Kumar, & P. Verma (Eds.), *Environmental Engineering and Waste Management: Recent Trends and Perspectives* (pp. 415–448). Springer Nature. https://doi.org/10.1007/978-3-031-58441-1_15

Su, G., Yu, C., Liang, S., Wang, W., & Wang, H. (2024). Multi-omics in food safety and authenticity in terms of food components. *Food Chemistry, 437*, 137943. https://doi.org/10.1016/j.foodchem.2023.137943

Vinestock, T., Short, M., Ward, K., & Guo, M. (2024). Computer-aided chemical engineering research advances in precision fermentation. *Current Opinion in Food Science, 58*. https://doi.org/10.1016/j.cofs.2024.101196

Visuña, L., Garcia-Blas, J., Ferrer-Bazaga, S., Lopez-Exposito, P., & Carretero, J. (2025). Automated monitoring of Mycobacterium population growth in time-lapse microscopy with Deep Learning pseudo-supervised algorithm. *Engineering Applications of Artificial Intelligence, 162*. https://doi.org/10.1016/j.engappai.2025.112503

Xi, N., Wang, S., Li, T., Wang, X., Xu, Y., Shen, Q., & Wei, Z. (2025). The synergistic effect of a phage-probiotic combination on suppressing bacterial wilt disease. *Acta Pedologica Sinica, 62*(4), 1185–1196. https://doi.org/10.11766/trxb202405160200

Xu, Z., Peng, X., Xie, S., Yi, Z., Yuan, L., Zhao, Q., Yin, C., Liu, Y., Li, Q., Yin, H., & Meng, D. (2025). Microbial community structures and potential functions in a food waste treatment facility. *Acta Microbiologica Sinica, 65*(10), 4444–4457. https://doi.org/10.13343/j.cnki.wsxb.20250181

Zhang, Y., Su, K., Munir, F., Wang, W., Liang, X., Wang, J., Hua, J., & Ma, W. (2025). Gut microbiome metagenomics in diarrheic and healthy Simmental cattle from Ningxia Province, China. *Research in Veterinary Science, 197*. https://doi.org/10.1016/j.rvsc.2025.105922

Index

C

D

E

F

G

O

P

Q

R

S

T

For Product Safety Concerns and Information please contact our EU representative GPSR@taylorandfrancis.com Taylor & Francis Verlag GmbH, Kaufingerstraße 24, 80331 München, Germany